PROPERTIES AND APPLICATIONS OF PEROVSKITE-TYPE OXIDES

CHEMICAL INDUSTRIES

A Series of Reference Books and Textbooks

Consulting Editor

HEINZ HEINEMANN
Berkeley, California

1. *Fluid Catalytic Cracking with Zeolite Catalysts,* Paul B. Venuto and E. Thomas Habib, Jr.
2. *Ethylene: Keystone to the Petrochemical Industry,* Ludwig Kniel, Olaf Winter, and Karl Stork
3. *The Chemistry and Technology of Petroleum,* James G. Speight
4. *The Desulfurization of Heavy Oils and Residua,* James G. Speight
5. *Catalysis of Organic Reactions,* edited by William R. Moser
6. *Acetylene-Based Chemicals from Coal and Other Natural Resources,* Robert J. Tedeschi
7. *Chemically Resistant Masonry,* Walter Lee Sheppard, Jr.
8. *Compressors and Expanders: Selection and Application for the Process Industry,* Heinz P. Bloch, Joseph A. Cameron, Frank M. Danowski, Jr., Ralph James, Jr., Judson S. Swearingen, and Marilyn E. Weightman
9. *Metering Pumps: Selection and Application,* James P. Poynton
10. *Hydrocarbons from Methanol,* Clarence D. Chang
11. *Form Flotation: Theory and Applications,* Ann N. Clarke and David J. Wilson
12. *The Chemistry and Technology of Coal,* James G. Speight
13. *Pneumatic and Hydraulic Conveying of Solids,* O. A. Williams
14. *Catalyst Manufacture: Laboratory and Commercial Preparations,* Alvin B. Stiles
15. *Characterization of Heterogeneous Catalysts,* edited by Francis Delannay
16. *BASIC Programs for Chemical Engineering Design,* James H. Weber
17. *Catalyst Poisoning,* L. Louis Hegedus and Robert W. McCabe
18. *Catalysis of Organic Reactions,* edited by John R. Kosak
19. *Adsorption Technology: A Step-by-Step Approach to Process Evaluation and Application,* edited by Frank L. Slejko
20. *Deactivation and Poisoning of Catalysts,* edited by Jacques Oudar and Henry Wise
21. *Catalysis and Surface Science: Developments in Chemicals from Methanol, Hydrotreating of Hydrocarbons, Catalyst Preparation, Monomers and Polymers, Photocatalysis and Photovoltaics,* edited by Heinz Heinemann and Gabor A. Somorjai
22. *Catalysis of Organic Reactions,* edited by Robert L. Augustine
23. *Modern Control Techniques for the Processing Industries,* T. H. Tsai, J. W. Lane, and C. S. Lin
24. *Temperature-Programmed Reduction for Solid Materials Characterization,* Alan Jones and Brian McNichol

25. *Catalytic Cracking: Catalysts, Chemistry, and Kinetics,* Bohdan W. Wojciechowski and Avelino Corma
26. *Chemical Reaction and Reactor Engineering,* edited by J. J. Carberry and A. Varma
27. *Filtration: Principles and Practices. Second Edition,* edited by Michael J. Matteson and Clyde Orr
28. *Corrosion Mechanisms,* edited by Florian Mansfeld
29. *Catalysis and Surface Properties of Liquid Metals and Alloys,* Yoshisada Ogino
30. *Catalyst Deactivation,* edited by Eugene E. Petersen and Alexis T. Bell
31. *Hydrogen Effects in Catalysis: Fundamentals and Practical Applications,* edited by Zoltán Paál and P. G. Menon
32. *Flow Management for Engineers and Scientists,* Nicholas P. Cheremisinoff and Paul N. Cheremisinoff
33. *Catalysis of Organic Reactions,* edited by Paul N. Rylander, Harold Greenfield, and Robert L. Augustine
34. *Powder and Bulk Solids Handling Processes: Instrumentation and Control,* Koichi Iinoya, Hiroaki Masuda, and Kinnosuke Watanabe
35. *Reverse Osmosis Technology: Applications for High-Purity-Water Production,* edited by Bipin S. Parekh
36. *Shape Selective Catalysis in Industrial Applications,* N. Y. Chen, William E. Garwood, and Frank G. Dwyer
37. *Alpha Olefins Applications Handbook,* edited by George R. Lappin and Joseph L. Sauer
38. *Process Modeling and Control in Chemical Industries,* edited by Kaddour Najim
39. *Clathrate Hydrates of Natural Gases,* E. Dendy Sloan, Jr.
40. *Catalysis of Organic Reactions,* edited by Dale W. Blackburn
41. *Fuel Science and Technology Handbook,* edited by James G. Speight
42. *Octane-Enhancing Zeolitic FCC Catalysts,* Julius Scherzer
43. *Oxygen in Catalysis,* Adam Bielański and Jerzy Haber
44. *The Chemistry and Technology of Petroleum: Second Edition, Revised and Expanded,* James G. Speight
45. *Industrial Drying Equipment: Selection and Application,* C. M. van't Land
46. *Novel Production Methods for Ethylene, Light Hydrocarbons, and Aromatics,* edited by Lyle F. Albright, Billy L. Crynes, and Siegfried Nowak
47. *Catalysis of Organic Reactions,* edited by William E. Pascoe
48. *Synthetic Lubricants and High-Performance Functional Fluids,* edited by Ronald L. Shubkin
49. *Acetic Acid and Its Derivatives,* edited by Victor H. Agreda and Joseph R. Zoeller
50. *Properties and Applications of Perovskite-Type Oxides,* edited by L. G. Tejuca and J. L. G. Fierro

ADDITIONAL VOLUMES IN PREPARATION

Computer Modeling of Catalysts, edited by E. Robert Becker and Carmo Pereira

PROPERTIES AND APPLICATIONS OF PEROVSKITE-TYPE OXIDES

edited by

L. G. Tejuca
J. L. G. Fierro
Instituto de Catálisis y Petroleoquímica
CSIC
Madrid, Spain

CRC Press
Taylor & Francis Group
Boca Raton London New York

CRC Press is an imprint of the
Taylor & Francis Group, an **informa** business

Taylor & Francis Group
6000 Broken Sound Parkway NW, Suite 300
Boca Raton, FL 33487-2742

First issued in paperback 2019

ISBN-13: 978-0-8247-8786-8 (hbk)
ISBN-13: 978-0-367-40252-5 (pbk)

Library of Congress Cataloging-in-Publication Data

Properties and applications of perovskite-type oxides / edited by L.G.
Tejuca, J.L.G. Fierro.
p. cm. -- (Chemical industries ; v. 50)
Includes bibliographical references and index.
ISBN 0-8247-8786-2
1. Perovskite. 2. Catalysts. I. Tejuca, L. G. (Luís G.), 1939-1989.
II. Fierro, J. L. G. (José L. G.). III. Series.
QD181.01P83 1992
541.3'95--dc20 92-36741
CIP

Visit the Taylor & Francis Web site at
http://www.taylorandfrancis.com

and the CRC Press Web site at
http://www.crcpress.com

In Memoriam

With my deepest regret, I must report the unexpected death of Dr. L.G. Tejuca, who first accepted the task of editing this book almost three years ago. Born on September 25, 1939 in Asturias, Dr. Tejuca received the master's degree in chemistry from Oviedo University in 1965 and the doctoral degree, also in chemistry, from the Complutense University in 1970. After serving as Associate Scientist from 1971 to 1975, he assumed a position as Staff Senior Scientist in 1976 at the Institute of Catalysis and Petrochemistry of the National Council for Scientific Research. He was also a postdoctoral research fellow at the University of Nottingham (Great Britain), Princeton University, and the University of Dundee (Great Britain).

Dr. Tejuca was best known for his pioneering research on the perovskite-type oxides, including synthesis, structural characterization, and application in several catalytic test reactions. His most important contribution was the establishment of the relationship between the local symmetry of the 3d transition cations in perovskite surfaces and the surfaces' adsorption and catalytic properties. The close parallelism he found between the extent of oxygen adsorption and the rate of carbon monoxide and low-molecular-weight olefin oxidations permitted him to discuss catalytic trends in terms of crystal field effects. Another major area of Dr. Tejuca's work involved the use of adsorption models to evaluate the number of surface sites on unsupported and supported metal oxides. These efforts of Dr. Tejuca resulted in over 130 original publications which have stimulated research in this paticular area of heterogeneous catalysis over the past decade.

As his friend and coworker, I decided to continue with the task of editing this pivotal book. All friends at the Institute, authors of the different chapters, the publisher, and readers of the book mourn the passing of Dr. Tejuca.

J.L.G. Fierro

In Memoriam

Preface

Many progress reports and a large body of publications in specialized surface science and catalysis journals have provided a convincing description of the processes taking place at the surface of perovskite-type structures. Accordingly, it has been our endeavor to offer the reader of this book an overview of the bulk and surface properties of these marvelous materials and how they can be related to the catalytic performance in many model and practical reactions.

Chapter 1 surveys the methods of preparation of bulk perovskites and offers the possibility of synthesizing these compounds on moderate- or high-surface-area carriers. Special emphasis is placed in Chapter 2 on the methods employed in the processing of high temperature superconductive cuprates. Chapter 3 provides a comprehensive review of the oxidative nonstoichiometric behavior of perovskites, including not only the excess oxygen but also the cation vacancies. This analysis is extended in Chapter 4 to several high-temperature superconductive cuprates for which a careful crystallographic analysis is made. Chapter 5 deals with the thermoelectric properties of various cuprates and the comparative analysis of the critical behavior of thermopower and electrical conductivity in the mean field region. Chapter 6 is concerned with the optical properties of perovskites and doped perovskites in which the ions may adopt a large variety of configurations.

Chapters 7 and 8 focus on methods used to evaluate the surface sites, including poisoning with sulfur dioxide, the study of the vibrational modes of adsorbed molecules, and the analysis of the desorption products of chemisorbed probe molecules. The role of surface impurities and phase segregation and the relevance of the surface structure on chemisorptive and catalytic properties are discussed in Chapter 9. Chapters 10 to 17 survey several practical and model catalytic reactions such as total oxidation of hydrocarbons, hydrogenation and hydrogenolysis of hydrocarbons, hydrogenation of carbon oxides, CO oxidation and NO reduction, partial oxidation of hydrocarbons and oxygenates, hydrogen evolution from water, reduction of sulfur dioxide, and decomposition of nitrous oxide. Finally, the use of these compounds as solid state chemical sensors is reviewed in Chapter 18.

The overall view depicted in this book indicates clearly how these model compounds provide basic information for the study of structure-activity relationships and also for catalytic design based on the flexible perovskite structure which can tolerate multiple ion substitution.

I acknowledge with gratitude the contributions made by the different authors, the technical staffs of the Institutes supporting this work, and the collaboration of the publisher, all of whose cumulative efforts and patience have made this book possible. Last but not least, my thanks go to Mrs. B. Pawelec whose indefatigable secretarial assistance has been invaluable.

J.L.G. Fierro

Contents

Contributors

Professor T. Arakawa Department of Industrial Chemistry, Faculty of Engineering, Kinki University in Kyushu, Iizuka, Fukuoka 820, Japan

Dr. M. Avudaithai Department of Inorganic and Physical Chemistry, Indian Institute of Science, Bangalore 560 012, India

Professor G. Blasse Debye Research Institute, University of Utrecht, 3508 TA Utrecht, The Netherlands

Dr. O. Cabeza Laboratorio de Física de Materiales (LAFIMAS), Universidad de Santiago de Compostela, 15706 Santiago de Compostela, Spain

Dr. M. T. Casais Instituto de Ciencia de Materiales, Consejo Superior de Investigaciones Científicas (CSIC), Serrano 113, 28006 Madrid, Spain

Dr. C. Cascales Instituto de Ciencia de Materiales, Consejo Superior de Investigaciones Científicas (CSIC), Serrano 113, 28006 Madrid, Spain

Dr. J. Christopher Research and Development Centre, Indian Oil Corporation, Ltd., Faridabad 121 007, India

Professor P. Duran Instituto de Cerámica y Vidrio, Consejo Superior de Investigaciones Científicas (CSIC), Electroceramics Department, 28500 Arganda del Rey, Madrid, Spain

Dr. J. L. G. Fierro Instituto de Catálisis y Petroleoquímica, Consejo Superior de Investigaciones Científicas (CSIC), Serrano 119, 28006 Madrid, Spain

Professor P. K. Gallagher Departments of Chemistry and Material Science and Engineering, The Ohio State University, Columbus, OH 43210

Professor M. Hervieu Laboratoire CRISMAT, Institut des Sciences de la Matière et du Rayonnement (ISMRA), Boulevard du Maréchal Juin, 14050 Caen Cédex, France

Professor D. Brynn Hibbert Department of Analytical Chemistry, University of New South Wales, Kensington, New South Wales 2033, Sidney, Australia

Professor K. Ichimura Department of Chemistry, Faculty of Science, Kumamoto University, Kurokami, Kumamoto 860, Japan

Professor Y. Inoue Department of Chemistry, Nagaoka University of Technology, Nagaoka, Niigata 940-21, Japan

Professor T. R. N. Kutty Department of Inorganic and Physical Chemistry, and Materials Research Centre, Indian Institute of Science, Bangalore 560 012, India

Professor Li Wan Department of Chemistry and Environmental Engineering, Beijing Polytechnic University, Beijing 100871, People's Republic of China

Dr. J. Maza Laboratorio de Física de Materiales (LAFIMAS), Universidad de Santiago de Compostela, 15706 Santiago de Compsotela, Spain

Professor C. Michel Laboratoire CRISMAT, Institut des Sciences de la Matière et du Rayonnement (ISMRA), Boulevard du Maréchal Juin, 14050 Caen Cédex, France

Professor C. Moure Instituto de Cerámica y Vidrio, Consejo Superior de Investigaciones Científicas (CSIC), Electroceramics Department, 28500 Arganda de Rey, Madrid, Spain

Professor I. Rasines Instituto de Ciencia de Materiales, Consejo Superior de Investigaciones Científicas (CSIC), Serrano 113, 28006 Madrid, Spain

Professor B. Raveau Laboratoire CRISMAT, Institut des Sciences de la Matière et du Rayonnement (ISMRA), Boulevard du Maréchal Juin, 14050 Caen Cédex, France

Professor T. Seiyama Graduate School of Engineering Sciences, Kyushu University, Kasuga, Fukuoka 816, Japan

Professor T. Shimizu Hakodate National College of Technology, Hakodate, Tokura-cho 14-1, 042 Japan

Professor D. M. Smyth Department of Materials Science and Engineering, Lehigh University, Bethlehem, PA 18015

Professor C. S. Swamy Department of Chemistry, Indian Institute of Technology, Madras 600 036, India

Dr. J. M. D. Tascón Instituto Nacional del Carbón y sus Derivados, Conjejo Superior de Investigaciones Científicas (CSIC), Apartado 73, 33080 Oviedo, Spain

Dr. L. G. Tejuca (deceased) Instituto de Catálisis y Petroleoquimica, Conjejo Superior de Investigaciones Científicas (CSIC), Serrano 119, 28006 Madrid, Spain

Dr. J. Twu Pharmaceutical R&D Laboratories, Development Center for Biotechnology, Taipei, Republic of China

Dr. J. A. Veira Laboratorio de Física de Materiales (LAFIMAS), Universidad de Santiago de Compostela, 15706 Santiago de Compostela, Spain

Professor F. Vidal Laboratorio de Física de Materiales (LAFIMAS), Universidad de Santiago de Compostela, 15706 Santiago de Compostela, Spain

Professor B. Viswanathan Department of Chemistry, Indian Institute of Technology, Madras 600 036, India

Dr. Y. Yadava Laboratorio de Física de Materiales (LAFIMAS), Universidad de Santiago de Compostela, 15706 Santiago de Compostela, Spain

Professor I. Yasumori Department of Applied Chemistry, Faculty of Engineering, Kanagawa University, Rokkakubashi, Yokohama 221, Japan

PROPERTIES AND APPLICATIONS OF PEROVSKITE-TYPE OXIDES

Chapter 1

Preparation of Bulk and Supported Perovskites

J. Twu[1] and P.K. Gallagher[2]

[1]Pharmaceutical R&D Laboratories, Development Center for Biotechnology, Taipei, Republic of China

[2]Departments of Chemistry and Materials Science & Engineering, The Ohio State University, Columbus, OH 43210

I. Introduction

Perovskite-related materials are represented by the general formula ABO_3 in which A ions can be rare earth, alkaline earth, alkali and other large ions such as Pb^{2+} and Bi^{3+} that fit into the dodecahedral site of the framework. The B ions can be 3d, 4d and 5d transitional metal ions which occupy the octahedral sites. A complete list of A and B ions is given in Table 1 (1). In general, the perovskite structure is formed if the tolerence factor, t, is in the range 0.8 - 1.0 (R is the ionic radius of the A, B, or oxide ion) (2):

$$t = (R_A + R_O)/\sqrt{2}(R_B + R_O)$$

Large classes of perovskite-like materials are comprised of basic perovskite cells separated by intervening layers such as alkaline earth oxides. There are many stacking sequences possible that give rise to such compounds as Sr_2FeO_4 or $Sr_3Fe_2O_7$ (2). In addition to the diversity of compositions mentioned above, perovskite materials can tolerate significant partial substitution and non-stoichiometry while still maintaining the perovskite structure. For example, metal ions having different valences can replace both A and B ions. This may generate a non-integral number of oxygen atoms. These flexible compositions can be represented by the related family of layered grossly oxygen deficient perovskites such as Sr_2FeO_{4-x} and $Sr_3Fe_2O_{7-x}$ (2), and the superconducting variety, $Y_1Ba_2Cu_3O_{6+x}$(3).

Because of their varied structure and composition, perovskite materials have attracted intense interest in many applied and fundamental areas of solid state chemistry, physics, advanced materials, and catalysis. A partial list of their potential applications is presented in Table 2 (4). Because of their diverse applications, the optimum preparation of these materials has proved challenging, requiring many synthetic approaches determined by the ultimate end use. For example, materials-oriented applications require densification by high temperature

Table 1. Radii of Cations A and B for ABO_3-Type Perovskite[a]

Dodecahedral A-site				Octahedral B-site			
Na^+	1.06	1.32?	(IX)	Li^+		0.68	0.74
K^+	1.45	1.60?		Cu^{+2}		0.72	0.73
Rb^+	1.61	1.73		Mg^{+2}		0.66	0.72
Ag^+	1.40	1.30	(VIII)	Ti^{+3}		0.76	0.67
Ca^{+2}	1.08	1.35		V^{+3}		0.74	0.64
Sr^{+2}	1.23	1.44		Cr^{+3}		0.70	0.62
Ba^{+2}	1.46	1.60		Mn^{+3}		0.66	0.65
Pb^{+2}	1.29	1.49		Fe^{+3}		0.64	0.64
La^{+3}	1.22	1.32?		Co^{+3}	(LS)	-	0.52
Pr^{+3}	1.10	1.14	(VIII)	Co^{+3}	(HS)	0.63	0.61
Nd^{+3}	1.09	1.12	(VIII)	Ni^{+3}	(LS)	-	0.56
Bi^{+3}	1.07	1.11	(VIII)	Ni^{+3}	(HS)	0.62	0.60
Ce^{+4}	1.02	0.97	(VIII)	Rh^{+3}		0.68	0.66
Th^{+4}	1.09	1.06	(VIII)	Ti^{+4}		0.68	0.60
				Mn^{+4}		0.56	0.54
				Ru^{+4}		0.67	0.62
				Pt^{+4}		0.65	0.63
				Nb^{+5}		0.69	0.64
				Ta^{+5}		0.69	0.64
				Mo^{+6}		0.62	0.60
				W^{+6}		0.62	0.58

[a] The coordination number is in parentheses if the radii given are not for 12-coordination. HS and LS refer to the spin states.

sintering to minimize both surface area and surface free energy in order to maximize mechanical strength. In contrast, catalytic materials have to maintain sufficiently high surface area in order to maximize their participation and activity in chemical reactions.

In this chapter, studies of the synthetic approach to produce high performance perovskite powders for both catalysis and advanced materials will be reviewed. Since other chapters are devoted to their applications and structures, the discussion here will be limited only to the synthetic aspects and related work. The literature on this topic is very extensive, and no attempt at an exhaustive coverage is attempted. Instead, a series of typical examples is selected to illustrate various approaches to the synthesis of perovskites and related materials.

II. Solution Preparation

Traditional ways of making perovskite materials usually adopt mixing the constituent oxides, hydroxides and/or carbonates. However, since these materials generally have a large particle size, this approach frequently requires repeated mixing and extended heating at high temperature to generate a homogeneous and single phase material. In order to overcome the disadvantages of low surface area and limited control of the micro-structure

Table 2. Applications of Perovskites in Materials

Multilayer Capacitor	$BaTiO_3$
Piezoelectric Transducer	$Pb(Zr,Ti)O_3$
P. T. C. Thermistor	$BaTiO_3$, doped
Electrooptical Modulator	$(Pb,La)(Zr,Ti)O_3$
Switch	$LiNbO_3$
Dielectric Resonator	$BaZrO_3$
Thick Film Resistor	$BaRuO_3$
Electrostrictive Actuator	$Pb(Mg,Nb)O_3$
Superconductor	$Ba(Pb,Bi)O_3$ layered cuprates
Magnetic Bubble Memory	$GdFeO_3$
Laser Host	$YAlO_3$
Ferromagnet	$(Ca,La)MnO_3$
Refractory Electrode	$LaCoO_3$
Second Harmonic Generator	$KNbO_3$

inherent in the high temperature process, precursors generated by sol-gel preparations or coprecipitation of metal ions by precipitating agents such as hydroxide, cyanide, oxalate, carbonate, citrate ions, etc., have been used.

These gel or coprecipitated precursors can offer molecular or near molecular mixing and provide a reactive environment during the course of subsequent heating and decomposition. Because of the improved solid state diffusion resulting from the improved mixing, they need a relatively lower temperature to produce similar materials compared to the traditional methods. These methods frequently offer additional advantages, such as better control of stoichiometry and purity, greater flexibility in forming thin films and new compositions, and an enhanced ability to control particle size. Consequently, they have opened new directions for molecular architecture in the synthesis of perovskites.

A convenient way of classifying the methods which start from solution is to consider the means used for solvent removal. Two basic classes exist. The first is based upon precipitation with subsequent filtration, centrifugation, etc., used to separate the solid and liquid phases. The second basic method depends upon thermal processes, e.g., evaporation, sublimation, combustion, etc., to remove the solvent. An additional consideration in the latter method is the possible simultaneous conversion of the residue into the desired product.

The principal advantages of starting from solution are better homogeneity and improved reactivity. No amount of mixing of solid particles will approach that obtained in solution and most of the other advantages are a direct result of this mixing on essentially an atomic scale. The necessary solid state reactions proceed more rapidly and at lower temperatures. As a consequence, the desired product can be obtained with a smaller particle size and greater reactivity.

Solubility is one of the most important considerations in designing solution techniques. Not only is it desirable that solubility be high in order to minimize the amount of solvent which must subsequently be removed, but also the particular components must be compatible, e.g., iron sulfates could not be combined with barium chloride in order to produce barium ferrite because barium sulfate would precipitate. Cost, purity and toxicity are other obvious factors.

A more subtle concern is the choice of the presumably inert anions, which will be determined by the pH values, ionic strength, degree of supersaturation and impurities. This is important firstly because of the tendency of these ions to become incorporated in the final product and secondly because their subsequent effects can vary dramatically. Conditions which favor large particle size, slow growth and equilibrium will generally produce the purest precipitate.

A. Precipitation

1. **Oxalate**-based preparations of ceramic and superconducting materials have been used extensively. However, due to the incorporation of extraneous ions by occlusion, adsorption and substitution, a special method has been applied to alleviate the deleterious effects. This method is based on digestion of the appropriate carbonates, hydroxides, or oxides with oxalic acid, and the only products of this process are the metal oxalates, water and carbon dioxide (5). Since the pH value of the resulting solution is close to 7, it minimizes the solubility problem. However, if solubility remains a problem, the solution could be reused to ensure a saturated solution as the starting point. Such recycling also reduces the potential enviromental impact by decreasing the disposal problem.

During calcination, an atmosphere that is sufficiently oxidizing has to be adopted in order to prevent carbon residues and carbide formation (6). In general, an oxygen atmosphere is maintained to prevent free carbon or metal formation by oxidation of CO and or any other metal formed from the more easily reduced oxides such as copper.

Clabaugh et al. precipitated from an aqueous chloride solution the unique and novel complex compound $BaTiO(C_2O_4)_2 \cdot 4H_2O$ as a precursor for the preparation of finely divided, stoichiometric $BaTiO_3$ (7). Modifications were made to this process to partially substitute Sr, Pb, or La for Ba in order to change the transition temperature and modify electrical conductivity (7-10). Gallagher et al. (11) reported the synthesis of $BaSnO_3$ by thermal decomposition of $BaSn(C_2O_4) \cdot 0.5H_2O$ or $Ba_2Sn(C_2O_4) \cdot 6H_2O$ to form $BaCO_3$ and SnO_2, which then subsequently react to form $BaSnO_3$. The intimate mixing resulting from the precursor method avoids the formation of Ba_2SnO_4 as an intermediate which is a problem during the conventional preparation. This avoidance of undesirable intermediates occurs because of the shorter solid state diffusion paths achieved during thermal decomposition of the precursor. The major use of an oxalate precursor for the synthesis of perovskites has been associated with the formation of alkaline earth titanates but $BaFeO_{3-x}$ and $SrFeO_{3-x}$ can also be generated from $Ba_3[Fe(C_2O_4)_2 \cdot 8H_2O$ and $Sr_3[Fe(C_2O_4)_3]_2 \cdot 2H_2O$ (12).

2. **Hydroxide** formation is often used because of its low solubility and the variety of precipitation schemes possible. Because of its advantages over traditional methods, e.g., chemical homogeneity, low temperature calcination, controlled hydrolysis for thin film formation, and room temperature deposition, the sol-gel process has been utilized for producing a wide range of new materials as well as improving existing materials. A general procedure can be divided into following steps: (i) preparations of metal alkoxides, (ii) conducting controlled hydrolysis and polymerization, (iii) drying, and (iv) firing. Blum et al. (13) reported the synthesis of $PbTiO_3$ using $Pb(CH_3COO)_2$ and $Ti(OC_3H_7)_4$ dissolved in methoxymethanol. The gel was formed by introducing water and nitric acid and then drying at room temperature for about three weeks. The dried gel was annealed at 873 K for several cycles. X-ray diffraction indicated that a tetragonal structure was formed at 673 K after sufficient time. However, absence of the ferroelectric anomaly was observed when heating below 823 K and the lack of extended crystallinity or long range order was proposed as an explanation. An alternative procedure is to start directly with a colloidal solution of one of the components. Commercial sols of many oxides are available, e.g., silica, alumina and titania. High T_c superconducting powders and thin films of $Y_1Ba_2Cu_3O_7$ have also been synthesized by sol-gel preparations (14).

Homogeneous and stoichiometric $LiNbO_3$ fibers were prepared by the sol-gel process (15). By controlling the concentration of the alkoxide solution and the amount of water used for partial hydrolysis, gel fibers having a uniform diameter were achieved. The gel fibers, prepared by drawing, crystallized directly to single-phase $LiNbO_3$ fibers by heating from 723 to 873 K. The density of the crystalline $LiNbO_3$ fibers was higher than 4.22 g/cm^3, and the dielectric constant at room temperature was about 10 at 10 Mhz.

Another approach via hydroxide formation involves hydrothermal synthesis. The hydrothermal synthesis of $BaTiO_3$-perovskite has been reported by Christensen (16). More recently, $Ba(Ti,Zr)O_3$, $SrTiO_3$ and $BaSnO_3$ have been reported by Kutty et al. (17). Reactive gels of hydrated TiO_2 and ZrO_2 were formed by the addition of NH_4OH to a mixed aqueous solution of $TiOCl_2$ and $ZrOCl_2$. The gel was washed free of chloride ions and suspended in a 0.2 M $Ba(OH)_2$ solution. Then the resulting slurry was placed in a sealed teflon container and heated to 358-403 K for 2-6 h under hydrothermal conditions. The product was filtered and washed with dilute acetic acid to remove excess $Ba(OH)_2$. The sample was air dried before being calcined. The perovskite structures of $BaTiO_3$ and $BaZrO_3$ can be detected by X-ray diffraction (XRD) for samples heated at 403 K for 6h. Adding polyvinyl alcohol during the hydrothermal treatment and LiF during the sintering process has been used to promote reactivity. Transmission electron microscopy (TEM) studies indicate that the hydrothermally prepared $BaTiO_3$ powder contains individual crystallite as well as aggregates of acicular crystallites. Mazdiyasni (18) also compared hydrothermally generated samples of fine (5 nm) of $BaTiO_3$ by heating at 323 K under vacuum. These are significantly smaller than the 30.5 nm formed at 973 K for 60 min.

3. **Acetate** ions have been used alone or together with nitrate ions to generate different perovskites and have shown some advantage over the traditional and oxalate methods. For instance, Nakamura et al. (19) reported the synthesis of $La_{1-x}Sr_xCoO_3$ compounds with x = to 0, 0.2, 0.4 and 0.6 by using acetate precursors followed by calcination at 1123 K in air for 5 h. From XRD, it was confirmed that the perovskite structure had been generated. Zhang et al. (20) reported the synthesis of $La_{1-x}Sr_xCo_{1-y}Fe_yO_3$ using lanthanum, strontium, and cobalt acetates and iron nitrate, then calcining at 1123 K in air between 5-10 h. XRD data showed that all of the oxides except $SrCoO_{2.5}$ showed a perovskite structure.

4. **Citrate** precursors have been investigated by Zhang et al. (21). The decomposition mechanism of these precursors was studied during the course of producing $LaCo_{0.4}Fe_{0.6}O_3$. By using thermogravimetry (TG), XRD and infrared (IR), it was found that citrate precursors undergo several decomposition steps before the perovskite structure is obtained. These steps correspond to the breakup of the citrate complexes, and the elimination of residual CO_3^{2-} and NO_3^- ions. The perovskite framework was detected by IR but was not observable by XRD, indicating that the perovskite was still lacking extended crystallinity. When the temperature was raised to 823 or 873 K, the diffractions due to perovskite were observed. Since the calcination temperature of citrate precursors

about 473 to 573 K lower than the acetate preparation for the same compositions, the resulting surface areas of both preparations were measured in order to compare them. Because of the lower calcination temperature, citrate preparations have a surface area increased by a factor between 3 and 7. It was proposed that the citrate complexes at the precursor stage have localized the metal ions more effectively resulting in a more homogeneous dispersion requiring a lower calcination temperature.

5. **Complex cyanides** have been used to prepare rare earth orthoferrites ($REFeO_3$) and the analogous cobalt compounds ($RECoO_3$) by the thermal decomposition of the appropriate rare earth ferricyanide and cobalticyanide compounds (22). These compounds such as $LaFe(CN)_6 \cdot 6H_2O$, $LaCo(CN)_6 \cdot 5H_2O$ and even ferrocyanides such as $NH_4LaFe(CN)_6 \cdot 5H_2O$ are readily precipitated from aqueous solution. A major advantage of such a process is that, since the excess of either component remains in solution, the precise control of stoichiometry is guaranteed. Because the mixing and stoichiometry are achieved on a atomic scale, the desired compounds are generated by relatively low temperature calcination and without any subsequent milling or contamination. Similar studies have been conducted for europium and other rare earth hexacyanoferrate compounds (23).

Monitoring the nature of the decompositions and identification of the intermediate and final products requires a wide range of techniques such as XRD, Mossbauer spectroscopy, IR, Raman, scanning electron microscopy (SEM), surface area, and TEM. The most convenient and powerful general method is thermal analysis, however. The use of TG, differential thermal analysis (DTA), and evolved gas analysis (EGA) techniques will determine weight loss, phase transitions, solid state reactions, oxidation/reduction, and the gas evolved or consumed. For example, the perovskite $BaPtO_3$ has been prepared by the thermal decomposition of a $BaPt(OH)_6$ precursor (24). TG and DTA were utilized to monitor the decomposition of $BaPt(OH)_6$ in both O_2 and vacuum. The decomposition pathway can be described by the reactions:

$$BaPt(OH)_6 . 0.5H_2O \text{ ----> } BaPt(OH)_6 + 0.5H_2O \quad (2)$$

$$BaPt(OH)_6 \text{ -----------> } BaPtO_3 + 3H_2O \quad (3)$$

$$BaPtO_3 \text{ --------------> } BaPtO_{3-x} + x/2\ O_2 \quad (4)$$

The degree of nonstoichiometry (value of x in eq. 4) could be readily determined by weight loss to form $Ba(OH)_2$ + Pt as determined by TG in H_2. Similar studies have characterized the decomposition or conversion for most of the precursors described herein.

B. Thermal Treatment

1. Combustion

Hydrolysis of aluminum, titanium and silicon in a flame around 1073-1673 K has long been used to prepare these oxides in a fine particulate form. Wenkus and Leavitt (25) first prepared a number of ferrites in a similar fashions by dissolving metal nitrates in alcohol, atomizing the resulting solution with oxygen through a nozzle, burning the alcohol, and finally collecting the resulting powder. The process, which is also referred to as self-propagating high temperature synthesis (SHS), provides energy- and cost-saving advantages over the more conventional processing routes for these materials (26). At the same time, the rapid heating and cooling rates provide a potential for the production of metastable materials with new and, perhaps, unique properties.

A novel combustion process for the synthesis of $YBa_2Cu_3O_7$ superconductor involves ignition of a metal nitrates-organic salts mixture and results in an intimate mixture of partially reacted oxides (27-29). Another modification uses a nitrate-urea mixture. The mixture is given a second combustion with NH_4NO_3 and urea followed by subsequent heat treatment and oxygenation. This produces a single phase superconducting $YBa_2Cu_3O_7$ powder (30).

Pankajavalli et al. discuss a procedure for the preparation of improved quality $YBa_2Cu_3O_{7-x}$ by the citrate pyrolysis method and optimization of parameters for annealing were presented (31). The characterization of the product of citrate pyrolysis and the quality control tests employed for the 1-2-3 compounds produced were discussed. Combustion of the citrate from the aqueous phase yielded uniform mixtures of Y_2O_3, $BaCO_3$, and CuO in the desired 1:4:6 molar ratio. The higher reactivity of this powder was observed in TG traces.

2. Freeze-Drying

The freeze-drying technique is conceptually simple. It involves: (i) dissolving salts in a suitable solvent, usually water; (ii) freezing this solution rapidly to preserve chemical homogeneity; (iii) freeze-drying the frozen solution to give dehydrated salts without involving a liquid phase; and (iv) decomposing the salts to give the desired oxide powder. The freezing and drying process can be comprehended easily with the aid of the schematic diagram in Fig. 1 (32). The most important criterion for the freezing step is the rate of heat transfer out of the solution. This should be as great as possible in order to minimize the ice-salt segregation and more importantly for multicomponent solutions, to prevent large scale segregation of the cation components. This can be accomplished by spraying the liquid through a small orifice into an immiscible liquid such as hexane chilled by dry ice-acetone, or directly into liquid nitrogen.

Early applications of this technique to forming mixed compounds involved ferrites (33,34) and lithium niobate (6). More recently freeze-drying was applied to prepare complex Bi-Pb-Sr-Ca-Cu-O powders which subsequently required sintering at 1113 K for 30 h to form the 110 K (2223) phase (35). Freeze-drying provided highly reactive, intimately mixed, and carbon-free precursors. The latter is important because the presence of carbonate in the uncalcined powders was the major cause of phase segregation and sluggishness of the 110 K phase formation.

The effect of the choice of precursor salts on the surface area of mixed metal oxides prepared by freeze-drying was studied

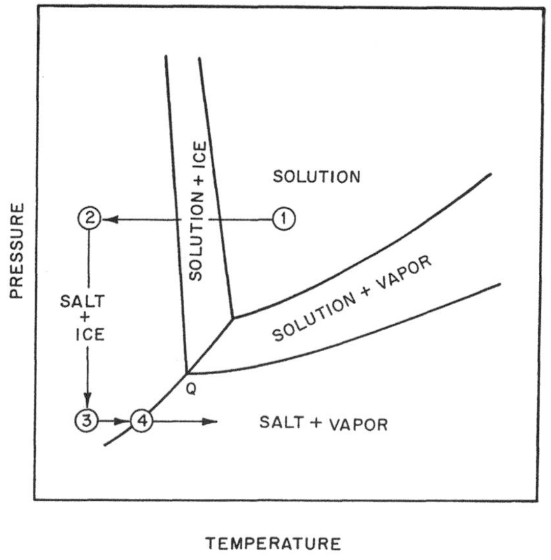

Fig. 1. Phase diagram of an aqueous solution illustrating the freeze-drying method (32).

using the nitrate of Ni or acetates and nitrates of La, Sr, and Co. The $NiCo_2O_4$ produced via the thermal decomposition of freeze-dried Ni Co nitrate had a greater surface area and lower decomposition temperature than $La0_{.5}Sr_{0.5}CoO_3$ prepared from the freeze-dried nitrates (36). The latter needed heating above 773 K to decompose the $Sr(NO_3)_2$, which fractionates from the mixed nitrates. The mist of a mixed solution of $Co(NO_3)_2$, $La(NO_3)_3$ and an additive was produced by supersonic atomization and treated in three temperatures to produce fine particles (37). These temperatures were adjusted to achieve evaporation of the water, decomposition of the mixed metal salts, and the crystallization of the mixed oxides, respectively. The surface area of these samples was increased from 11 m^2/g to 55 m^2/g when NH_4Cl and polyvinyl alcohol were added.

3. Plasma Spray-Drying

Plasma techniques can be divided into two main stages: the injection of reactants and the generation and interaction of the molten droplets (either with the substrate or with the previous generated droplets). They have been applied to various precursors which include gaseous, liquid and solid materials. Since this process offers improved performance in various aspects such as economy, purity, particle size distribution and reactivity, it has been applied to various ceramic, electronic and catalytic materials. This technique has recently been applied to produce thick $YBa_2Cu_3O_x$ films covering large areas (38). The deposition conditions, such as plasma parameters and substrate temperatures, as well as post-deposition treatment of films, were varied in order to obtain the optimum superconducting oxide phase.

III. Solid State Reactions

Conventional processing of perovskite-related materials uses solid state reactions between metal-carbonates, hydroxides, and oxides. A typical case is represented by $BaTiO_3$ (39). Impurities are introduced from raw materials, milling media, and the calcination container. Because of the high temperature required for complete reaction and coarse particles, problems such as multiphases have to be minimized in order to generate homogeneous high performance $BaTiO_3$. Low temperature approaches described in the previous sections have been tested by various groups. The formation of the $Ba_2YCu_4O_8$ superconducting compound initally required a high pressure oxygen process (40). It was later found that the reactivity could be promoted and the desired product obtained at atmospheric pressure by adding an extra (0.2-0.5) molar ratio of an alkaline metal carbonate (41,42). The 124 perovskite formed in this way was more nearly single phase compared to the sample without the addition of alkaline metal carbonates. The formation of a low melting eutectic phase by the addition facilitates reactivity. The additive volatilizes during subsequent processing at higher temperatures.

Rather than starting with the individual components, it may be advisable to begin with prereacted powders (43). This approach has been applied to the synthesis of $Pb(FeW)O_3$ and $Pb(FeNb)O_3$ in order to obtain materials having high dielectric constants with a low temperature dependence. Presumably, the prereacted phases facilitate initial homogeneity leading to reduced diffusion paths during the formation of product.

Oxidation reactions during solid state reactions can be facilitated by using highly oxidized starting materials, e.g., BaO_2 and Fe_2O_3 were used to synthesize $BaFeO_3$ by utilizing the oxidizing ability of the peroxide ion (44). Fe_2O_3 was prepared by thermal decomposition of $FeC_2O_4 \cdot H_2O$ and then mixed with the BaO_2. The cubic structure of $BaFeO_3$ was observed at 633 K and the tetragonal structure of $BaFeO_{2.75}$ at 848 K by XRD. High pressure oxygen systems have also been used to prepare these difficult materials.

A float zone technique was used to prepare single crystal and textured ceramics of incongruently melting perovskites (45). A CO_2 laser produced a stable float zone of $Bi_2CaSr_2Cu_2O_x$ in air or in O_2 pressure up to 2.6 atm. In the best case, nearly 100% single crystal can be obtained. This technique shows great promise for preparing ceramic materials even when the detailed phase diagrams are not available.

Perovskite-related compounds have demonstrated a wide range of oxygen non-stoichiometry. This property determines the vacancy content of the materials which can affect the temperature of melting, electrical properties and catalytic performance (46). The degree of non-stoichiometry depends upon several parameters such as oxygen partial pressure, calcination temperature and substitution of metal ions with different valences. As an example Lindemer et al. reported the influence on oxygen stoichiometry exhibited by variations in P_{02} and calcination temperature for YBa_2Cu_3O superconducting materials, see Fig. 2 (47). The transi-

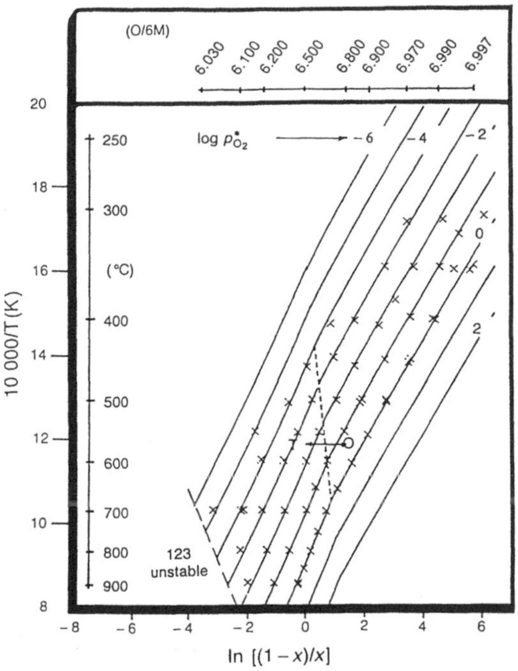

Fig. 2. Oxygen stoichiometry in $YBa_2Cu_3O_{7-x}$ perovskite-related structure (47). The orthorhombic-to-tetragonal transition (O <—> T, dashed line) is from Specht et al. (48).

tion from orthorhombic to the tetragonal phase is indicated by the dashed line based on the data of Specht (48). Since high temperature superconductivity has been shown to depend upon a high oxygen content, the need for a low anneal temperature is clearly evident. This low temperature anneal can be accomplished using oxidizing agents other than the conventional molecular oxygen, e.g., ozone, oxygen plasmas, etc.

IV. Gas Phase Preparations and Reactions

The deposition of perovskite films with a specific thickness and composition generally requires gas phase reaction or transport. Many physical techniques have been developed for gas phase deposition such as laser ablation (49), molecular beam epitaxy (50), dc sputtering (51), magenetron sputtering (52), electron beam evaporation (53) and thermal evaporation (54). In general they can be divided into two categories based on the target they use. The first type uses separate targets where a different speed of deposition for each element has to be determined. The second method uses the preformed perovskite material itself as target and the stoichiometric phase is transported to the substrate by sputtering or ablation techniques. Gas phase depositions can be divided further into three categories: (i) deposition at a low substrate temperature followed by a post-annealing at elevated temperatures, (ii) deposition at an intermediate temperature of 873 to 1073 K followed by a post-annealing treatment, and (iii) deposition at the crystallization temperature under an appropriate atmosphere.

Films of $YBa_2Cu_3O_7$ have been made by co-evaporation of Y, Cu, and BaF_2 and subsequent annealing in wet O_2 (54). Addition of water vapor to the annealing gas at high temperatures greatly reduces the annealing time and thus the substrate interaction. Transition temperatures (zero resistance) of 89 to 92 K are obtained on $SrTiO_3$ and cubic ZrO_2 substrates. Critical current densities on $SrTiO_3$ as high as 10^6 A/cm^2 at 81 K were obtained.

PZT thin films were deposited by radio frequency magnetron sputtering of a multi-element metal target (55). Control of the composition of PZT film was accomplished by adjusting the relative surface areas of the individual metallic targets. The perovskite structure was obtained by a post-annealing treatment. $Pb(BaPb_{1-X}Bi_X)O_3$ and $(Ba0_{.95}Sr_{0.05}Pb0_{.75}Bi_{0.25})O_3)$ were prepared. The original films were deficient in Pb and Bi but this tendency was reduced by a higher partial pressure of O_2.

V. Support

Because of the ease in adjusting the composition of perovskites over a wide range, these materials can be investigated in a systematic way and correlation made between their performance and structure. This unique feature has made perovskites an ideal system for both fundamental studies in solid state chemistry and physics as well as useful electronic and catalytic materials. On the other hand, catalytic materials require a combination of advantages to achieve maximum activity and stability, such as high surface area, redox reactivity, product selectivity, and a stable framework. Some perovskites, such as cobaltate and manganite, exhibited competitive activity for the oxidation of CO and reduction of NO compared with noble metal catalysts (56). However, due to their poisoning by SO_2, low surface areas, and weak mechanical strength, most perovskites have not been able to compete successfully in many catalytic applications (57).

In order to overcome these problems, developing a well dispersed pervoskite on support has been a attractive strategy to optimize the performance except for poisoning. In the preparations of a supported perovskite from solution, the method of drying can be extremely important. Gallagher et al. reported the preparation of supported $LaO_{.5}Pb_{0.5}MnO_3$ using different supports with various impregnation and drying techniques and then compared the catalytic performance of these materials (58). Various drying techniques were applied to maximize the homogeneity and stoichiometry on different supports and their conversion activity examined for the oxidation of CO (59). The immobility of the solution during the freeze-drying process can be invaluable to uniformly distribute the catalyst throughout a porous support, as shown clearly in Figs. 1 and 2 of reference 58.

At this point the discussion will be divided into sections based upon the choice of supports.

A. Cordierite

Mizuno et al. (60) reported the preparation of $La_{1-x}Sr_xCoO_3$ on a cordierite surface ($2Al_2O_3 \cdot 5SiO_2 \cdot 2MgO$) starting from metal citrates and nitrates. Attempts to use Al_2O_3 have not been successful because of the incorporation of Co atoms into the aluminate framework to form spinel. This precludes their participation in the catalytic reactions. Since the generation of the pervoskite depends upon the loading of lanthanum ions, it was proposed that lanthanum oxide was formed at the interface between the pervoskite and support and consequently reduced the diffusion of cobalt ions. Partial substitution of La^{3+} by Sr^{2+} ions results in a similar pervoskite structure and enhanced propane oxidation.

B. $La_2O_3 \cdot 19Al_2O_3$ (LA)

Zhang et al. (61) reported the preparation of $La_{1-x}Sr_xMnO_3$ or MnO_2 supported on LA by pre-treating LA with citrate precursors. XRD of 20 wt% samples supported on LA was performed from 873 to 1173 K and the pervoskite reflections were observed only for calcinations above 1073 K. Comparing this result with that for an unsupported pervoskite generated by citrate precursors indicated a difference of 473 K for the appearance of pervoskite reflections. This was attributed to the monolayer dispersion (20 wt%) of the pervoskite on the supports. When LA was treated with 10 % Mn_2O_3, a new phase was observed by XRD when calcined above 1273 K. This was proposed to be a new phase of La. These supported perovskites show better stability and activity in methane oxidation when compared with a Pt/Al_2O_3 catalyst.

C. ZrO_2

Fujii et al. (62) studied the structure and catalytic activity of $LaCoO_3$ and $La_{1-x}Sr_xCoO_3$ when supported on ten metal oxides. Acetate precursors were used and the supported pervoskite were calcined in air at 1123 K for 5 h. Formation of the pervoskite phases was confirmed by XRD. Catalytic activity for propane oxidation of these supported perovskites is shown in Fig. 3. Because of the outstanding activity on ZrO_2, additional pyridine adsorption, TEM and loading analysis were employed to study the structure developed on ZrO_2 surface. These studies indicated that both $LaCoO_3$ and $La_{1-x}Sr_xCoO_3$ had developed a well dispersed pervoskite on the ZrO_2 surface which were responsible

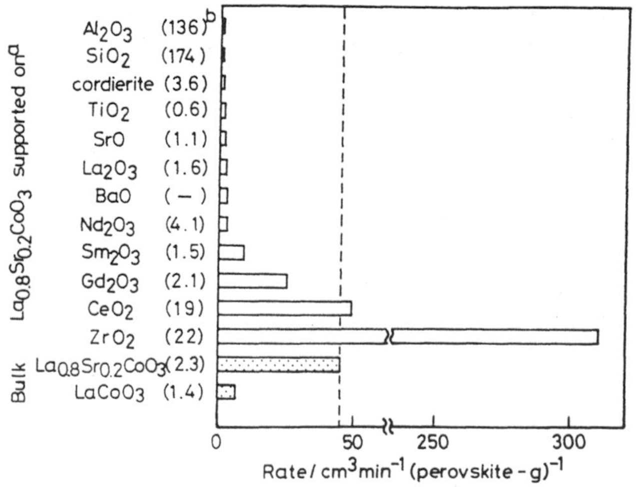

Fig. 3. Catalytic activities of perovskite supported on oxides (62). The amount of $La_{0.8}Sr_{0.2}CoO_3$ loaded was 5.8 wt%. Surface areas of the supports (m^2/g) are given in parentheses.

for the unusual activity. A similar observation was made using a ZrO_2 supported pervoskite electrode.

D. Perovskite as a Support

Perovskites have been tested extensively for CO oxidation and NO_x reduction since these have the most impact for the control of automobile emissions. However, one of the serious obstacles in developing these catalysts to replace the noble metal catalysts is the poisoning produced by the SO_2 reaction to form sulfide and/or sulfate ions which incapacitate the catalytic activity. Addition of Pt to the supported perovskite can help sustain the activity in the presence of sulfur poisoning significantly. From ESCA studies, it can be concluded that the Pt can exist in tetravalent and +4 state when added to the perovskites (63). However, for a similar amount of Pt, the $BaPtO_3$ shows much less activity for CO oxidation. This implies that pervoskite may be critical in developing a unique Pt structure. More studies have to be done to understand and optimize these interactions.

VI. Conclusion and Perspectives

The synthesis of new perovskite materials has been an ongoing interest and challenge of material scientists interested in catalysis and solid state chemistry or physics. Since the discovery of high T_c superconductors, however, there has been even greater interest in perovskites and related materials and brought about many new studies on processing these materials. In the past decade, molecular architecture and micro-structural control have been advanced by using molecular precursor and low temperature treatments. In addition, a combination of perovskites with other materials has proved to be an efficient approach to enhance the desired properties and develop new applications. Based on this new understanding and development, several promising directions emerge for further studies, such as oxygen sensors (64), NO decomposition (65), and emission control catalyst. Novel methods of chemical synthesis and processing will undoubtedly play an ever more important role in the future.

References

1. Voorhoeve, R.J.H., D.W. Johnson, Jr., J.P. Remeika and P.K.

Gallagher, Perovskite oxides: materials science in catalysis, *Science*, **195:** 827 (1977).
2. Gallagher, P.K., J.B. MacChesney and D.N.E. Buchanan, Mössbauer effect in the system $Sr_3Fe_2O_{6-7}$, *J. Chem. Phys.*, **45:** 2466 (1966).
3. Gallagher, P.K., H.M. O'Bryan, S.A. Sunshine and D.W. Murphy, Oxygen stoichiometry in $Ba_2YCu_3O_x$, *Mat. Res. Bull.*, **22:** 995 (1987).
4. Newnham, R.E., Structure-propertity relationship in perovskite electroceramics, in *Perovskite: a structure of great interest to geophysics and materials science*, A. Navrotsky and D.J. Weidner (Eds.), American Geophysical Union, Washington, D.C. (1989).
5. Gallagher, P.K. and D.A. Fleming, Influence of oxygen partial pressure on the synthesis of $Ba_2YCu_3O_7$ from a novel oxalate precursor, *Chem. Mat.*, **1:** 659 (1989).
6. Gallagher, P.K. and F. Schrey, The thermal decomposition of freeze-dried tantalum and mixed lithium-niobium oxalates, *Thermochim. Acta*, **1:** 465 (1970).
7. Clabaugh, W.S., E.M. Swiggard and R.J. Gilchrist, Preparation of barium titanyl oxalate tetrahydrate for conversion to barium titanate of highly purity, *J. Res. Natl. Bur. Std.*, **56:** 289 (1956).
8. Gallagher, P.K., F. Schrey and F.V. DiMarcello, Preparation of semiconducting titanates by chemical methods, *J. Am. Ceram. Soc.*, **46:** 359 (1963).
9. Northover, W.R., *J. Am. Ceram. Soc.*, **48:** 173 (1965).
10. Shrey, F., Effect of pH on the chemical preparation of barium-strontium titanate, *J. Am. Ceram. Soc.*, **48:** 359 (1965).
11. Gallagher, P.K. and D.W. Johnson, Kinetics of the formation of $BaSnO_3$ from barium carbonate and tin(IV) oxide or oxalate precursors, *Thermochim. Acta*, **4:** 283 (1972).
12. Gallagher, P.K., Thermal decomposition of barium and strontium trisoxalatoferrates(III), *Inor. Chem.*, **4:** 965 (1965).
13. Blum, J. B. and S.R. Gurkovich, Sol-gel-derived lead titanate, *J. Mater. Sci.*, **20:** 4479 (1985).
14. Ravindranathan, P., S. Komarneni, S.C. Choi, A.S. Bhalla and R. Roy, Sol-gel process for the preparation of fine electroceramic powders, *Ferroelectrics*, **87:** 133 (1988).
15. Hirano, S., T. Hayashi, K. Nosaki and K. Kato, Preparation of stoichiometric crystalline lithium niobate fibers by

sol-gel processing with metal alkoxides, *J. Am. Ceram. Soc.*, **72:** 707 (1989).

16. Christensen, A.N., Hydrothermal preparation of barium titanate by transport reactions, *Acta Chem. Scand.*, **24:** 2447 (1979).
17. Kutty, T.R.N., R. Vivenkanandan and P. Murugrraji, Precipitation of rutile and anatase (TiO_2) fine powders and their conversion to metal titanate ($MTiO_3$) (M = barium, strontium, calcium) by the hydrothermal method, *Mater. Chem. Phys.*, **19:** 533 (1978).
18. Mazdiyasni, K.S., Fine particle perovskite processing, *Am. Ceram. Soc. Bull.*, **63:** 591 (1984).
19. Nakamura, T., M. Misono and Y. Yoneda, Reduction-oxidation and catalytic properties of perovskite-type mixed oxide catalysts ($La_{1-x}Sr_xCoO_3$), *Chem. Lett.*, 1589 (1981).
20. Zhang, H.M., Y. Shimizu, Y. Teraoka, N. Miura and N. Yamazoe, Oxygen sorption and catalytic properties of lanthanum strontium cobalt iron oxide ($La_{1-x}Sr_xCo_{1-y}Fe_yO_3$) perovskite-type oxides, *J. Catal.*, **121(2):** 432 (1990).
21. Zhang, H.M., Y. Teraoka and N. Yamazoe, Preparation of perovskite-type oxides with large surface area by citrate process, *Chem. Lett.*, 665 (1987).
22. Gallagher, P.K., A simple method for the prepararion of (RE)FeO_3 and (RE)CoO_3, *Mat. Res. Bull.*, **3:** 225 (1967).
23. Gallagher, P.K. and B. Prescott, Further studies of the thermal decomposition of europium hexacyannoferrate (III) and ammonium europium hexacyanoferrate (II), *Inor. Chem.*, **9:** 2510 (1970).
24. Gallagher, P.K., D.W. Johnson, Jr., E.M. Vogel, G.K. Wertheim and F.J. Schnetter, Synthesis and structure of $BaPtO_3$, *J. Catal.*, **21:** 277 (1977).
25. Wenkus, J.F. and W.Z. Leavitt, Papers of the Conf. on Mag. and Mag. Mater., *AIEE Publication T-91*, 526 (1957).
26. Munir, Z.A., Synthesis of high-temperature materials by self-propegating combustion methods, *Am. Ceram. Soc. Bull.*, **67:** 342 (1988).
27. Kourtakis, K., M. Robbins and P.K. Gallagher, Synthesis of $Ba_2YCu_3O_7$ powder by anionic oxidation of (NO_3^{-1}) - reduction ($RCOO^{-1}$, where R is H, CH_3, and CH_3CH_2), *J. Solid State Chem.*, **83:** 230 (1989).
28. Kourtakis, K., M. Robbins and P.K. Gallagher, A novel synthetic method for the preparation of oxide superconduc-

tors: anionic oxidation-reduction, *J. Solid State Chem.*, **82:** 390 (1989).
29. Kourtakis, K., M. Robbins, P.K. Gallagher and T. Tiefel, Synthesis of $Ba_2YCu_4O_8$ by anionic oxidation-reduction, *J. Mater. Res.*, **4:** 1289 (1989).
30. Varma, H., K.G. Warrier and A.D. Damodaran, Metal nitrate-urea decomposition route for Y-Ba-Cu-O powder, *J. Am. Ceram. Soc.*, **73:** 3103 (1990).
31. Pankajavalli, R., J. Janaki, O.M. Sreedharan, J.B. Gnanamoorthy, G.V.N. Rao, V. Sankarasastry, M. Janawadkar, Y. Hariharan, T.S. Radhakrishnan, Synthesis of high quality 1-2-3 compound through citrate combustion, *Physica C*, **156(5):** 737 (1988).
32. Schnettler, F.J., F.R. Monforte and W.W. Rhodes, A cryochemical method of prparing ceramic materials, *Science in Ceramics*, Vol. 4, G.H. Stewart (Ed.), H. Blacklock Co., Manchester, 1968.
33. Johnson, D.W., P.K. Gallagher, D.J. Nitti and F. Schrey, Effect of preparation technique and calcination temperature on the densification of lithium ferrites, *Am. Ceram. Soc. Bull.*, **53:** 163 (1974).
34. Gallagher, P.K., D.W. Johnson, E.M. Vogel and F. Schrey, Microstructure of some freeze dried ferrites, in *Ceramic Microstructures'76*, R.M. Fulrath and J.A. Pask (Eds.), Westview Press, Bouler, CO, 1976, p. 423.
35. Song, K.H., H.K. Liu, S.X. Dou and C.C. Sorrell, Rapid formation of the 110 K phase in bismuth lead strontium calcium copper oxide through freeze-drying powder processing, *J. Am. Ceram. Soc.*, **73(6):** 1771 (1990).
36. Hibbert, B.D., J. Lovegrove and A.C.C. Tseung, A critical examination of a cryochemical method for the preparation of high-surface-area semiconducting powders. Part 3. Factors which determine surface area, *J. Mater. Sci.*, **22(10):** 3755 (1977).
37. Imai, H., K. Takami and M. Naito, Preparation of $CoLaO_3$ catalyst fine particles by mist decomposition method II effect of additives for the increase of surface area, *Mat. Res. Bull.*, **19:** 1293 (1984).
38. Cuomo, J.J., C. Guarnieri, M. Richard, S.A. Shivashankar, R.A. Roy, D.S. Yee and R. Rosenberg, Large area plasma spray deposited superconducting $YBa_2Cu_3O_7$ thick films, *Adv. Ceram. Mater.*, **2(3B):** 422 (1989).

39. Phule, P.P. and S.H. Risbud, Low-temperature synthesis and processing of electronic materials in the BaO-TiO_2 system, (1990).
40. Karpinski, J., E. Kaldis, E. Jilek, S. Rusiecki and B. Bucher, Bulk synthesis of the 81-K superconductor yttrium barium copper oxide ($YBa_2Cu_4O_8$) at high oxygen pressure, *Nature*, **336:** 660 (1988).
41. Cava, R.J., J.J. Krajewski, W.F. Peck, B. Batlogg, L.W. Rupp, R.M. Fleming, A.C.W.P. James and P. Marsh, Synthesis of bulk superconducting $Ba_2YCu_4O_8$ at one atmosphere oxygen pressure, *Nature*, **338:** 328 (1989).
42. Pooke, D.M., R.G. Buckley, R. Presland and J.L. Tallon, Bulk superconducting $Y_2Ba_4Cu_7O_{15-}$ and $YBa_2Cu_4O_8$ prepared in oxygen 1 atm, *Phys. Rev. B*, **41:** 6616 (1990).
43. Takahara, H., Preparation and dielectric properties of mixed-sintering ceramics, *Jpn. J. Appl. Phys. Part 1*, **24:** 427 (1985).
44. Martinez, E., E. Sanchez Viches, A. Beltran-Porter and D. Beltran-Porter, Low temperature synthesis of Ba-Fe mixed oxides having perovskite structures, *Mat. Res. Bull.*, **21:** 511 (1985).
45. Brody, H.D., J.S. Haggerty, M.J. Cima, M.C. Flemings, R.L. Barns, E.M. Gyorgy, D.W. Johnson, W.W. Rhodes, W.A. Sunder and R.A. Laudise, Highly textured and single crystal bismuth calcium strontium copper oxide ($Bi_2CaSr_2Cu_2O_X$) prepared by laser-heated float zone crystallization, *J. Crystal. Growth.*, **96:** 225 (1989).
46. Vogel, E.M., D.W. Johnson, Jr. and P.K. Gallagher, Oxygen stoichiometry in $LaMn_{1-X}Cu_XO_{3+Y}$ by thermogravimetry, *J. Am. Cer. Soc.*, **60:** 31 (1977).
47. Lindemer, T.B., J.F. Hunley, J.E. Gates, A.L. Sutton, Jr., J. Byrnestad, C.R. Hubbard and P.K. Gallagher, Experimental and thermodynamic study of nonstoichiometry in $YBa_2Cu_3O_{7-X}$, *J. Am. Ceram. Soc.*, **72:** 1775 (1989).
48. Specht, E.D., C.J. Sparks, A.G. Dhere, J. Brynestad, O.B. Cavin, D.M. Kroeger and H.A. Oye, Effect of oxygen pressure on the orthorhombic-tetragonal transition in the high-Temperature superconductor $YBa_2Cu_3O_X$, *Phys. Rev. B, Condens. Matter.*, **37:** 7426 (1988).
49. Dijkkamp, D., T. Venkatesan, X.D. Wu, S.A. Shaheen, N. Jisrawi, Y.H. Min-Lee, W.L. McLean and M. Croft, Preparation of Y-Ba-Cu-O oxide superconductor thin films

using pulsed laser evaporation from high T_c bulk material, *Appl. Phys. Lett.*, **51:** 24 (1987).
50. Webb, C., S.L. Weng, J.N. Eckstein, N. Missert, K. Char, D.G. Schlom, E.S. Hellman, M.R. Beasley, A. Kapitulink and J.S. Harris, Growth of high T_c Superconducting thin films using molecular beam epitaxy techniques, *Appl. Phys. Lett.*, **51:** 12 (1988).
51. Croteau, A., S. Matsubara, Y. Miyasaka and N. Shohata, Ferroelectric lead zirconate titanate ($Pb(Zr,Ti)O_3$) thin films prepared by metal target sputtering, *Jpn. J. Appl. Phys., Part 1*, **26:** 18 (1990).
52. Han, Z., L. Bourget, H. Li, M. Ulla W.S. Millman, H.P. Baun, M.F. Xu, B.K. Sarma, M. Levy and B.P. Tonner, Single-target deposition, post-processing and electron spectroscopy of perovskite superconductor thin films, *AIP Conf. Proc.*, **165:** 66 (1988).
53. Naito, M., R.H. Hammond, B. Oh, M.R. Hahn, J.W.P. Hsu, P. Rosenthal, A.F. Marshall, M.R. Beasley, T.H. Geballe and A. Kapitulnik, Thin-film synthesis of the high-Tc oxide superconductor yttrium barium copper oxide ($YBa_2Cu_3O_7$) by electron-beam codeposition, *J. Mater. Res.*, **2:** 713 (1987).
54. Mankiewich, P.M., J.H. Scofield, W.J. Skocpol, R.E. Howard, A.H. Dayem and E. Good, Reproducible technique for fabrication of thin films of high transition temperature superconductors, *Appl. Phys. Lett.*, **51(21):** 1753 (1987).
55. Suzuki, M., Y. Enomoto, T. Murakami and T. Inamura, Thin film preparation of superconducting perovskite-type oxides by rf sputtering, *Jpn. J. Appl. Phys.*, **20:** 13 (1981).
56. Voorhoeve, R.J.H., J.P. Remeika, L.E. Trimble, A.S. Cooper, F.J. Disalvo and P.K. Gallagher, Perovskite-like $La_{1-x}K_xMnO_3$ and related compounds: solid state chemistry and the catalysis of the reduction of NO by CO and H_2, *J. Catal.*, **14:** 395 (1990).
57. Gallagher, P.K., D.W. Johnson, J.P. Remeika, F. Schrey, L.E. Trimble, E.M. Vogel and R.J.H. Voorhoeve, The activity of $La_{0.7}Sr_{0.3}MnO_3$ without Pt and $La_{0.7}Pb_{0.3}MnO_3$ with varying Pt contents for the catalytic oxidation of CO, *Mater. Res. Bull.*, **10:** 529 (1975).
58. Johnson, D.W., P.K. Gallagher, F.J. Schnettler and E.M. Vogel, Novel preparative technique for supported oxide catalysts, *Am. Ceram. Soc. Bull.*, **56:** 786 (1977).
59. Gallagher, P.K., J.W. Johnson, Jr. and E.M. Vogel,

Preparation, structure and selected catalytic properties of the system $LaMn_{1-x}Cu_xO_3$, *J. Am. Ceram. Soc.*, 60: 28 (1977).
60. Mizuno, N., H. Fujii and M. Misono, M., Preparation of perovskite-type mixed oxide ($La_{1-x}Sr_xCoO_3$) supported on cordierite. An efficient combustion catalyst, *Chem. Lett.*, 1333 (1986).
61. Zhang, H. M., Y. Teraoka and Y. Yamazoe, Preparation of supported $La_{1-x}Sr_xMnO_3$ catalysts by the citrate process, *Appl. Catal.*, **41:** 137 (1988).
62. Fujii, H., N. Mizuno and M. Misono, Pronounced catalytic activity of $La_{1-x}Sr_xCoO_3$ highly dispersed on ZrO_2 for complete oxidation of propane, *Chem. Lett.*, 2147 (1987).
63. Johnson, D.W. Jr., P.K. Gallagher, G.K., Wertheim and E.M. Vogel, The nature and effects of plantium in perovskite catalysts, *J. Catal.*, **48:** 87 (1977).
64. Shimizu, Y., Y. Fukuyama, H. Arai and T. Seiyama, Oxygen sensor using perovskite-type oxides. Measurements of electrical characteristics, *ACS Symp. Ser.*, **309:** 83 (1986).
65. Shimada, H., S. Miyama and H. Kuroda, Decomposition of nitric oxide over Y-Ba-Cu-O mixed oxide catalysts, *Chem. Lett.*, 1797 (1988).

Chapter 2

High T_C Superconductive Cuprates: Processing and Properties

C. Moure and P. Duran

Instituto de Cerámica y Vidrio, CSIC, Electroceramics Department, 28500 Arganda del Rey, Madrid, Spain

I. Introduction

Since the discovery of the superconducting properties of some cuprate compounds with perovskite or related structures (1-4) and unexpected high transition temperatures (T_c), an exceptionally large number of papers have been devoted to the study of these new materials. Most of them focus on the synthesis and ceramic processing of the different compounds and attempt to establish the relations between the ceramic parameters (density, porosity, microstructure, nature of grain boundaries) and the superconducting properties.

Although in the beginning it was said that the cuprate compounds could be very easily processed, actually their preparation shows serious difficulties due to their complex crystal chemistry. Because of this, numerous discrepances about the most relevant features and numerical parameters exist between the different authors. The lack of a perfect knowledge of the appropriate ceramic processing leads to the lowering of the performance of the actual superconducting materials in comparison with the parameters calculated from theoretical models.

In the present chapter, we attempt to expose briefly the state-of-the-art of the synthesis and ceramic processing of one of the most important superconducting ceramic materials, the $Y_1Ba_2Cu_3O_x$ compound and their more relevant features.

II. Synthesis

A. Solid State Synthesis

The $Y_1Ba_2Cu_3O_x$ is a point compound of the ternary oxide system (Fig.1). It decomposes peritectically at a temperature of ≈ 1270 K to give Y_2BaCuO_5 (the "green phase" isolating), and a liquid. This temperature is only slightly above the processing temperatures (1273-1248 K). In the ternary system also exist other low melting binary eutectics. Moreover, other binary and ternary compounds are formed in the system: Ba_2CuO_3, $BaCuO_2$, $YBa_3Cu_2O_y$, $Y_2Cu_2O_5$, BaY_2O_4, and others (5-8). Small deviations of stoichiometry, lead to the formation of second phases and/or liquid phases which can strongly alter the microstructure and final properties of the materials.

The most common procedure for superconducting powder synthesis via solid state reactions is the calcining of a homogeneous mixture of the corresponding oxides and carbonates or nitrates (9-12). The yttrium and copper oxides are easily disposable materials. The more reliable source of barium is the barium carbonate. It decomposes to give BaO at high temperatures (≈ 1720 K), far above the liquidus temperature of the ternary system. Although the presence of other compounds acts as a catalyst, it is necessary, in general, to lower the decomposition temperature to attain temperatures above 1173 K for complete reaction of components and elimination of CO_2.

The decomposition rate of $BaCO_3$ is, therefore, the controlling process for the $Y_1Ba_2Cu_3O_x$ synthesis. The sluggish decomposition of the carbonate allows the formation of other phases, such as $Y_2Cu_2O_5$, $BaCuO_2$, and Y_2BaCuO_5, which are stable phases in the ternary system. Some of these phases are compatible with the YBaCuO compound. Therefore the reactions between them to give the 123 phase is often slower than the direct reaction between the three oxides. As a consequence, that reaction is not complete and small amounts of the other phases remain along with the $Y_1Ba_2Cu_3O_x$ phase. Several steps of regrinding and recalcining are necessary for a more complete reaction.

The use of other crystalline compounds as a source of Ba has shown some advantages: BaO of $Ba(OH)_2.nH_2O$ allowed lowering the synthesis temperature, and to complete the reaction more quickly than the carbonate. Nitrate and peroxide were not so advantageous (13). Nevertheless, handling of these compounds is more harder because of their stoichiometry deviations, which can lead to formulating incorrectly the compositions. Moreover, a strict control of the synthesis atmosphere is necessary for avoiding exposure to CO_2 and moisture which is normally in the air. In fact, the unreacted BaO, present during the initial step of the synthesis at relatively low temperatures, reacts very quickly with CO_2 and moisture to form $BaCO_3$, leading to the several problems posed when using $BaCO_3$ as starting material.

On the other hand, use of crystalline materials leads to powders with micronic sizes, and wide, non-homogeneous, size distributions. It makes it very difficult to control the grain size of sintered bodies and their homogeneity, and favors the existence of localized non-stoichiometries. As a consequence, a liquid phase could appear promoting exaggerated grain growth and, therefore, large microstructural anisotropies.

B. Wet Chemical Synthesis

To avoid the problems pointed out above, many approaches to synthesis have been attempted. The most widely studied have been the chemical processing methods, which make up several procedures already applied in the synthesis of other high technology ceramics. Table 1 shows some of those chemical procedures, along with solid state procedures and other less common processes.

The oxalate coprecipitation route is a relatively simple method for the synthesis of YBaCuO powders (14-16). Starting from salt solutions, addition of oxalic acid or ammonium oxalate in basic medium leads to precipitation of a complex cationic oxalate or a mixture of several oxalates. Figure 2 shows a flow sheet of the process. The solvent must be organic rather than water, because the solubility of oxalates is much lower in that medium, thus assuring a correct stoichiometry. The oxalate route allows obtaining overall stoichiometric precipitates, but fails in obtaining individual precipitated particle stoichiometry (14). An additional problem is that the formed barium oxalate transforms to barium carbonate at temperatures at which the formation reaction progresses slowly or does not even begin. Although the reactivity of the emerging barium carbonate is very high and the synthesis proceeds rapidly, it is necessary to

Table 1. Processing Methods for Preparing High-T_c Superconducting Ceramics

Main reaction feature	Reaction types	Applications
Solid State Reaction	Calcining of oxides carbonates and or nitrates	Powders
	Intermediate phase reactions	Powders
Wet Chemical Reactions	Oxalate coprecipitation	Powders
	Carbonate coprecip.	Powders
	Citrate synthesis	Powders, thin and thick films
	Sol-gel synthesis	Thin and thick films
	Alkoxidehydrolysis	Powders, films, fibers
	Aerosol decomposition	Powders, films
	Freeze drying	Powders
Other Methods	Oxidation of metallic precursors	Thin and thick films, fibers
	Glass ceramic approach	Films,bulk shapes
	Molten oxide reactions	Bulk textured shapes,fibers

attain relatively high temperatures for accomplishing the reaction (15).

The citrate process is a synthesis route which can lead to the YBaCuO formation without the presence of $BaCO_3$ for any of the processing steps (17-20). It is a particular type of sol-gel method, and offers the advantage of relative simplicity and the feasibility of the chemical compounds used in the development of the process. Figure 3 shows a typical flow sheet of the citrate process. The method assures a great local and overall stoichiometry. Nevertheless, it is necessary to raise the synthesis temperature to values above certain critical ones to complete the reaction avoiding the carbonate formation. Another common problem associated with this method is the precise control of the self-spontaneous combustion of citrates (19). Other sol-gel methods which have also been developed, although in some lower extension, are: hydroxide precipitation, gelation of ethylenglycol solutions, liquid mixing techniques (21-23), etc.

There is a versatile class of methods for fabricating complex metal oxide powders beginning with molecular precursors: the aerosol processes are methods in which particles are formed in the gas phase and are deposited onto surfaces to form thin and thick films, or to be recollected as powder particles (24-27). The aerosol process allows the preparation of powders with a combination of properties that in many cases cannot be obtained by other process, such as: high purity in single or polycrystalline systems, preparation of multicomponent materials, control of porosity of the individual particles, sizing from few nanometers to > 5 μm, unagglomerated, soft or hard agglomerated powders, and others.

Several aerosol methods are described in the literature, but only one has demonstrated ability to produce chemically homogeneous complex superconducting oxides: aerosol decomposition. This procedure consists in passing a precursor solution such as nitrate solution through an aerosol to form fine, size-variable droplets, and then to heat them for chemical synthesis reaction. The resulting synthesized powders are collected on a filter. The method leads to obtaining particles with controlled size and morphology. Particularly, the aerosol decomposition of a solution of the barium, copper, and yttrium nitrates has allowed obtaining very pure, fully reacted superconducting powders at temperatures below 1170 K (25). Alternative methods based on very similar principles are the spray-drying and the freeze-drying methods, in which the droplet generation and the decomposition-reaction steps are completely separated. By means of these methods it is possible to prepare prereacted particles which can be stored for posterior heat-treatment and synthesis of the superconductor powders.

Furthermore, in the literature are cited a series of chemical methods, such as carbonate precipitation alkoxide hydrolysis, spray pyrolysis, etc. (26-28). They are variations of coprecipitation, sol-gel, or aerosol methods. Only the alkoxide hydrolysis seems to have a promising future, although the high cost of the metal alkoxides can be a severe handicap for commercial development of the method. The carbonate precipitation has the advantage of simplicity and good precipitate stoichiometry, but the decomposition and synthesis steps are very similar to the corresponding solid state synthesis.

By means of these wet chemical methods, it is possible to obtain monophasic, pure, fully reacted powders with submicronic

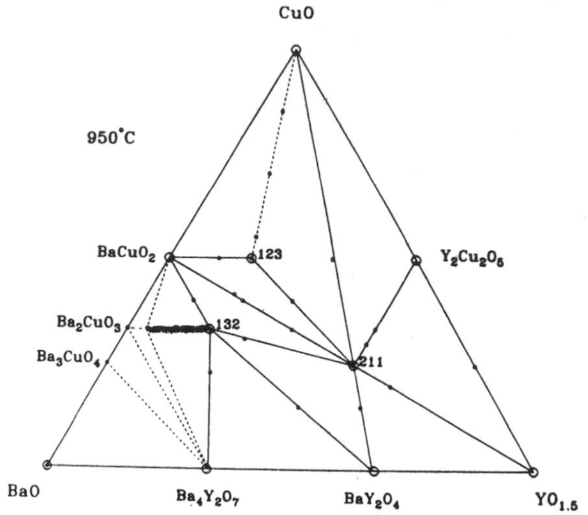

Fig. 1. Compatibility relations in the ternary phase diagram of the Y_2O_3-BaO-CuO system (From ref. 5).

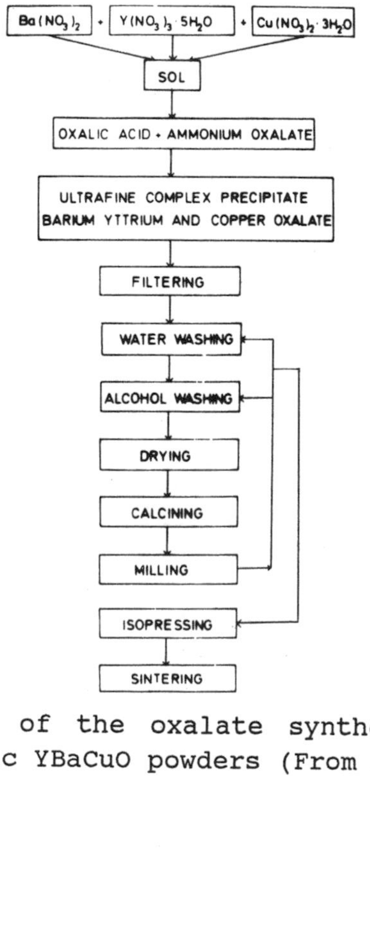

Fig. 2. Flow sheet of the oxalate synthesis processing for obtaining submicronic YBaCuO powders (From ref. 15).

and even nanometric sizes, homogeneous and narrow size distribution, and very reactive characteristics. All of them need a careful control of the reaction steps to prevent barium carbonate formation from environmental CO_2, particularly when the YBaCuO formation begins and it is not complete, and free BaO could exist in the powder. The presence of barium carbonate traces in synthesized YBaCuO powders adversely affects the ceramic and superconducting properties. When the carbonate traces have not been fully eliminated during synthesis, they must be eliminated in the sintering step. If the sintering rates are high enough, the CO_2 could be trapped in the solid, creating an additional porosity or remaining as a second phase in the grain boundaries. As we will see later, the carbon or the CO_2 affects very negatively the electrical properties of the superconducting YBaCuO.

C. Other Synthesis Methods

Other less common methods exist for synthesizing YBaCuO, avoiding the presence or formation of barium carbonate in the synthesis steps. One of them is the controlled oxidation of metallic precursors: a metallic alloy of the 1:2:3 ratio is melted and slowly cooled. This alloy is carefully oxidized, in absence of CO_2, to give a fully oxygenated $Y_1Ba_2Cu_3O_x$ compound (28).

Another explored possibility is to prepare compounds of barium yttrium and/or barium copper as starting materials for synthesizing YBaCuO. Synthesis temperatures of 1170 K or below and one-step firing procedure have allowed the obtaining of YBaCuO powders with small or no presence of secondary phases (Fig. 4). The main problem posed by these methods lies with the instability of some yttrium barium compounds, and with the difficulty of maintaining the stoichiometry of the barium copper oxides. The advantage of this procedure is to have the barium oxide always combined, avoiding the possible carbonation (29).

III. Influence of the Oxygen Content on the Superconducting Properties

The $Y_1Ba_2Cu_3O_x$ compound exhibits superconductivity at 91 K when x is close to the value of 7. The oxygen stoichiometry

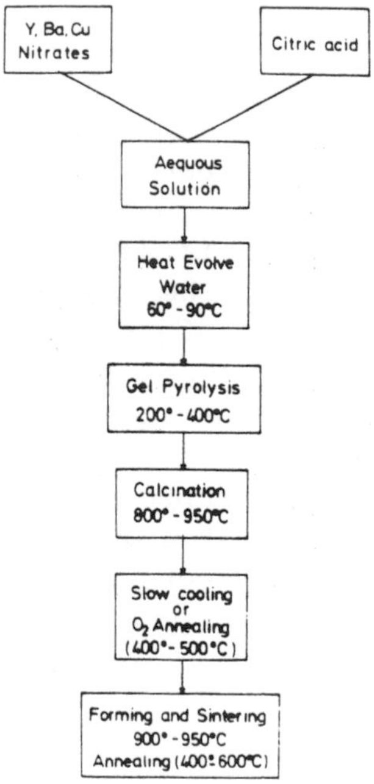

Fig. 3. Flow sheet of the citrate synthesis processing of $Y_1Ba_2Cu_3O_x$ powders.

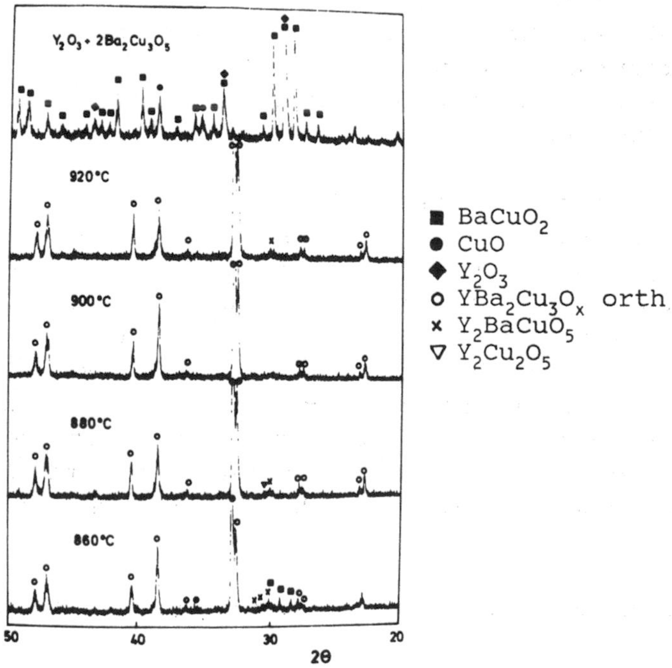

Fig. 4. XRD patterns of YBaCuO powders synthesized at several temperatures from a mixture of Y_2O_3 and previously reacted $Ba_2Cu_3O_5$. The common time was 12 h (From ref. 29).

strongly depends on the temperature and oxygen partial pressure. On the other hand, the superconductivity behavior is directly associated with the stoichiometry. Three phases are found with the 1:2:3 atom ratios, all of them stable at room temperature. Figure 5 shows a diagram of T_c vs the oxygen content (30). Table 2 summarizes some of the most representative values of lattice parameters against oxygen content.

Starting from the orthorhombic phase, the O_2 evolves when the temperature rises. At 983 K, x < 6.4 and a normal superconducting transition takes place for an oxygen partial pressure of 1 atm (10^5 Pa)(31). This transition temperature lowers as oxygen pressure lowers, i.e. for air temperature close to 863 K. Phase II is formed when the 123 compound is quenched from high temperatures in pure O_2 atmospheres (32).

The oxygen release is a reversible process: the tetragonal phase picks up O_2 during cooling and transforms to the orthorhombic one. The kinetics of the reoxidation process have been widely studied (33-37). The powder oxidation is a simple process with an adequate cooling cycle (38). On the other hand, the sintering and densification of the YBaCuO material is carried out at temperatures higher than 1170 K, for which the stable phase is the tetragonal one even in very rich O_2 environments. Because of the dense nature of sintered bodies the oxygen pickup is a more difficult process (39). It progresses through an O_2 diffusion mechanism. It has been proved that the oxygen diffusion is quicker along the grain boundaries than through the crystal lattice (40) in ceramic materials.

Moreover, the activation energy for oxygen diffusion depends

Table 2. Lattice Parameters of the $Y_1Ba_2Cu_3O_x$ Compound Against the Oxygen Content

x Value	Symmetry	a_o	b_o	c_o
6.98	Orthorhombic I	3.8170	3.8830	11.6350 (33)
		3.8200	3.8822	11.0814 (36)
6.92	Orthorhombic I	3.8216	3.8860	11.6865 (36)
6.80	Orthorhombic I	3.8236	3.8828	11.6937 (36)
6.64	Orthorhombic I	3.8270	3.8900	11.7000 (80)
6.51	Orthorhombic II	3.8377	3.8772	11.7363 (36)
6.39	Tetragonal	3.8595	3.8595	11.7939 (36)
6.00	Tetragonal	3.8630	3.8630	11.8300 (43)

on the oxygen stoichiometry in such a way that when the x value rises, the diffusion rates lower (41). As a consequence, fully densified bodies are very difficult to completely oxidize, whereas the less densified materials can be adequately oxidized by means of single cooling cycles in O_2 or in air (42).

Owing to the predominant grain boundary diffusion mechanism, the nature of these boundaries plays a very important role for attaining a complete phase transition and, therefore, good superconducting properties. The location of secondary phases and/or trapped CO_2 are detrimental for the correct oxygen pickup (44). On the other hand, a uniformly distributed porosity along the boundaries, and not located exclusively in the triple points, is beneficial for easy oxidation. Nevertheless, this porosity could be a suitable path for corrosive agents, such as water or CO_2, which degrade the materials. The low lattice diffusion coefficient value leads to a minimization of the grain size for rapid and complete oxidation (45).

IV. Electrical and Magnetic Properties

The superconducting oxide compounds are type II superconducting materials, with a magnetic field penetration depth larger than the coherence length (46). For the $Y_1Ba_2Cu_3O_x$ compound the $Hc_2(0)$ is very high (> 200 T), and it varies with a 2T/K rate, remaining relatively high at 77 K, [40T] (47). The crystalline anisotropy leads to the existence of strong magnetic anisotropy. For single crystals the ratio $Hc_2 /Hc_2^{\perp} >7$, with Hc_2 the critical field parallel to the ab plane and $Hc_2^{\perp}$ the perpendicular one to the same plane (48).

The critical current density (J_c) is also very high. Theoretical considerations have led to a value of 3×10^8 A/cm^2 at O K and zero magnetic field (49). The variation with temperature is not very large. At 77 K, $J_c \approx 10^7$ A/cm^2. In a manner similar to the magnetic field anisotropy, critical current anisotropy has been noticed for the YBaCuO compound (50). The magnetic field dependence of J_c in single crystals is not very pronounced because of the large Hc_2 value (51).

Thin films prepared by several methods have shown J_c values close enough to the theoretical one. Particularly, epitaxial-growth thin films with preferred oriented ab planes showed J_c values above 10^6 A/cm^2 when measured in those directions at zero magnetic field, and temperatures near the liquid N_2 temperature

Table 3. Critical Current Values (J_c) of $Y_1Ba_2Cu_3O_x$ for Several Shapes and Processes

Shape	Processing	J_c values (A/cm^2) For zero magnetic field (At 77 K except as indicated)
Thin films	Sputtering	3.5×10^4 (81)
	Epitaxial growth	2.6×10^6 (at 4.2 K) (52)
Thick films	Screen printing	7.5×10^2 (55)
Bulk Materials	Pressureless sintering	4.7×10^2 (58)
	Ag_2O doping	2.25×10^2 (77)
	Melt-textured growth	$3.1\text{-}7.4 \times 10^3$ (68)
	Liquid phase processing	$1{,}85 \times 10^4$ (73)
	Magnetic aligned crystals	1.0×10^7(at 4.3K) (75)

(52-55). On the contrary, J_c falls spectacularly when it is measured on ceramic polycrystalline samples. J_c values well below 10^3 A/cm^2 are normally measured on high-quality ceramic superconducting samples. Moreover, J_c depends markedly on the magnetic field and deteriorates exponentially for weak fields (56-59). The explanation of such a behavior lies with the nature of ceramic materials: they consist in superconducting grains coupled by weak or non-superconducting grain boundaries (Josephson junctions). These regions are known as weak-links, and are very sensitive to magnetic fields (60). The total current transport is controlled by the weak-links and not by the grain superconductivity. Table 3 shows some of the J_c values determined on different geometries and several preparation procedures.

Several possibilities exist for the appearance of weak-links in grain boundaries: (i) the formation of microcracking from residual thermal stresses and volumetric changes accounted by tetragonal-orthorhombic transition; (ii) the formation of insulating impurity layers, as green phase layers, carbon or carbonate, arising from unreacted BaO or $BaCO_3$, or CO_2-post firing contamination; (iii) the electrical mismatching between grains because of the conductivity anisotropy, and (iv) the lack of correct oxygen stoichiometry.

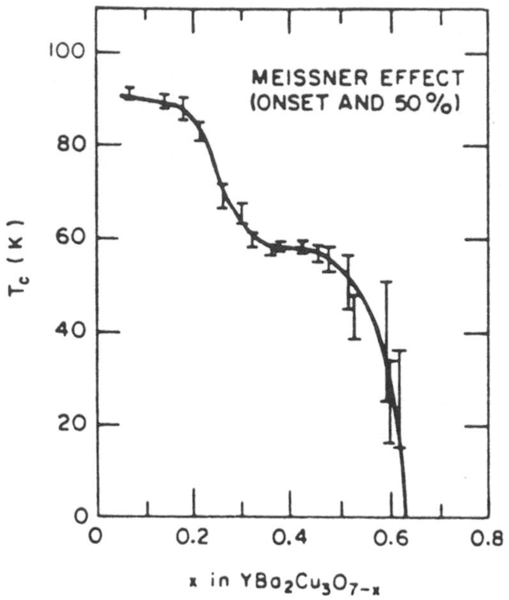

Fig. 5. Transition temperature vs. oxygen content (From ref. 30).

Fig. 6. ROM micrograph obtained from polished surface showing the crack formation through a superconducting grain.

The thermal residual stresses arise from the anisotropic thermal expansion, and it is proved that the c axis contracts more than the a_0 and b_0 axis (31,61) on cooling. These stresses can induce fracture along the grain boundaries or by cleaving the grains (Fig. 6). Nevertheless the anisotropic changes caused by the tetragonal-orthorhombic transition during the oxidation process is more pronounced than the thermal anisotropy. Through the transition there is a volume change > 1 % along with a differential contraction of the c_0 axis against the a_0 and b_0 ones (moreover, the b_0 axis suffers an increasing value whereas a_0 decreases). Furthermore, as indicated above, the lattice oxygen diffusion is very slow. The grain surface forms quickly but the inner grain changes more slowly or remain unchanged. This is another residual stress source. For these forms of cracking, the crack lengths can be greater than several grain lengths whereas the former effect is a typically a one-grain length (40).

These problems can be avoided by means of suitable thermal processes: slow cooling and the presence of interconnected porosity diminish the effects of the oxidation gradients (62). On the other hand, the anisotropic thermal expansion effect can be minimized by decreasing the grain size (45). The grain growth control is achieved by using submicronic powder obtained from wet chemical processes, or by the presence of second phases (as the "green phase") which avoid the liquid phase formation and retard the grain boundaries migration (38,63) (Fig.7). However, it is possible that the presence of these second phases negatively influences the critical current values as a consequence of their insulating nature, and could act as barriers for oxygen diffusion, leading to stoichiometric-deficient materials with poor superconducting properties.

A lot of works have shown that the presence of carbon or CO_2 in the grain boundaries adversely affects the critical current values by trapping mobile holes and lowering their concentration (40). The CO_2 can be a residual product from $BaCO_3$ starting material. It is necessary to control the total decarbonatation during the YBaCuO synthesis or to employ other barium sources, as BaO, BaO_2, etc. However, if adequate care is not taken during the process, it is possible to have a recarbonation of free barium in the intermediate stages of the synthesis and sintering (64). But the most important contamination arises when the sintered bodies which, having an adequate pore structure for facilitating oxygen pickup through cooling, are exposed to environment media. Especially in the presence of moisture, the

Fig. 7. ROM micrograph of polished surface of small grained $Y_1Ba_2Cu_3O_x$ compound sintered with minor amounts of Y_2BaCuO_5 phase (From ref. 38).

ambient CO_2 can be fixed onto the surfaces of YBaCuO and it can progress towards the interior via porosity structure (65). Careful control must be taken with respect to the absence of CO_2 in the gas flow used in the annealing step of ceramic YBaCuO materials (44).

Hitherto, the exposed causes of the limiting values of Jc are extrinsic to the material, and could be resolved by adequate processing, leading to fully oxidated, crack-free and carbon-free ceramics. However, such materials so prepared even show relatively low values of Jc. On the other hand, oriented, epitaxially growth thin films, as mentioned before (Table 3), show high values of J_c, (> 10^6 A/cm^2). The strong electrical current anisotropy of crystals along with the above considerations led to the conclusion that one of the most important causes of these low J_c values in polycrystalline materials is the relative misorientation between individual crystals separated by grain boundaries (66). Experiments carried out on bicrystals showed that the J_c across the grain boundary falls quickly when the relative angle between basal plane orientations increases above 15° (67). These considerations led to the preparation of highly textured microstructures in polycrystalline materials. Several methods have been developed for obtaining such a microstructure: melt-textured growth from supercooled melts (68, 69), hot pressing (70), hot-forging of sintered materials (71), liquid phase processing (72,73), magnetic field orientation, (74,75), etc. Most of these procedures lead to materials with high J_c values (10^3 -10^4 A/cm^2) in preferred directions along with large conductivity anisotropy. Textured samples have shown small J_c dependence with respect to the magnetic field. The procedures based on the cooling of total or partially molten batches allow a more complete decarbonation and full Ba combination. Moreover the microstructure is dense, the grain boundaries are cleaner, and the grain boundary area lowers significantly. On the other hand, these processes have as a principal inconvenience the great size of grains, which makes oxidation difficult during cooling. Combination of solid state sintering or hot-pressing with magnetic field alignment allows a better grain size control, and an easier stoichiometry preservation (76).

The incorporation of metallic Ag or Ag_2O as dopant of the YBaCuO also has been shown to enhance the critical current values. The improvement lies with the increase of texture that the Ag_2O promotes (77,78) and with the increment in the oxidation kinetics caused by this oxide when it precipitates in the grain boundaries (79).

References

1. Bednorz, J.G. and K.A. Muller, Possible high-T_c superconductivity in the La-Ba-Cu-O system, *Z. Phys. B*, **6**: 189 (1986).
2. Wu, W.K., J.R. Ashburn, C.J. Torng, P.H. Hor, R.L. Meng, L. Gao, Z.J. Huang, Y.Q. Wang and C.W. Chu, Superconductivity at 93 K in a new mixed-phase Y-Ba-Cu-O compound system at ambient pressure, *Phys. Rev. Lett.*, **58**: 908 (1987).
3. Maeda, M., Y. Tanaka, M. Fukutom and A. Asano, A new high T_c oxide superconductor without a rare earth element, *Jpn. J. Appl. Phys.*, **27**: L 209 (1988).
4. Sheng, Z.Z. and A.M. Hermann, A.M., Bulk superconductivity of 120 K in the Tl-Ca/Ba-Cu-O system, *Nature*, **332**: 138 (1988).
5. Frase, K.G. and D.R. Clarke, Phase compatibilities in the system Y_2O_3-BaO-CuO, *Adv. Ceram. Mater. 2 Special Issue*: 295 (1987).
6. Aselage, T. and K. Keeler, Liquidus relations in Y-Ba-Cu-O

oxides, *J. Mater. Res.*, **3**: 1279 (1988).

7. Roth, R.S., C.J. Rawn, F. Beech, J.D. Whitler and J.O. Anderson, Phase equilibria in the system Ba-Y-Cu-O-CO_2 in air, *Ceramic Superconductors II, M.F.Yan (Ed.)*, Am. Ceram. Soc. Inc., Westerville, OH, 1988, p. 13.
8. De Leeuw, D.M., C.A.H.A. Mutsaers, C. Langereis, H.C.A. Smoorenburg and P.J. Rommers, Compounds and phase compatibilibilities in the system Y_2O_3-BaO-CuO at 1223 K, *Physica C*, **152**: 39 (1988).
9. Takao, I., M. Uzawa and A. Uchikawa, Formation of superconducting ceramic $YBa_2Cu_3O_x$ from a mixture of Y_2O_3, $BaCO_3$, and CuO, *J. Mater. Sci. Lett.*, **7**: 130 (1988).
10. Lay, K.W., Formation of yttrium barium cuprate powder at low temperatures, *J. Am. Ceram. Soc.*, **72**: 696 (1988).
11. Alford, N. NcN., J.D. Birchall, W.J. Clegg, M.A. Harmer, K. Kendall, D.J. Eaglesham, L.J., Humphreys and D.R. Jones, Processing and properties of high T_c superconductors, *Superconducting Ceramics, R. Freer (Ed.)*, The Institute of Ceramics, Stoke-on-Trent, 1988, p.149.
12. Tarascon, J.M., P.B. Barboux, B.G. Bagley, L.H. Greene and G.W. Hull, On synthesis of high T_c superconducting perovskites, *Mater. Sci. Eng.*, **B1**: 29 (1988).
13. Gadalla, A.M. and T. Hegg, Optimum conditions for synthesizing $YBa_2Cu_3O_x$, *Private communication* (1988).
14. Schüler, C.Chr., P. Kluge-Weiss and F. Breuter, Processing of $Y_1Ba_2Cu_3O_{7-x}$ powders by oxalate coprecipitation, *Ceramic Powder Processing Science, H. Hausner, G.L. Messing and S. Hirano (Eds.)*, Deuts. Keram. Ges. e.v., Köln 90, 1988, p. 121.
15. Moure, C., P. Duran, S. Vieira, S. Bourgeal, R. Villar, A. Aguiló and M.A. Ramos, Low temperature thermal expansion of a high T_c superconductor phase $Y_1Ba_2Cu_3O_{7-x}$ prepared from oxalate precursors, *Superconducting Ceramics, R.Freer (Ed.)*, The Institute of Ceramics, Stoke-on-Trent, 1988, p. 237.
16. Wang, X.Z., M. Henrt, J. Livage and I. Rosenndu, The oxalate route to superconducting $YBa_2Cu_3O_{7-x}$, *Solid State Commun.*, **64**: 881 (1987).
17. Blank, D.H.A., H. Kruidhof and J. Flokstra, J., Preparation of $YBa_2Cu_3O_{7-x}$ by citrate synthesis and pyrolisis, *J. Phys. D, Appl. Phys.*, **21**: 226 (1987).
18. Sale, F.R. and F. Mahloojchi, Citrate gel processing of oxide superconductors, *Ceram. International.*, **14**: 229 (1988).

19. Pankajavalli, R., J. Janaki, D.M. Sreedharan, J.B. Gnanamoorthy, G.V.N. Rao, V. Sankarasastry, M.P. Janawadkar, Y. Hariharan and T.S. Radhakrishnan, Synthesis of high quality 1-2-3 compound through citrate combustion, *Physica C*, **156**: 737 (1988).
20. Thomson, W.J., H. Wang, D.B. Parkman, D.X. Li, M. Strasik, T.S. Luhman, Ch. Han and I.A. Aksay, Reaction sequencing during processing of the 123 superconductors, *J. Am. Ceram. Soc.*, **72**: 1987 (1989).
21. Bunker, B.C., J.A. Woigt, D.H. Doughty, D.L. Lamppa and K.M. Kimball, Precipitation of superconductor precursor powders, *High-temperature superconducting materials, W.E. Haffield and J.H. Miller Jr. (Ed.)*, Marcel Dekker, Inc., N.Y., 1988, p. 121.
22. Nagano, M. and M. Greenblatt, High temperature superconducting films by sol-gel preparation, *Processing and applications of high Tc superconductors, W.E. Mayo (Ed.)*, Metallurgical Society, N.Y., 1988, p. 43.
23. González, E.G., M. Vallet, J.M. González-Calbet, E. Morán, R. Sáez-Puche, M.A. Alario, M.T. Pérez-Frias and J.L. Vicent, $Ba_2YCu_3O_7$ prepared by liquid-mix technique, *Proc. Eur. Workshop on High Tc Superconductors and Potential Applications*, EEC, Genoa, 1987, p. 451.
24. Kodas, T.T., Generation of complex metal oxides by aerosol processes: superconducting ceramic particles and films, *Adv. Mater.*, **6**: 180 (1987).
25. Kodas, T.T., E.M. Engler, V.Y. Lee, R. Jacowiti, T.H. Bacum, K. Roche, S.S.P. Parkin, W.S. Young, S. Hughes, J. Kleder and W. Auser, Aerosol flow reactor production of fine $Y_1Ba_2Cu_3O_7$ powder: fabrication of superconducting ceramics, *Appl. Phys. Lett.*, **52**: 1622 (1988).
26. Kodas, T.T., E.M. Engler and V.Y. Lee, Generation of thick $Ba_2YCu_3O_7$ films by aerosol deposition, *Appl. Phys. Lett.*, **54**: 1923 (1989).
27. Hoste, S., H. Vlaeminck, P.H. DeRyck, F. Persyn, R. Mouton and G. van der Kelen, A simple apparatus for the preparation ofspray-dried $YBa_2Cu_3O_{7-x}$ ceramic superconductors, *Supercond. Sci. Technol.*, **1**: 239 (1988).
28. YureK, G.J., J.B. van der Sander, D.A. Rudman and Y.M. Chiang, Superconducting microcomposites by oxidation of metallic precursors, *J. Metals*, **40**: 16 (1988).
29. Moure, C., J.F. Fernández, P. Recio and P. Durán, A new route for synthesizing YBaCuO superconducting powders,

Silicates Ind., **55:** 127 (1990).
30. Jorgensen, J.D., Shaked, H., Kinks, D.G., Dabrowsky, B., Veal, B.W., Paulikas, A.P., Nowicki, L.J., Crabtree, G.W., Kwok, W.K., Nunez, L.H. and Claas, H., Oxygen Vacancy or Leving and Superconductivity in $YBa_2Cu_3O_{7-x}$, *Physica C*, **153:** 578 (1988).
31. Sprecht, E.D., C.J. Sparks, A.G. Dhere, J. Brynestad, O.B. Cavin, D.M. Kroeger and H.A. Dye, The effect of oxygen pressure on the orthorhombic-tetragonal transition in the high temperature superconductor $YBa_2Cu_3O_x$, *Phys. Rev. B*, **37:** 7426 (1988).
32. Kubo, Y., T. Ichihash, T. Manako, K. Baba, J. Tabuchi and H. Igarashi, Orthorhombic (II) superstructure phase in oxygen- defficient $YBa_2Cu_3O_{7-x}$ prepared by quenching, *Phys. Rev. B*, **37:** 7858 (1988).
33. O'Bryan, H.M. and P.K. Gallagher, P.K., Kinetics of the oxidation of $Ba_2YCu_3O_x$ ceramics, *J. Mater. Res.*, **3:** 619 (1988).
34. O'Bryan, H.M., P.K. Gallagher, R.A. Laudise, A.J. Caporaso and R.C. Sherwood, R.C., Oxidation of $Ba_2YCu_3O_x$ at high PO_2, *J. Amer. Ceram. Soc.*, **72:** 1298 (1988).
35. Verweij, H., Thermodynamics of the oxidation reaction of $YBa_2Cu_3O_x$, *Ann. Phys. Fr.*, **13:** 349 (1988).
36. Verweij, H. and W.H.M. Bruggink, Constant stoichiometry cooling of $YBa_2Cu_3O_x$, *J. Phys. Chem. Solids*, **50:** 75 (1989).
37. Tu, K.N., C.C. Tsuei, S.I. Oark and A. Levu, Oxygen diffusion in superconducting $YBa_2Cu_3O_{7-x}$ oxides in ambient helium and oxygen, *Phys. Rev. B*, **38:** 772 (1988).
38. Moure, C., J.F. Fernández,P. Recio and P. Durán, Densification of $Y_1Ba_2Cu_3O_x$ superconducting ceramics, *Euro-Ceramics, Vol. 2, G. de With, R.A. Terpstra and R.Metselaar, (Ed.)*, Elsevier Publis., NL, 1989, p. 2415.
39. Blendell, J.E. and L.L. Stearns, Densification, susceptibility and microstructure of $Ba_2YCu_3O_{6+x}$, *Ceramic Powder Processing II, G.L. Messing, E.R. Muller and H. Hauser Eds.)*, Am. Ceram. Soc. Inc., Westerville, OH, 1988, p. 1146.
40. Clarke, D.R., T.M. Shaw, T.M. and D. Dimos, Issues in the processing of cuprate ceramic superconductors, *J. Am. Ceram.* Soc., **72:** 1103 (1989).
41. Tu, K.N., N.C. Yech, S.I. Park and C.C. Tsuei, Diffusion of oxygen in superconducting $YBa_2Cu_3O_{7-x}$ ceramic oxides, *Phys. Rev. B*, **39:** 304 (1989).
42. Chen, I. W., S. Keating, C.Y. Keating, X. Wu, J. Xu, P.E.

Reyes-Morel and T.Y. Tien, Superconductivity and the tailoring of lattice parameters of the compound $YBa_2Cu_3O_x$, *Adv. Ceram. Mater. (Special Issue)*, **2**: 457 (1987).
43. Swinnea, J.S. and H. Steinfink, The crystal structure of $YBa_2Cu_3O_6$, the x =1 phase of the superconductor $YBa_2Cu_3O_{7-x}$, *J. Mater. Res.*, **2**: 424 (1987).
44. Zhang, L., J. Chen, H.M. Chan and M.P. Harmer, Formation of grain boundary carbon-containing phase during annealing of $YBa_2Cu_3O_{6+x}$, *J. Amer. Ceram. Soc.*, **72**: 1997 (1989).
45. Shaw, T.M., S.L. Shinde, D. Dimos, R.F. Cook, P.R. Duncome and C. Kroll, C., The effect of grain size on microstructure and stress relaxation in polycrystalline $Y_1Ba_2Cu_3O_{7-x}$, *J. Mater. Res.*, **4**: 248 (1989).
46. Küpfer, H., I. Apfelstedt, W. Schauer, R. Flükiger, R. Meier- Hirmer and H. Wühl, Critical currents and upper critical field of sintered and powdered superconducting $YBa_2Cu_3O_7$, *Z. Phys. B, Condensed Matter*, **69**: 159 (1987).
47. Cava, R.J., B. Batlogg, R.B. van Dover, D.W. Murphy, S. Sunshine, T. Siegrist, J.P. Remeika, E.A. Rietman, S. Zahurak and G.P. Espinosa, Bulk superconductivity at 91 K in single phase oxygen-deficient perovskite $Y_1Ba_2Cu_3O_{9-x}$, *Phys. Rev. Lett.*, **58**: 408 (1989).
48. Worthington, T.K., W.J. Gallagher and T.R. Dinger, Anisotropic nature of high temperature superconductivity in single crystal $Y_1Ba_2Cu_3O_{7-x}$, *Phys. Rev. Lett.*, **59**: 1160 (1987).
49. Malozemoff, A.P., W.J. Gallagher and R.E. Schawall, Applications of high temperature superconductivity, *Chemistry of High Temperature Superconductors, D.H. Nelson, M.S. Whittinghan and T.F. George (Ed.)*, Amer. Chem. Soc., Washington D.C., 1987, p. 280.
50. Ausloos, M. and Ch. Laurent, Thermodynamic fluctuations in the superconductor $Y_1Ba_2Cu_3O_{9-y}$. Evidence for two dimensional superconductivity, *Phys. Rev. B*, **37**: 611 (1988).
51. Dinger, T.R., T.K. Worthington, W.J. Gallagher and R.L. Sandstrom, Direct observation of electronic anisotropy in single crystal $Y_1Ba_2Cu_3O_{7-x}$, *Phys. Rev. Lett.*, **58**: 2687 (1987).
52. Kwo, J., M. Hong, R.M. Fleming, T.C. Hsich, S.H. Liou and B.A. Davidson, Single crystal superconducting $Y_1Ba_2Cu_3O_{7-x}$ oxide films by molecular beam epitaxy, *Novel Superconductivity, S.A. Wolf and V.Z. Kresined (Ed.)*, Plenum Publishing Co, N.Y.,1987, p. 699.
53. Laibowitz, R.B., R.H. Koch, P. Chandari and R.J. Gambino,

Thin superconducting oxide films, *Phys. Rev. B*, **35**: 8821 (1987).

54. Heintze, G.N., R. McPherson, D. Tolino and C. Andrikidis, The structure of thermally sprayed $YBa_2Cu_3O_{7-x}$ superconducting coating, *J. Mater. Sci.Lett.*, **7**: 253 (1988).

55. Yamamoto, T., Terao, G., Okazaki, K., Natsashewa, H., Koshimura, M., Ueyama, T. and Ono, M., Macrostructure of $YBa_2Cu_3O_{7-x}$/metal Ag Hy and thick film, *Superconductivity and Ceramic Superconductors, K.M. Nair and E.A. Giess (Eds.)*, Am. Ceram. Soc. Inc., Westerville, OH, 1990, p. 489.

56. Thompson, J.R., D.K. Christen, S.T. Sekula, J. Brynestad, Y.C. Kim, Magnetization studies of the high-T_c compound $Y_1Ba_2Cu_3O_x$, *J. Mater. Res.*, **2**: 779 (1987).

57. Camps, R.A., J.E. Evetts, B.A. Glowacki, S.B. Newcomb, R.E. Somekh and W.M. Stubbs, Microstructure and critical current of superconducting $YBa_2Cu_3O_{7-x}$, *Nature*, **329**: 229 (1987).

58. Alford, N. McN., W.J. Clegg, M.A. Harmer, J.D. Birchall, K. Kendall and D.H. Jones, The effect of density on critical current and oxygen stoichiometry of $YBa_2Cu_3O_x$ superconductors, *Nature*, **332**: 58 (1988).

59. Babic, E., D. Drobac, J. Horvat, Z. Marohnic and M. Prester, Critical currents in high-quality YBaCuO ceramics investigated by magnetization, a.c. susceptibility and electrical transport, *J. Less-Comm. Metals*, **151**: 89 (1989).

60. Ekin, J.W., A.I. Braginski, A.J. Panson, M.A. Janocko, D.W. Capone, N.J. Zaluzec, B. Flandermeyer, B., O.F. de Lima, M. Hong, J. Kwo, J. and S.A. Liou, Evidence for weak link and anisotropic limitations on the transport critical current in bulk polycrystalline $YBa_2Cu_3O_x$, *J. Appl. Phys.*, **62**: 4821 (1987).

61. Jorgensen, D., M.A. Beno, D.G. Hinks, L. Soderholm, K.J. Volin, R.L. Hitterman, J.D. Grace, I.K. Schuller, C.U. Segre, K. Zhang, K. and M.S. Kleefisch, Oxygen ordering and the orthorhombic to tetragonal phase transition in $YBa_2Cu_3O_7$, *Phys. Rev. B*, **36**: 3608 (1987).

62. Cheng, Ch.-W., A.C. Rose-Irnes, N.McN. Alford, M.A. Harmer and J.D. Birchall, The effect of porosity on the superconducting properties of $Y_1Ba_2Cu_3O_x$ ceramic, *Superconductor Science and Technology*, **1**: 113 (1988).

63. No, K., J.D. Verhoeven, R.W. McCallum and E.D. Gibson, Grain size control in powder processed $Y_1Ba_2Cu_3O_x$, *IEEE Trans. on Magnetics*, **25**: 2184 (1989).

64. Yarnoff, J.A., D.R. Clarke, W. Drube, V.O. Karlson, A. Taleb Ibrahimi and F.J. Himpsel, Valence electronic structure of $YBa_2Cu_3O_x$, *Phys. Rev. B*, **36**: 3967 (1987).
65. Verhoeven, J.D., A.J. Bevolo, R.W. McCallum, E.D. Gibson and M.A. Noack, Auger study of grain boundaries in large grained $Y_1Ba_2Cu_3O_x$, *Appl. Phys. Lett.*, **52**: 745 (1988).
66. Laurent, Ch., M. Laguesse, S.K. Patapis, H.W. Vanderschueren, G.V. Lecomte, A. Rulmont, P. Tarte and M. Ausloos, Dimensionality crossover regimes due to microcrystalline anisotropy, in the resistivity of granular $Y_1Ba_2Cu_3O_{7-y}$ in weak magnetic fields at low temperature, *Z. Phys.B, Cond. Matt.*, **69**: 435 (1988).
67. Dimos, D., P. Chaudhari, J. Mannhart and F.K. Le Goues, Orientation dependence of grain boundary critical currents in $YBa_2Cu_3O_{7-x}$ bicrystals, *Phys. Rev. Lett.*, **61**: 219 (1988).
68. Jin, S., T.H. Tiefel, R.C. Sherwood, R.B., van Dover, M.E. Davis, G.W. Kammlott and R.A. Fastnacht, Melt-textured growth of polycrystalline $YBa_2Cu_3O_{7-x}$ with high transport J_C at 77 K, *Phys. Rev. B*, **37**: 7850 (1988).
69. Hojaji, H., K.A. Michael, A. Barbatt, A.N. Thorpe, M.F. Ware, I.G. Talmy, D.A. Haught and S. Alterescu, A comparative study of sintered and melt-grown recrystallized $YBa_2Cu_3O_x$, *J. Mater. Res.*, **4**: 28 (1989).
70. Chen, I.-W., X. Wu, S.J. Keating, C.Y. Keating, P.A. Johnson and T.-Y. Tien, Texture development in $YBa_2Cu_3O_x$ by hot extrusion and hot pressing, *J. Amer. Ceram. Soc.*, **70**: C-388 (1987).
71. Robinson, Q., P. Georgopoulos, D.L. Johnson, H.O. Marcy, C.R. Kannewurf, S.J. Hwu, T.J., Marks, K.R. Poeppelmeier, S.N. Song and J.B. Ketterson, Sinter-forged $Y_1Ba_2Cu_3O_x$, *Adv. Ceram. Mater., Special Issue*, **2**: 380 (1987).
72. Ginley, D.S., E.L. Venturin, J.F. Kwak, R.J. Baughman and B. Morosin, Improved superconducting $Y_1Ba_2Cu_3O_{6.9}$ through high temperature processing, *J. Mater. Res.*, **4**: 496 (1988).
73. Salama, K., V. Selvamanickam, L. Gao and K. Sun, High current density in bulk $YBa_2Cu_3O_x$ superconductor, *Appl. Phys. Lett.*, **54**: 2352 (1989).
74. Farrel, D.E., B.S. Chandrasekhar, M.R. de Guire, M.M. Fang, V.G. Kogan, J.R. Clen and D.K. Finnemore, Superconducting properties of aligned crystalline grains of $YBa_2Cu_3O_{7-x}$, *Phys. Rev. B*, **36**: 4025 (1987).
75. Seuntgens, J., X. Cai and D.C. Larbalester, Preparation and

characterization of magnetically textured Y and Dy 123 compounds, *IEEE Trans. Magn.*, **25**: 2 (1989).
76. Lusnikov, A., L.L. Miller, R.W. McCallum, S. Mitra, W.C. Lee and D.C. Johnston, Mechanical and high temperature (920 C), magnetic field grain alignment of polycrystalline (Ho,Y) $Ba_2Cu_3O_{7-x}$, *J. Appl. Phys.*, **65**: 3136 (1989).
77. Malik, M.K., V.D. Nair, A.R. Biswas, R.V. Raghavan, P. Chaddah, P.K. Mishra, G. Ravikumar and B.A. Dasannacharya, Texture formation and enhanced critical currents in $YBa_2Cu_3O_7$, *Appl. Phys. Lett.*, **52**: 1525 (1988).
78. Tiefel, T.H., S. Jin, R.C. Sherwood, M.E. Davis, G.W. Kammott, P.K. Gallagher, D.W. Johnson, R.A. Fastnacht and W.W. Rhodes, Grain-growth enhancement in the $YBa_2Cu_3O_{7-x}$ superconductor by silver-oxide doping, *Matter. Lett.*, **7**: 363 (1989).
79. Song, Y., Y. Cao, A. Misra and J.R. Gaines, Superconducting (YBaCuO)-(CuO)-Ag composites, *J. Mater. Res.*, **4**: 802 (1989).
80. Pietraszko, A., M. Wolcyrz, R. Horyn, Z. Bukowski, K. Lukaszewicz and J. Klamnt, Orthorhombic-tetragonal phase transition and oxygen index of $YBa_2Cu_3O_{6+x}$, *Cryst. Res. Technol.*, **23**: 351 (1988).
81. Kwasnick, R.F., F.E. Lublorsky, E.L. Hall, M.F. Garbauskas, K. Borst and M.J. Curran, Microstructure and properties of superconducting sputter deposited Y-Ba-Cu-O films, *J. Mater. Res.*, **4**: 257 (1989).

Chapter 3

Oxidative Nonstoichiometry in Perovskite Oxides

D. M. Smyth

Department of Materials Science and Engineering,
Lehigh University,Bethlehem, PA 18015

I. Introduction

"Oxidative nonstoichiometry" can be defined in crystallographic terms, i.e. oxygen in excess of the number of sites in the oxygen sublattice, or a cation content that is less than the number of sites in the cation sublattices; or in stoichiometric terms, i.e. an oxygen content in excess of that needed to balance the cationic constituents in their normal oxidation states. It is not possible to achieve a crystallographic excess of oxygen in the ideal perovskite structure under any readily achievable experimental conditions. The combined array of large cations and oxide ions approximates a cubic-close-packed sublattice, and there is no space available for additional large anions. Thus the presence of a crystallographic excess of oxygen requires a modification of the structure, as in the series of compositions with the generic formula $A_mB_mO_{3m+2}$, e.g. $Sr_2Nb_2O_7$, the reduced formula for a member of the family with m = 4. In these structures, two-dimensional slabs of the perovskite structure, m unit cells thick, are separated by layers containing the excess oxygen. Such materials can be considered to have perovskite-related structures (1,2). In contrast with the case of a crystallographic excess, it is quite common to have a stoichiometric excess of oxygen in the perovskite structure, in the sense of more oxygen than corresponds to a combination of the stoichiometric binary constituents. Moreover, in the case of donor-doped materials, when the positively charged impurity centers are compensated by cation vacancies, the ratio of occupied oxygen sites to occupied cation sites exceeds that of the ideal perovskite structure. All of these types of "oxidative nonstoichiometry" will be discussed in this chapter. Although there are halides that have the perovskite structure, e.g. $KNiF_3$, the discussion will be limited to the much more important oxides. For the purpose of self-consistency, all of the oxidation and reduction reactions have been written in terms of the gain or loss of a single oxygen atom.

II. Reference States

It is essential to distinguish between two different kinds of reference states. There is the chemical or thermodynamic reference, usually referred to as the stoichiometric composition, and there is the reference structure, the ideal, defect-free structure. For a simple, pure, binary oxide, such as MgO, the compositional and structural reference states are the same, but the situation is more complicated for solid solutions or more complex compositions.

For a ternary or higher oxide, the stoichiometric composition corresponds to a combination of the stoichiometric binary constituents, i.e. A_2O + B_2O_5 ($KTaO_3$), AO + BO_2 ($BaTiO_3$), or A_2O_3 + B_2O_3 ($LaCrO_3$) for perovskites with monovalent, divalent, and trivalent large cations, respectively. For these compositions, the ideal perovskite structure is completely filled, and the stoichiometric composition and the reference structure are identical. The materials are electronically balanced such that the valence band is nominally completely filled and the conduction band is nominally completely empty. The only source of charge carriers is by ionization across the band gap to give equal numbers of electrons and holes. When the materials contain substitutional aliovalent impurity ions, the stoichiometric composition still corresponds to the combination of stoichiometric binary oxides including that of the impurity oxide, e.g. BaO + $(1-x)TiO_2$ + $x/2Al_2O_3$ or $BaTi_{1-x}Al_xO_{3-x/2}$ for acceptor-doped $BaTiO_3$ with compensating oxygen vacancies (3,4), and BaO + $(1-5x/4)TiO_2$ + $x/2Nb_2O_5$ or $BaTi_{1-5x/4}Nb_xO_3$ for donor-doped $BaTiO_3$ with compensating titanium vacancies (5,6). In these thermodynamic reference states the primary chemical bonds are completely satisfied and the electronic structure corresponds to nominally full and empty bands. Phenomenologically, the stoichiometric composition separates a region of n-type conduction that increases with decreasing equilibrium values of $P(O_2)$, from a region of p-type conduction that increases with increasing $P(O_2)$. It thus corresponds to a minimum value in the equilibrium electronic conductivity, assuming equal mobilities for electrons and holes. The reference structure for both doped and undoped materials, i.e. the structure from which lattice defects are defined, is the ideal perovskite structure and hence differs in composition from the stoichiometric composition in the case of the doped materials.

III. Undoped Perovskites

The ideally pure material does not exist, and the perovskite oxides are particularly good examples of systems that invariably have enough impurity content to have a strong influence on the defect chemistry (4,7). Nevertheless, this fictional state is always a convenient starting point for discussions of defect chemistry. $BaTiO_3$ will be taken as an example that has been studied in considerable detail. The stoichiometric composition will be the combination of stoichiometric binary constituents

$$BaO + TiO_2 \longrightarrow Ba_{Ba} + Ti_{Ti} + 3O_o \equiv BaTiO_3 \qquad (1)$$

For the ideally pure, stoichiometric material, the most abundant defects are electrons and holes that result from direct ionization across the band gap

$$nil \rightleftharpoons e' + h^{\bullet} \qquad (2)$$

where nil refers to the defect-free reference state. For $BaTiO_3$, the standard enthalpy for this reaction, sometimes known as the band gap at absolute zero, is 3.4 eV (7), which results in intrinsic carrier concentrations of about 2×10^{16} cm^{-3} at 1273 K. Under normal equilibration conditions, no form of intrinsic ionic disorder creates defects in comparable concentrations, but under crystal growth or sintering conditions there may be a small contribution from Schottky disorder,

$$nil \rightleftharpoons V_{Ba}'' + V_{Ti}'''' + 3V_o^{\bullet\bullet} \qquad (3)$$

Computation of relative defect energies have shown that this is the preferred type of ionic disorder in $BaTiO_3$, with an enthalpy of formation of 2.3 eV per defect (8). This yields a value of about 10^{-9} at 1273 K for the exponential enthalpy term per defect in the mass-action constant for Eq (3), this being the term that primarily sets the scale for defect concentrations. Assuming that the entropy term in the mass-action expression contributes no more than a factor of 100, the defect concentrations are expected to be less than 0.1 ppm (per formula unit). As a result, impurity effects invariably determine the ionic defect content in real, stoichiometric materials (4).

The perovskite titanates reduce readily with the formation of oxygen vacancies, $V_o^{\bullet\bullet}$.

$$O_o \rightleftharpoons 1/2O_2 + V_o^{\bullet\bullet} + 2e' \quad (4)$$

This is facilitated by the presence of a reducible species, Ti^{+4}, and the formation of the easily tolerated anion vacancy. As indicated, the vacancies seem to be always present in the fully ionized state, at least under the equilibration conditions (7). However, this situation corresponds to "reductive nonstoichiometry" relative to both the stoichiometric composition and the reference structure, so it is outside the scope of this discussion.

There are two possible oxidation reactions for ideally pure $BaTiO_3$, depending on whether unit cells are conserved or increased

$$1/2O_2 \rightleftharpoons O_I'' + 2h^{\bullet} \quad (5)$$

$$1/2O_2 \rightleftharpoons 1/3V_{Ba}'' + 1/3V_{Ti}'''' + O_o + 2h^{\bullet} \quad (6)$$

Neither one of these occurs in practice. The first involves the highly unfavorable anion interstitial, and the second involves a combination of highly charged cation vacancies. Oxidation is also suppressed because the material contains no chemically oxidizable species. Because of these restrictions, ideally pure perovskites, including $BaTiO_3$, have no behavior that is of interest to this chapter.

IV. Acceptor-Doped Perovskites

The formation of an acceptor-doped perovskite, e.g. the

partial substitution of Al^{+3} for Ti^{+4} in $BaTiO_3$, can be expressed as follows

$$BaO + (1-x)TiO_2 + x/2Al_2O_3 \longrightarrow Ba_{Ba} + (1-x)Ti_{Ti} + xAl'_{Ti} + (3-x/2)O_o + x/2V_o^{\bullet\bullet} \quad (7)$$

or, alternatively as

$$xBaO + x/2Al_2O_3 \xrightarrow{(xTiO_2)} xBa_{Ba} + xAl'_{Ti} + 5x/2O_o + x/2V_o^{\bullet\bullet} \quad (8)$$

where the entity in parentheses above the arrow is replaced by the dopant oxide. These incorporation reactions are not written as equilibrium reactions since the impurity is usually incorporated into the material during its chemical preparation, rather than by equilibration with some external source of fixed activity. The impurity content is subsequently taken as invariant. Exceptions to this include cases where one of the constituent binary oxides is volatile, as in the case of $PbTiO_3$ and other lead-containing compounds.

Eqs. (7) and (8), as written, yield the stoichiometric composition, $BaTi_{1-x}Al_xO_{3-x/2}$, since the product derives from a conbination of stoichiometric binary oxides. They represent the equilibrium composition for which n = p, and at which the sum of the electron and hole concentrations has its minimum value at any given equilibration temperature. Thus it will correspond to the composition at the minimum in the equilibrium conductivity measured as a function of oxygen activity, except insofar as the electron and hole mobilities are not equal. However, while chemically stoichiometric, this composition is oxygen deficient relative to the ideal perovskite reference structure. The incomplete oxygen sublattice gives an easy mechanism for the accommodation of a stoichiometric excess of oxygen

$$yV_o^{\bullet\bullet} + y/2O_2 \rightleftharpoons yO_o + 2yh^{\bullet} \quad (9)$$

This reaction is greatly enhanced by the availability of normal lattice sites for the added oxygen, and by the fact that the number of lattice defects is actually being decreased by the oxidation reaction, whereas in the case of the reduction reaction, Eq. (4), the number of defects is increased. These factors can counterbalance the absence of a chemically oxidizable species. In $BaTiO_3$, the enthalpy for the oxidation reaction is 0.92 eV per added oxygen, which is six times smaller than the enthalpy of reduction per oxygen removed (7).

The product of oxidation has the composition $BaTi_{1-x}Al_xO_{3-x/2+y} + 2yh^{\bullet}$, and is oxygen-excess relative to the stoichiometric composition. The equilibrium p-type conductivity increases from the stoichiometric value with increasing oxygen activity. The material will remain oxygen-deficient relative to the reference structure until all oxygen sites become filled at the composition $BaTi_{1-x}Al_xO_3 + xh^{\bullet}$, when $y = x/2$. This saturation of the oxygen sublattice does not occur for acceptor-doped $BaTiO_3$ for any equilibrium $P(O_2)$ up to 1 atm, but it can occur for compounds that contain easily oxidizable cations, such as Fe^{+3} and Mn^{+3}, as will be discussed. At the point of saturation, further oxidation becomes highly unfavorable for the same reasons discussed for the case of ideally pure $BaTiO_3$ in the preceding section. When the oxygen sublattice becomes filled, the condition of charge neutrality is primarily determined by the expression $[Al'_{Ti}] = p$, whereas at the stoichiometric composition it is $[Al'_{Ti}] = 2[V_O^{\bullet\bullet}]$. In other words, the oxidation reaction results in the gradual replacement of vacancies by holes as the compensating defect. The equilibrium compositions can be visualized in the form of a Kroger-Vink diagram as shown in Fig. 1. All of the region to the high $P(O_2)$ side of the intersection of the electron and hole concentrations at $n = p$ corresponds to a stoichiometric excess of oxygen. The extent of this region below the usual experimental limit of 1 atm of oxygen is highly dependent on the chemical nature of the cations in the host perovskite. In the alkaline earth titanates, which do not contain an oxidizable cation, the oxidation reaction never results in the consumption of a significant fraction of the available oxygen vacancies.

When the perovskite contains an easily oxidizable cation, as in the case of $LaMnO_3$ or $LaFeO_3$, the oxygen-excess, p-type region may extend over a wide range of oxygen activities, and the oxygen sublattice may become completely filled at high oxygen activities. This is demonstrated in Fig. 2, in which the equilibrium electrical conductivities of $LaFeO_3$ and $La_{0.9}Sr_{0.1}FeO_{2.95}$ are compared (9,10). Since oxygen-excess, p-type compositions are assumed to be possible only when the stoichiometric composition contains oxygen vacancies that can be filled by a reaction such as Eq. (9), the undoped material is assumed to contain either a small net excess of acceptor impurities because of its naturally occurring impurity content (4), or, as suggested by the original authors, a small deficiency of La_2O_3 because of the difficulty of controlling the cation ratio precisely. The results show the rapid saturation of the oxygen vacancy content in the undoped

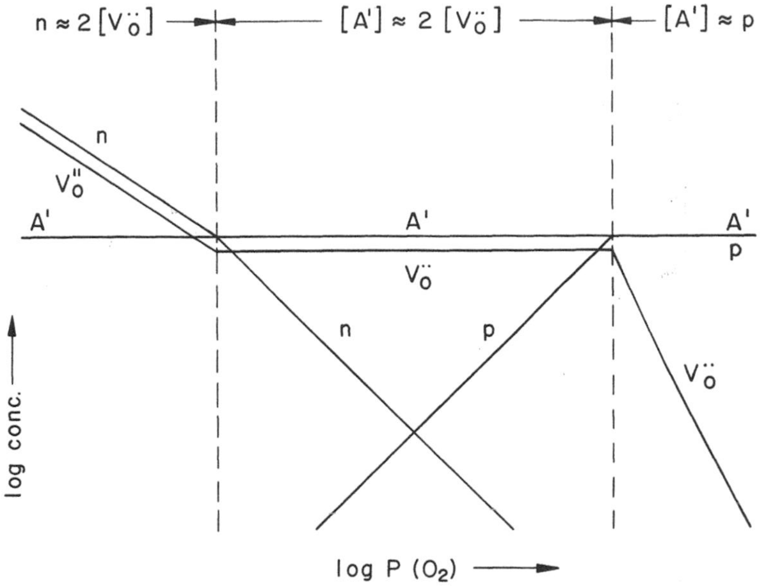

Fig. 1. Kroger-Vink diagram for a perovskite oxide containing a singly-charged acceptor center A', under equilibrium conditions at constant temperature.

sample, and the higher hole concentrations achieved in the acceptor-doped sample because of its larger concentration of compensating oxygen vacancies that can be filled by the stoichiometric excess of oxygen. The conductivity minimum, i.e. the stoichiometric composition, is located at lower values of $P(O_2)$ for the more highly doped material, as is the usual case for acceptor-doped oxides (3). Below the conductivity minima, the materials are reduced by the loss of oxygen from the stoichiometric composition according to Eq (4). This region of reduction is quite extensive in $LaFeO_3$ because of the easy reduction of Fe^{+3} to Fe^{+2}. In a material such as $LaCrO_3$, where the cations can only be oxidized, the oxygen-excess, p-type region extends to much lower values of $P(O_2)$ (11). In the undoped material the plateau at high $P(O_2)$ becomes temperature dependent, because the intrinsic electronic ionization reaction, Eq (2), becomes a larger source of carriers than the nonstoichiometry, as shown schematically in Fig. 3. This is to be expected when the material contains both reducible and oxidizable cations, in this case Fe^{+3} that can be reduced to Fe^{+2} and oxidized to Fe^{+4}. This chemical behavior is reflected in the electronic structure of the solid compound as a relatively small band gap, 2.0 eV in this case. Ionization across the band is related to the chemical

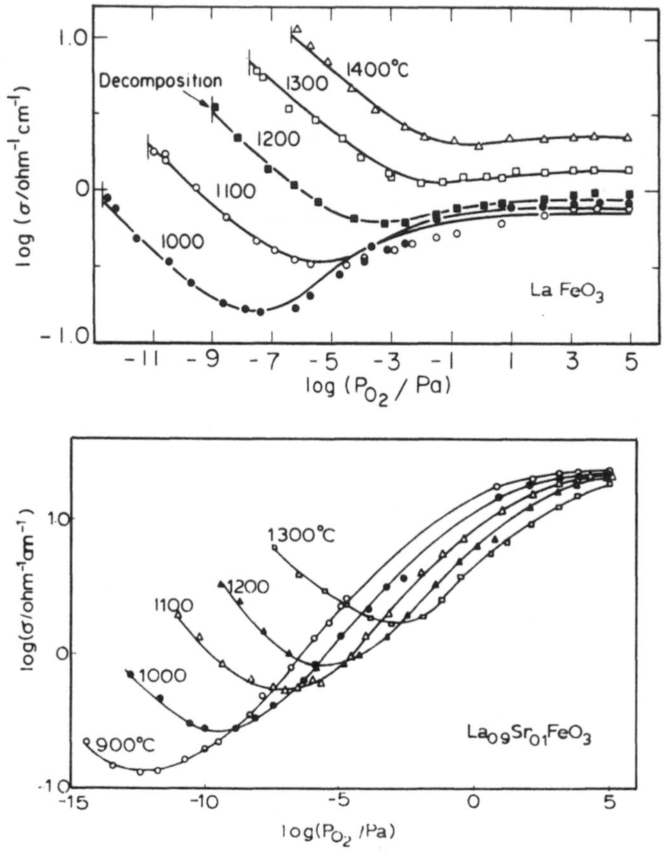

Fig. 2. The equilibrium electrical conductivities of $LaFeO_3$ and $La_{0.9}Sr_{0.1}FeO_{2.95}$ as a function of $P(O_2)$. (Reproduced from Refs. (9) and (10) with permission from the authors and from the American Ceramic Society).

disproportionation of the Fe^{+3}

$$2Fe^{+3} \rightleftharpoons Fe^{+2} + Fe^{+4} \tag{10}$$

The authors of this investigation view the electrons and holes as being localized on the Fe to give Fe^{+2} and Fe^{+4}, respectively (9). Conduction then takes place by a polaron mechanism.

Both the relative saturation levels of the conductivities at high $P(O_2)$, and the shift in the conductivity minima, are consistent with $La_{0.9}Sr_{0.1}FeO_{2.95}$ having a concentration level of oxygen vacancies that is 17 times that of the undoped material. Oxidation effectively ceases when the oxygen sublattice becomes filled, even though the material contains additional oxidizable cations, because one of the highly unfavorable oxidation reactions of the type represented by Eqs (5) or (6) would have to come into play.

The presence of an oxidizable cation not only influences the balance between oxygen-excess and oxygen-deficient non-stoichiometry, it also has a strong effect on the resulting electrical properties near room temperature. The occupied atomic d states that contribute to the structure of the valence band can easily lose additional electrons, i.e. undergo oxidation, and this band is thus hospitable to the presence of holes. Thus in the oxygen-excess material the holes remain in the valence band down to quite low temperatures, and the materials are black and semiconducting after having been cooled from equilibration in an oxygen-rich atmosphere. This means that the electronic states associated with the acceptor impurities must lie very close to

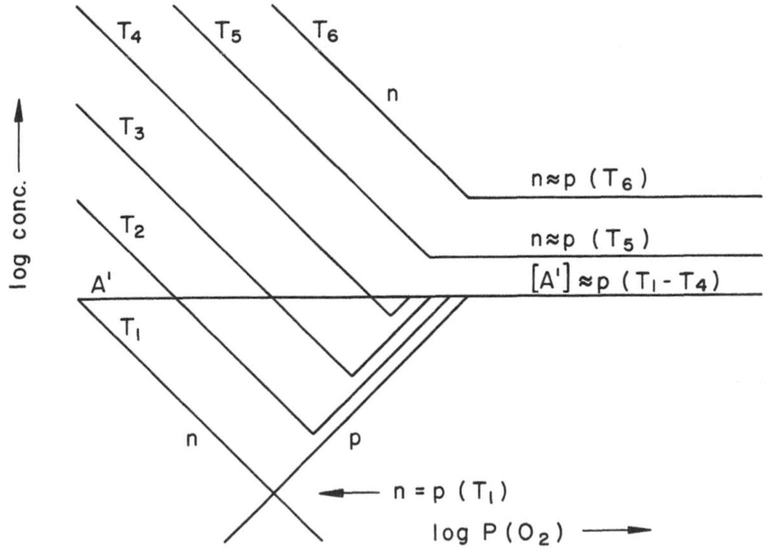

Fig. 3. Partial Kroger-Vink diagram at several temperatures for a material such as $LaFeO_3$ containing a small net acceptor excess, A' ; $T_6 > T_5 > T_4 > T_3 > T_2 > T_1$, $[A'] > (np)^{1/2}$ for $T_1 - T_4$, $n = p > [A']$ for T_5 and T_6.

the top of the valence band so that they do not act as effective traps for the holes as the material is cooled down. This is the usual situation for any oxide that contains an oxidizable cation and that has been equilibrated to an oxygen-excess, p-type state. Such materials are not useful for applications that require an insulating behavior after having been processed in air, but they can be useful when conductivity is desired in an oxidizing environment, as in the cathode of a high temperature fuel cell.

While the presence of an oxidizable cation enhances the tendency to form oxygen-excess nonstoichiometry, modest amounts of excess oxygen can be achieved even if there are no oxidizable cations present. The principle requirement is that there be unoccupied oxygen sites in the perovskite lattice that can accommodate the stoichiometric excess of oxygen. Thus in $SrTiO_3$ and $KTaO_3$, the alkali, alkaline earth, and transition metal ions have the rare gas electronic configuration, there are no electrons in their d shells, and they can only be reduced to lower oxidation states by normal chemical means. Nevertheless, these compounds can take up a small stoichiometric excess of oxygen, as shown in Fig. 1, by the filling of oxygen vacancies that are present due to acceptor impurities, either naturally occurring or deliberately added, or to a nonstoichiometric ratio of the cation content (4,12). Apparently the presence of such vacant sites is a more stringent requirement for accommodating excess oxygen than is the presence of an oxidizable cation. However, in these cases, the oxygen excess is limited to the sub-percent level, and only a small fraction of the available oxygen vacancies are ever filled.

In the absence of an oxidizable cation, there are no d electrons to occupy the valence band, which is therefore made up entirely by filled oxygen 2p states. Putting a hole into the valence band is thus equivalent to oxidizing O^{2-} to O^{-}, and this is chemically very unfavorable. Thus as these oxygen-excess materials are cooled toward room temperature, the holes become trapped by the acceptor centers, and the materials become insulators

$$A' + h^{\bullet} \rightleftharpoons A^{x} \tag{11}$$

where A' is a singly-charged acceptor center, and A^{x} is an acceptor center that has become neutral by trapping a hole. On the other hand, the presence of electrons in the conduction band, which is made up of empty d states derived from the transition metal ion, is the equivalent of reducing those ions to lower oxidation states. That is compatible with the chemistry of these elements, and the conduction band states readily accept electrons. As a result, electrons are not trapped out by positively charged defects as the reduced materials are cooled, and the materials are black and semiconducting at room temperature (4). In this case, donor levels lie very close to the bottom of the conduction band. This behavior is obviously just the mirror image of that of the compounds that contain oxidizable cations. The titanates, tantalates, niobates, tungstates, etc., are normally colorless and insulating after having been processed in oxidizing atmospheres, usually air. Perovskite oxides of this type are widely used as capacitor dielectrics, piezoelectric transducers, electrooptic materials, and other applications that require electrically insulating or optically transparent materials.

V. Cation-Cation Nonstoichiometry

Oxygen excesses can also be readily achieved in perovskites that are structurally deficient in oxygen because of cation nonstoichiometry. For example, it has been shown that $CaTiO_3$ can accept excess TiO_2 in solid solution up to one percent level, and it is assumed that this is accommodated by the formation of calcium and oxygen vacancies (12)

$$xTiO_2 \rightleftharpoons xV_{Ca}'' + xTi_{Ti} + 2xO_o + xV_o^{\bullet\bullet} \qquad (12)$$

The resulting oxygen vacancies enhance the oxidation of the material in the same way that was discussed above for acceptor-doped perovskites, i.e. by way of Eq. (9). The minimum value in the equilibrium conductivity shifts to lower oxygen activities with increasing excess TiO_2 in the same way that it shifts with increasing acceptor content (3,12).

The TiO_2-rich compositions can be considered to be stoichiometric with respect to oxygen content, i.e. with respect to oxidation-reduction equilibria, when they are composed of a combination of the stoichiometric binary oxides, $(1-x)CaO + TiO_2$, or $Ca_{1-x}TiO_{3-x}$. On the other hand, they are obviously nonstoichiometric with respect to cation ratio. This further emphasizes the care that must be used in the definition of stoichiometry. As these materials oxidize according to Eq. (9), the resulting oxygen-excess compositions can be expressed as $Ca_{1-x}TiO_{3-x+y} + 2yh^{\bullet}$. If a full oxygen sublattice could be achieved, which it cannot in this system, the limiting composition would become $Ca_{1-x}TiO_3 + 2xh^{\bullet}$, when y = x. As in the acceptor-doped case, oxygen vacancies are gradually replaced by holes during oxidation as the compensating defect for the negatively charged cation vacancy, which formally behaves the same as an acceptor center; although, in this case, it is a doubly-charged center, and would be more directly analogous to a doubly-charged acceptor center such as Mg_{Ti}.

There is experimental evidence for significant solubility for excess CaO in $CaTiO_3$, in that the conductivity minima have been observed to move to lower oxygen activities for such compositions (12). It has been proposed that this results from some occupation of Ti-sites by Ca^{+2} to give an acceptor-like center Ca_{Ti}, as has been quite conclusively shown for the substitution of Ca^{+2} for Ti^{+4} in $BaTiO_3$ (13). This would result in an incorporation reaction of the type

$$xCaO \longrightarrow x/2Ca_{Ca} + x/2Ca_{Ti}'' + xO_o + x/2V_o^{\bullet\bullet} \qquad (13)$$

The resulting oxygen vacancies would behave in the same way as described above, and the materials could oxidize according to Eq. (9). The stoichiometric composition (relative to oxygen content) would be $CaTi_{1-x/2}Ca_{x/2}O_{3-x/2}$, and would become $CaTi_{1-x/2}Ca_{x/2}O_{3-x/2+y} + 2yh^{\bullet}$ when oxidized, with a limiting composition of $CaTi_{1-x/2}Ca_{x/2}O_3 + xh^{\bullet}$ when the oxygen sublattice becomes filled for the condition that y = x/2.

The occupation of Ti-sites by Ca+2 in $CaTiO_3$ has been questioned on the basis of theoretical calculations of the energy of the incorporation process, even though such calculations for the same process in $BaTiO_3$ is compatible with the model (14). However, no alternative explanation has been offered for the observed shift in the minimum value of the equilibrium conductivity, and for the observed increase in the ionic conductivity that has been assumed to result from the increase in oxygen vacancy content.

The chemical properties of the constituent cations play the same role as described for acceptor-doped perovskites in determining the ease and extent of oxidation, the trapping levels of acceptor and donor states, and the resulting equilibration conditions for obtaining insulating or semiconducting materials.

VI. Donor-Doped Perovskites

A donor impurity has a higher cationic charge than the host cation that it replaces, and thus the donor impurity oxide is trying to bring more oxygen into the system than the host oxide, e.g. Nb_2O_5 substituted for $2TiO_2$ in $BaTiO_3$. The stoichiometric composition for the donor-doped material will, as usual, correspond to a combination of the stoichiometric binary constituents, $BaO+(1-x)TiO_2+x/2Nb_2O_5$, or $BaTi_{1-x}Nb_xO_{3+x/2}$. The properties of the donor-doped material are critically dependent on the fate of the extra oxygen. The incorporation of the excess oxygen has the same problem as the oxidation of pure, stoichiometric $BaTiO_3$; it must be either a structural excess in the form of interstitial oxygen, or it must be incorporated into the oxygen sublattice with the formation of cation vacancies. These alternatives can be viewed as the cases for the conservation of unit cells, or the addition of unit cells, respectively. The corresponding incorporation reactions are

$$BaO + (1-x)TiO_2 + x/2Nb_2O_5 \longrightarrow Ba_{Ba} + (1-x)Ti_{Ti} + xNb^{\bullet}_{Ti} + 3O_o + x/2O_I'' \quad (14)$$

$$Ba + (1-x)TiO_2 + x/2Nb_2O_5 \longrightarrow Ba_{Ba} + (1-x)Ti_{Ti} + xNb^{\bullet}_{Ti} + x/6V_{Ba}'' + x/6V_{Ti}'''' + (3+x/2)O_o \quad (15)$$

The conditions for charge neutrality at the stoichiometric compositions are $[Nb^{\bullet}_{Ti}] = 2[O_I'']$ or $[Nb^{\bullet}_{Ti}] = 2[V_{Ba}''] + 4[V_{Ti}'''']$. Eq. (14) gives a stoichiometric material that is oxygen-excess with respect to the perovskite reference structure, while Eq. (15) gives a stoichiometric material that is metal-deficient relative to the perovskite reference structure. There are no known cases of compensation by interstitial oxygen, but there have been reports of compensation by mixtures of cation vacancies (5), comparable to incorporation by Eq. (15). Such materials fall under the jurisdiction of this chapter, since the oxygen/metal ratio is greater than that of an ideal, stoichiometric perovskite because of the cation vacancies. Their chemical reduction would be of interest to this discussion, but will be delayed to a later part of this section, because, in the opinion of this author, donor-compensation by a cation vacancy is more likely to involve primarily a single vacancy species, rather than a mixture of two species. This is based on experimental evidence (6), as well as the consideration that for the two different cation vacancies to have sufficiently comparable lattice energies to be both significantly involved would be quite fortuitous.

The alternative incorporation reaction involves the loss of the extra oxygen from the lattice, with a chemically-equivalent number of electrons being left behind. This can be viewed as the loss of the interstitial oxygen in Eq. (14) by reduction

$$BaO + (1-x)TiO_2 + x/2Nb_2O_5 \longrightarrow Ba_{Ba} + (1-x)Ti_{Ti} + xNb^{\bullet}_{Ti} + 3O_o + xe' + x/4O_2 \tag{16}$$

This gives an ideally filled perovskite lattice, and is clearly a reduced material, since it contains excess electrons that are the chemical equivalent of a reduction in the average oxidation state of the cations. Such materials are semiconducting and are of practical importance in such devices as boundary-layer capacitors and PTC (positive temperature coefficient) devices, where the bulk of the grains in a ceramic structure are conducting. It is the mode of incorporation of donors into $BaTiO_3$ for small concentrations, typically less than 0.5 mol%, in oxygen-rich atmospheres, or for any concentration in highly reducing atmospheres. In $SrTiO_3$ it appears to occur only in reducing atmospheres. In any case, this situation does not fall within the province of this chapter.

Oxidation of any of the donor-doped compositions does not occur, because of the previously-stated reasons involving unfavorable defects. However, reduction can be of interest to us. If the donor centers were compensated by oxygen interstitials, i.e. by a lattice excess of oxygen, reduction could occur by the subtractive loss of oxygen

$$yO_I'' \rightleftharpoons y/2O_2 + 2ye' \tag{17}$$

giving compositions with general formula $BaTi_{1-x}Nb_xO_{3+x/2-y} + 2ye'$. When y = x/2 at the composition $BaTi_{1-x}Nb_xO_3 + xe'$, electrons have completely replaced the interstitial oxygens as the charge-compensating species, and the oxygen sublattice has become exactly filled in an ideal perovskite structure. Further reduction will give compositions that are oxygen-deficient in every respect. The analogous reduction reaction for the case of compensation by cation vacancies involves the elimination of the vacancies by the loss of unit cells

$$y/3V''_{Ba} + y/3V''''_{Ti} + yO_o \rightleftharpoons y/2O_2 + 2ye' \tag{18}$$

Once again, the reduced compositions are $BaTi_{1-x}Nb_xO_{3+x/2-y} + 2ye'$, the ionic defects are replaced by electrons as the compensating species, and the ideal perovskite is reached when y = x/2 at the composition $BaTi_{1-x}Nb_xO_3 + xe'$.

Probably of greater practical interest is the situation where only one type of cation vacancy exists in the perovskite lattice to compensate for the charge of the donor centers. For an ideal perovskite composition, where the number of A-site and B-site occupants are equal, e.g. Ba/(Ti + Nb) = 1, this will require either the separation of a phase rich in the oxide of the species whose vacancy is the compensating defect, or the retention of this species in solid solution to give a mixed situation involving both the impurity substitution and cation-cation nonstoichiometry. We will deal only with the more common case of phase separation. For the case that the titanium vacancy is the preferred defect, the situation can be derived by elimination of the barium vacancies from Eq. (15) by the addition of the reaction

$$x/2O_o + x/6V_{Ba}'' + x/6V_{Ti}'''' + x/4Ti_{Ti} \longrightarrow x/4V_{Ti}'''' + x/4TiO_2 \qquad (19)$$

By canceling the V_{Ti}'''' common to both sides, this can be rewritten

$$x/2O_o + x/6V_{Ba}'' + x/4Ti_{Ti} \longrightarrow x/12V_{Ti}'''' + x/4TiO_2 \qquad (20)$$

The addition of either of these to Eq. (15) gives for the incorporation reaction

$$BaO + (1-x)TiO_2 + x/2Nb_2O_5 \longrightarrow Ba_{Ba} + (1-5/4x)Ti_{Ti} + xNb_{Ti}^{\bullet} + x/4V_{Ti}'''' + 3O_o + x/4TiO_2 \qquad (21)$$

or, if the starting composition is adjusted to avoid the formation of the second phase

$$BaO + (1-5/4x)TiO_2 + x/2Nb_2O_5 \longrightarrow Ba_{Ba} + (1-5/4x)Ti_{Ti} + xNb_{Ti}^{\bullet} + x/4V_{Ti}'''' + 3O_o \qquad (22)$$

The composition of the perovskite phase is $BaTi_{(1-5/4x)}Nb_xO_3$, and is chemically stoichiometric but structurally deficient in cations due to the titanium vacancies. The condition of charge neutrality is $[Nb_{Ti}^{\bullet}] = 4[V_{Ti}'''']$. If the Ti-rich second phase is formed as in Eq. (21), it has been shown that its actual composition is $Ba_6Ti_{17}O_{40}$ (15,16), formerly thought to be a trititanate, $BaTi_3O_7$.

It has been shown by transmission electron microscopy that the only composition that is single phase in the $BaO-TiO_2-Nb_2O_5$ system after processing in air is $BaTi_{(1-5/4x)}Nb_xO_3$, for $x > 0.005$, in accord with the case for preferred compensation by titanium vacancies (6). It will be recalled that for smaller donor concentrations in $BaTiO_3$, compensation is primarily by electrons, according to Eq. (16).

If the barium vacancy is the preferred species, the incorporation reaction can be derived from Eq. (15) by addition of the reaction

$$x/2O_o + x/6V_{Ba}'' + x/6V_{Ti}'''' + x/2Ba_{Ba} \longrightarrow x/2V_{Ba}'' + x/2BaO \qquad (23)$$

or, by elimination of the V_{Ba}'' common to both sides

$$x/2O_o + x/6V_{Ti}'''' + x/2Ba_{Ba} \longrightarrow x/3V_{Ba}'' + x/2BaO \qquad (24)$$

The addition of either of these to Eq. (15) gives the incorporation reaction

$$BaO + (1-x)TiO_2 + x/2Nb_2O_5 \longrightarrow (1-x/2)Ba_{Ba} + x/2V_{Ba}'' + (1-x)Ti_{Ti} + xNb_{Ti}^{\bullet} + 3O_o + x/2BaO \qquad (25)$$

or, if the starting composition is adjusted to avoid the formation of a second phase

$$(1-x/2)BaO + (1-x)TiO_2 + x/2Nb_2O_5 \longrightarrow (1-x/2)Ba_{Ba} + x/2V_{Ba}'' + (1-x)Ti_{Ti} + xNb_{Ti}^{\bullet} + 3O_o \qquad (26)$$

The composition of the perovskite phase is $Ba_{1-x/2}Ti_{1-x}Nb_xO_3$, and the Ba-rich phase, if present, would be expected to be Ba_2TiO_4 (17). While this has been shown not to be the preferred mechanism of incorporation for donor-doped $BaTiO_3$ (6), there is evidence that it does occur in the case of donor-doped $SrTiO_3$ (18).

Since these materials have oxygen-excess compositions by virtue of their cation vacancy content, their reduction back to the ideal perovskite is of interest to us. If the starting composition has the ideal perovskite ratio, such that a second phase remains because of the exsolution of one binary oxide species in order to leave vacancies of that cation, then reduction will result in the gradual loss of the second phase. For the case that titanium vacancies are the favored defect, and there is a Ti-rich phase present, the reduction reaction is then

$$y/2V_{Ti}'''' + y/2TiO_2 \rightleftharpoons y/2Ti_{Ti} + y/2O_2 + 2ye' \quad (27)$$

The reduced compositions are $BaTi_{15/4x+y/2}Nb_xO_3 + (x/4-y/2)TiO_2 + 2ye'$. When y = x/2, the Ti vacancies have been completely replaced by electrons as the charge-compensating species, the material is single phase, and it has the ideal perovskite structure with the composition $BaTi_{1-x}Nb_xO_3 + xe'$. The corresponding reduction reaction for the case of compensation by barium vacancies with a Ba-rich second phase is

$$yV_{Ba}'' + yBaO \rightleftharpoons yBa_{Ba} + y/2O_2 + 2ye' \quad (28)$$

The reduced compositions are $Ba_{1-x/2+y}Ti_{1-x}Nb_xO_3 + (x/2-y)BaO + 2ye'$, with an ideal single phase perovskite with the composition $BaTi_{1-x}Nb_xO_3 + xe'$ being achieved when x = 2y. In both cases, the reduced, ideal perovskite has the same composition as if electrons were the main compensating defect in the first place, i.e. Eq. (16). While this situation can be achieved by sintering a polycrystalline ceramic in a reducing atmosphere, it is dubious that it can be attained by reduction of material which has been processed in an oxidizing atmosphere so that cation vacancies are the compensating defect; cation diffusion is too slow in the perovskite structure to allow the necessary long-range diffusion to occur in reasonable reduction times. The reactions written here assume that equilibrium has been reached, without concern for the kinetic problems involved in getting there.

If the starting composition of the donor-doped titanate is adjusted to give single phase material with compensation by cation vacancies, reduction will result in the separation of a second phase, and a loss of unit cells, as the vacancies are replaced by electrons as the compensating defects. For the case of titanium vacancies as the preferred defect, the reduction reaction is

$$y/2V_{Ti}'''' + y/2Ba_{Ba} + 3y/2O_o \rightleftharpoons y/2O_2 + 2ye' + y/2BaO \quad (29)$$

The reduced compositions are $BaTi_{1-5/4x+y/2}Nb_xO_3 + 2ye' + y/2BaO$, and reach the ideal perovskite structure at $BaTi_{1-x}Nb_xO_3 + xe' + x/4BaO$, when y = x/2. If barium vacancies are the preferred cation vacancy, the reduction reaction is

$$yV_{Ba}'' + yTi_{Ti} + 3yO_o \rightleftharpoons y/2O_2 + 2ye' + yTiO_2 \quad (30)$$

The reduced compositions are $Ba_{1-x/2+y}Ti_{1-x}Nb_xO_3 + 2ye' + yTiO_2$, and reach the ideal perovskite structure at $BaTi_{1-x}Nb_xO_3 + xe' + x/2TiO_2$, when $y = x/2$. As before, the systems may be kinetically hindered by slow cation diffusion from achieving these equilibrium compositions for reasonable diffusion times.

VII. Cation Vacancies in Perovskite Structures

It does not appear possible to add oxygen to a nominally pure perovskite oxide that already has a filled oxygen sublattice by exposure to oxygen partial pressures of one atmosphere or less. As described earlier, the two options of forming either oxygen interstitials or cation vacancies has a prohibitive energetic cost. However, donor impurities are known to be compensated by cation vacancies in many cases. How do we reconcile this apparent inconsistency? It involves the energetics of the electronic structure of the material.

If an oxygen atom is added to pure, stoichiometric $BaTiO_3$ with the formation of cation vacancies, Eq. (6), there is an energetic cost of forming 1/3 of a vacancy on each cation sublattice, and of putting two holes in the valence band. However, if the starting material is a donor-doped perovskite with compensation by electrons, e.g. Nb-doped $BaTiO_3$ processed in a reducing atmosphere to give $BaTi_{1-x}Nb_xO_3 + xe'$, according to Eq. (16), the situation is quite different. Oxidation of this material by the same oxidation reaction, Eq. (6), to give $Ba_{1-x/3}Ti_{1-4x/3}Nb_xO_3$, again yields 1/3 of a vacancy on each cation sublattice and two holes, but, in this case, the two hole states can be filled by two of the compensating electrons from the conduction band. For each added oxygen, two electrons drop from the conduction band to the valence band, and each such recombination gives back the value of the band gap in energy. Thus for donor-doped $BaTiO_3$, the enthalpy of oxidation per added oxygen is 3.4 x 2 = 6.8 eV less than for the undoped material. This is a very substantial energetic advantage, and is apparently enough to make the oxidation of donor-doped $BaTiO_3$ with the formation of cation vacancies (of either or both types) a feasible process. Once the donor content has become fully compensated by cation vacancies, and there are no more electrons in the conduction band to compensate, there is no more energetic advantage to the addition of more oxygen, and the oxidation process stops.

VIII. Structural Accommodation

The perovskite structure is marvelously resilient, and can accommodate large nonstoichiometry or dopant concentrations by a variety of methods. This has become particularly apparent with the study of the complex family of superconducting oxides (1,2). While the discussion in this chapter has so far been limited to classical point defect chemistry, it is only one of the ways that the perovskites react to compositional abuse. It is particularly pertinent for us to note that a structural excess of oxygen can also be accommodated by a modification of the structure. In effect, the extra oxygen associated with donor impurities is sometimes built into the structure in the form of oxygen-rich layers that separate two-dimensional slabs of perovskite structure of various thicknesses (19-23). This has been described in detail in a previous review (1), and will only be described briefly here for the sake of completeness of our survey of "oxidative nonstoichiometry".

The excess oxygen in these structurally modified perovskites comes from a very high concentration of donor impurities. Thus the $SrO-TiO_2-Nb_2O_5$ system is a particularly good model system for this effect. If the TiO_2 component of $SrTiO_3$ is completely replaced by Nb_2O_5, i.e. if the composition is $2SrO + Nb_2O_5$, the product is $Sr_2Nb_2O_7$, sometimes referred to as the pyroniobate in analogy with aqueous anion chemistry. This single phase material consists of two-dimensional slabs of the perovskite structure, four unit cells thick, separated by oxygen-rich layers. As shown schematically in Fig. 4, it is as if an ideal perovskite structure were cut at regular intervals along (100) planes of shared oxygen atoms. When the slabs are separated, the oxygens that lie on the plane of separation can go with only one of the two NbO_6 octahedra with which they had previously been shared, and when extra oxygen is brought in to complete the other octahedra, the $Sr_2Nb_2O_7$ structure results. For lesser amounts of donor dopant, the oxygen-rich planes can be spaced further apart and the perovskite slabs become stepwise thicker. Thus there is a series of homologous structures having the generic composition $Sr_mNb_4Ti_{m-4}O_{3m+2}$ (20-22). $Sr_2Nb_2O_7$ is the reduced formula of the first member of the series with m = 4, and members with successively higher values of m are $Sr_5Nb_4TiO_{17}$, $Sr_3Nb_2TiO_{10}$, $Sr_7Nb_4Ti_3O_{23}$, etc. The perovskite slabs are m octahedra thick. As the value of m increases, the precision of the repeated spacing decreases, and perovskite layers of various thicknesses may be seen by lattice imaging. For compositions between those of the ideal members of the homologous series, intergrowths of different spacings are observed, leading to a rich variety of very complex structures that may have very long repeat distances (24,25). Similar structures can be obtained by substituting acceptor impurities on the A-sites of the perovskite structure. Thus the compounds $Sr_4NaNb_5O_{17}$ and $Sr_4Na_2Nb_6O_{20}$ are members of the homologous series $Sr_4Na_{m-4}Nb_mO_{3m+2}$.

Since the A/B site occupation ratio is unchanged in these series of compounds, the sequence of structural members with excess oxygen can be obtained by reduction as well as by a decrease in the net donor content. Both should result in a move toward more ideal perovskite structures as the amount of excess oxygen is reduced, and the spacing between oxygen-rich layers increases. The reduction of $Sr_2Nb_2O_7$ can be viewed as the

Fig. 4. Schematic structures in the homologous series $A_mB_mO_{3m+2}$. (Reproduced from Ref. (20) with permission from the authors and from Academic Press).

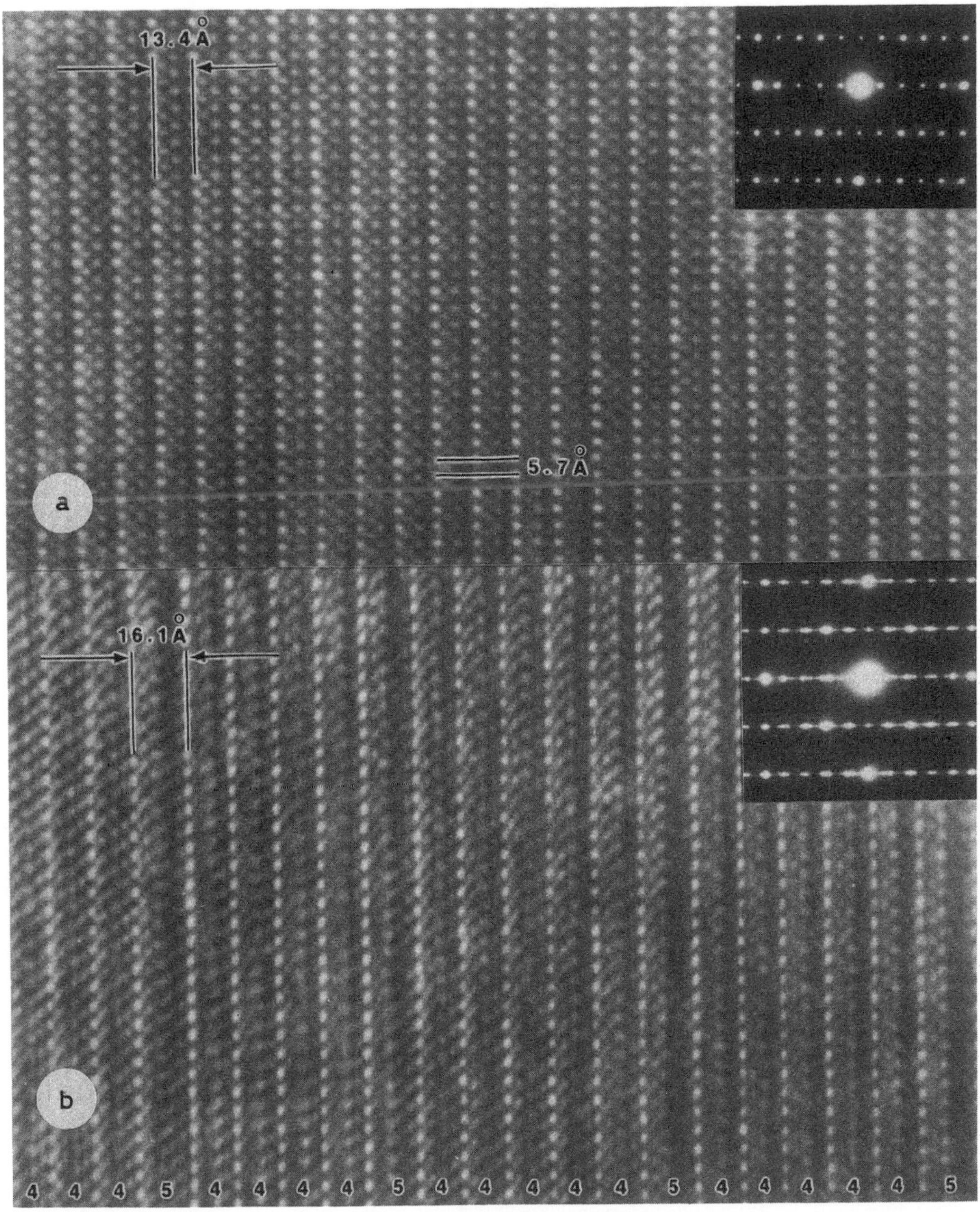

Fig. 5. Lattice images of $Sr_2Nb_2O_7$ in [100] projection. a) Equilibrated in pure O_2 for 3 hours at 1523 K; uniform spacing of 13.4 A corresponds to m = 4 in $Sr_mNb_mO_{3m+2}$. b) Equilibrated in CO/CO_2 mixture with $P(O_2) = 10^{-8}$ atm for 10 hours at 1523 K; intergrowth of 1.61 nm spacings corresponds to m = 5.

replacement of Nb^{+5} by Nb^{+4}, analogous to the case of the replacement of Nb^{+5} by Ti^{+4} described above. Recent studies in our laboratory have shown that the reduction of $Sr_2Nb_2O_7$ soon results in an intergrowth of perovskite layers with thicknesses characteristic of the next member of the series with m = 5. Fig. 5 shows lattice images for stoichiometric and reduced $Sr_2Nb_2O_7$, where spacings corresponding to perovskite layers five units thick are seen to occur randomly in the reduced sample (26). This implies that the perovskite blocks very quickly saturate with oxygen vacancies, and that further reduction results in structural adjustment. The latter phase of reduction can be written as

$$5Sr_2Nb_2O_7 \longrightarrow 2Sr_5Nb_5O_{17} + 1/2O_2 + 2e' \qquad (31)$$

The apparent charge imbalance in this reduction reaction results from a mixture of defect and conventional chemical notation. The stoichiometric formula for the m = 5 member with all cations in their normal oxidation states is $Sr_5Nb_4TiO_{10}$, so that in $Sr_5Nb_5O_{17}$ there has been a substitution of Nb^{+5} for the Ti^{+4}, and the perovskite layers contain $Nb_{Ti}^{\bullet}$ that compensate for the negative charge of the electrons created by reduction. From another point of view, 1/5 of the Nb^{+5} has been reduced to Nb^{+4}, and the latter dissociates into Nb^{+5} and electrons. When all of the Nb has been reduced to the +4 state, the formula is $SrNbO_3 + e'$, and the material should have the ideal perovskite structure.

The superlattice structures of these highly donor-doped materials can be viewed approximately as ordered arrays of interstitial oxygen ions. Disordered versions would be the equivalent of point defect compensation by dispersal of oxygen interstitials throughout a perovskite structure, or by an expansion of the number of unit cells to create compensating cation vacancies. No such disorder reactions have been observed, and this is in accord with the unfavorable nature of interstitial oxygen in this structure, and the kinetic restraints on the formation of cation vacancies because of the long-range cation diffusion that would be required.

IX. Summary

This review has described a number of ways that "oxidative nonstoichiometry" can be incorporated into the perovskite structure. Oxygen excess compositions have been discussed for the following cases: (i) Oxygen contents in excess of that of the stoichiometric composition where the oxygen sublattice is not completely filled. This is the situation for acceptor-doped perovskites with compensation by oxygen vacancies. These materials oxidize toward the ideal perovskite structure as the vacancies are replaced by holes. (ii) Oxygen contents in excess of the stoichiometric composition (in terms of oxygen/metal ratios) in undoped compositions with unfilled oxygen sublattices that result from cation-cation nonstoichiometry. At the stoichiometric composition the nonstoichiometry involves a combination of cation and oxygen vacancies. These materials oxidize toward the cation-deficient perovskite structures as holes replace oxygen vacancies as the compensating defects for the cation vacancies. (iii) Oxygen/metal ratios in excess of the ideal value because of the presence of cation vacancies. This is the case for most donor-doped compositions. These materials reduce toward the ideal perovskite structure as electrons replace cation vacancies as the compensating defects. (iv) Structural accommodation where ordered layers of excess oxygen separate two-dimensional layers of the perovskite structure of various thicknesses. This is the case for highly donor-doped compositions in which structural modification has replaced compensating point defects. These materials reduce toward the ideal perovskite structure as electrons replace the structural modification as the mode of accommodation for the donor content.

Throughout the discussion emphasis has been placed on the distinction between two different reference states: the stoichiometric composition that results from a combination of the stoichiometric binary oxide constituents and corresponds to the condition $n = p$, and the reference structure from which lattice defects are defined. Variable ratios of the binary constituents can also result in cation-cation nonstoichiometry. The compositions described herein represent electrical properties that range from insulating to superconducting. The materials have found use in a wide variety of applications, including capacitors, electrochemical electrodes, catalysts, electro-optic devices, and mechanical sensors and transducers. They combine a fascinating combination of basic science and practical applications.

Acknowledgment

Long-term support from the Division of Materials Research of the National Science Foundation has been an essential ingredient in the development of the concepts described in this chapter, and is gratefully acknowledged. Important contributions to an understanding of the world of perovskites have also been made by my many students, postdocs, and colleagues.

References

1. Smyth, D.M., Composition characterization of dielectric oxides, *Ann. Rev. Mater. Sci.* **15:** 329 (1985).
2. Smyth, D.M., Defects and structural changes in perovskite systems: from insulators to superconductors, *Cryst. Lattice Defects Amorph. Mat.* **18:** 355 (1989).
3. Chan, N.H., R.K. Sharma and D.M. Smyth, Nonstoichiometry in acceptor-doped barium titanate, *J. Am. Ceram. Soc.* **65:** 167 (1982).
4. Smyth, D.M., The role of impurities in insulating transition metal oxides, *Prog. Solid St. Chem.* **15:** 145 (1984).
5. Jonker, G.H. and E.E. Havinga, The influence of foreign ions on the crystal lattice of barium titanate, *Mater. Res. Bull.* **17:** 345 (1982).
6. Chan, H.M., M.P. Harmer and D.M. Smyth, Compensating defects in highly donor-doped barium titanate, *J. Am. Ceram. Soc.* **69:** 507 (1986).
7. Chan, N.H., R.K. Sharma and D.M. Smyth, Nonstoichiometry in undoped barium titanate, *J. Am. Ceram. Soc.* **64:** 5 56 (1981).
8. Lewis, G.V. and C.R.A. Catlow, Computer modeling of barium titanate, *Radiation Effects* **73:** 307 (1983).
9. Mizusaki, J., T. Sosamoto, W.R. Cannon and H.K. Bowen, Electronic conductivity, Seebeck coefficient, and defect structure of lanthanum iron oxide ($LaFeO_3$), *J. Am. Ceram. Soc.* **65:** 363 (1982).
10. Mizusaki, J., T. Sosamoto, W.R. Cannon and H.K. Bowen, Electronic conductivity, Seebeck coefficient, and defect structure of lanthanum strontium iron oxide ($La_{1-x}Sr_xFeO_3$) (x = 0.1, 0.25), *J. Am. Ceram. Soc.* **66:** 247 (1983).
11. Flandermeyer, B.K., M.M. Nasrallah, D.H. Sparlin and H.U. Anderson, in *Transport in Nonstoichiometric Compounds*, G. Simkovich

and V. S. Stubican (Eds.), Plenum, New York, 1985, p. 17.
12. Han, Y.H., M.P. Harmer, Y.H. Hu and D.M. Smyth, in *Transport in Nonstoichiometric Compounds, G. Simkovich and V. S. Stubican, Eds.*, Plenum, New York (1985), pp. 73-85.
13. Han, Y.H., J.B. Appleby and D.M. Smyth, Calcium as an acceptor impurity of barium titanate ($BaTiO_3$), *J. Am. Ceram. Soc.* **70:** 96 (1987).
14. Udayakumar, K.R. and A. Cormack, Nonstoichiometry in alkaline earth-excess alkaline earth titanates, *J. Phys. Chem. Solids* **50:** 55 (1989).
15. O'Bryan, H.M., Jr. and J. Thompson, Jr., Phase equilibriums in the titanium oxide-rich region of the system barium oxide-titanium oxide, *J. Am. Ceram. Soc.* **57:** 522 (1974).
16. Sharma, R.K., N.H. Chan and D.M. Smyth, Solubility of titanium dioxide in barium titanate (IV), *J. Am. Ceram. Soc.* **64:** 448 (1981).
17. Han, Y.H., M. Lal, X.W. Zhang and D.M. Smyth, Dielectric properties of doped barium titanate, *Mater. Sci. Monogr.* **28A:** 369 (1985).
18. Burn, I. and S. Neirman, Dielectric properties of donor-doped polycrystalline strontium titanate (IV), *J. Mat. Sci.* **17:** 3510 (1982).
19. Brandon, J.K. and H.D. Megaw, Crystal structure and properties of $Ca_2Nb_2O_7$ "calcium pyroniobate", *Phil. Mag.* **21:** 189 (1970).
20. Nanot, M., F. Queyroux and J.C. Gilles, Etude cristallographique des termes n = 4.5, 5 et 6 des séries $(La,Ca)_nTi_nO_{3n+2}$, $(Nd,Ca)_nTi_nO_{3n+2}$ et $Ca_n(Ti,Nb)_nO_{3n+2}$, *J. Solid State Chem.* **28:** 137 (1979).
21. Carpy, A., P. Amestoy and J. Galy, Calcium niobate-sodium niobate ($Ca_2Nb_2O_7$-$NaNbO_3$) system. Synthesis and radiocrystallography of members of the series $A_nB_nX_{3n+2}$ (n = 4, 5, 6), *C.R. Acad. Sci. Ser. C* **77:** 501 (1973).
22. Nanot, M., F. Queyroux, J.C. Gilles, A. Carpy and J. Galy, Phases multiples dans les systèmes $Ca_2Nb_2O_7$-$NaNbO_3$ et $La_2Ti_2O_7$-$CaTiO_3$: Les séries homologues de formule $A_nB_nO_{3n+2}$, *J. Solid State Chem.* **11:** 272 (1974).
23. Ishizawa, N., F. Marumo, T. Kawamura and M. Kimura, Compounds with perovskite-type slabs.II. The crystal structure of strontium tantalum oxide $Sr_2Ta_2O_7$, *Acta Crystallogr.* **1332:** 2564 (1976).
24. Portier, R., M. Fayard, A. Carpy and J. Galy, Electron

microscopic study of some members of the (sodium, calcium) niobium oxide ($Na,Ca_nNb_nO_{3n+2}$) series, *Mater. Res. Bull.* **6:** 371 (1974).

25. Portier, R., A. Carpy, M. Fayard and J. Galy, Perovskite-like compounds ABO_{3+x} ($0.44 \leq x < 0.5$). Electron microscopy survey of the sodium calcium niobate $(NaCa)_nNb_nO_{3n+2}$ ($4 < n \leq 4.5$), *Phys. Status Solidi* **30:** 683 (1975).
26. Liu, D.H., Ph.D. dissertation, Xian Jiaotong University, Peoples' Republic of China (1989).

Chapter 4

High T_C Superconductive Cuprates: Classification and Nonstoichiometry

B. Raveau, C. Michel and M. Hervieu

Laboratoire CRISMAT-ISMRA, Bd. du Maréchal Juin, 14050 Caen, France

I. Introduction

A tremendous number of researchers have investigated the copper based oxides since the discovery in 1986 of superconductivity at 40 K in the Ba-La-Cu-O system (1). The critical temperature has rapidly increased in several months from 40 K to 125 K, suggesting a promising field for applications. Although the compounds can be easily obtained they are most times synthesized in the form of mixtures and are not well characterized making it difficult to interpret the superconducting properties of those materials. This difficulty results from the complex crystal chemistry of these oxides which is so far not understood. We present here the structural classification of these oxides and the main rules which govern the nonstoichiometry of these phases.

II. The Different Layered Copper Oxides

Six series of superconductive cuprates have been successively discovered: (i) the La_2CuO_4 type oxides (2-5) which superconduct at 35-40 K; (ii) the $YBa_2Cu_3O_{7-\delta}$ family (6-12) which exhibits T_c's ranging from 92 K to 40 K; (iii) the bismuth cuprates (13-29)

whose T_c's range from 22 K to 108 K; (iv) the thallium cuprates (30-46) for which the highest T_c's of 125 K was observed; (v) the lead cuprates (47-50) characterized by T_c's ranging from 46 K to 90 K; (vi) the Nd_2CuO_4-type oxides (51,52) doped with cerium whose T_c's are close to 25 K.

The five first series of oxides belong in fact to the same structural family and are hole superconductors whereas the sixth series belongs to a different structural type and is an electron superconductor. All these compounds have a common feature, the bidimensional character of the structure, which has been shown to be an important factor for the existence of superconductivity at high temperature (53-55).

The hole superconductors exhibit a mixed valence of copper Cu(II)-Cu(III). They are characterized by the presence of alkaline earth ions in order to stabilize the mixed valence of copper. Their structure can be described as an intergrowth of oxygen deficient perovskite slabs containing m copper layers with rock salt-type slabs containing n AO layers, so that they can be represented by the formulation $(ACuO_{3-x})_m (AO)_n$. Thus the different members of the series can be symbolized [m,n] where m and n are integral numbers in the case of single intergrowths, i.e. containing only one kind of perovskite slab and one kind rock salt-type layers. Non-integral m or n values are also possible; they correspond to the formation of multiple intergrowths, i.e. to the coexistence in the structure of several sorts of perovskite or rock salt-type layers. For instance m = 1.5 results from the existence in the structure of two sorts of perovskite slabs built up of double copper layers (m = 2) and single copper layers (m = 1). Table 1 summarizes the different compounds with such a layered structure which have been isolated up to now.

This table can be analyzed in terms either of horizontal periods or of vertical columns. Considering the different columns, one can propose for each of them the following general formulations :

(i) [1,n] compounds formulated $A_{n+1}CuO_{3+n}$ are characterized by single perovskite layers (Fig. 1) built up of corner-sharing CuO_6 octahedra.

(ii) [2,n] compounds whose formulation is $A_{n+1}A'Cu_2O_{5+n}$ all exhibit double layers of corner-sharing CuO_5 pyramids (Fig. 2) interleaved with A' cations (A'= Ca, Y or Ln, and rarely Sr).

(iii) [3,n] oxides which are formulated $A_{n+1}A'_2Cu_3O_{7+n}$ correspond to the presence of triple copper layers (Fig. 3) involving two pyramidal layers and one layer of corner-sharing

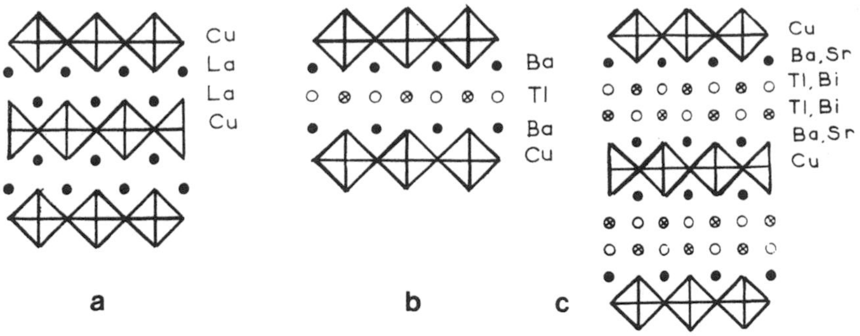

Fig. 1. Structures of the [1,n], $A_{n+1}CuO_{3+n}$ oxides. a) La_2CuO_4 (n = 1); (b) $Tl_1Ba_2CuO_5$ (n = 2); (c) $Bi_2Sr_2CuO_6$ and $Tl_2Ba_2CuO_6$ (n = 3).

CuO_4 square planar groups interleaved with A' cations (A' = Y, Sr or Ca).

(iv) [4,n] compounds, which correspond to the formula $A_{n+1}A'_3Cu_4O_{9+n}$ are only obtained in the case of thallium (A = Tl, Ba) and for A' = Ca; they are characterized by quadruple copper layers of corner-sharing pyramids and of corner-sharing CuO_4 square planar groups (Fig. 4).

(v) [1.5,n] oxides correspond to the intergrowth of [1,n] oxides and [2,n] oxides. Thus they would be formulated $A_{2n+2}A'$

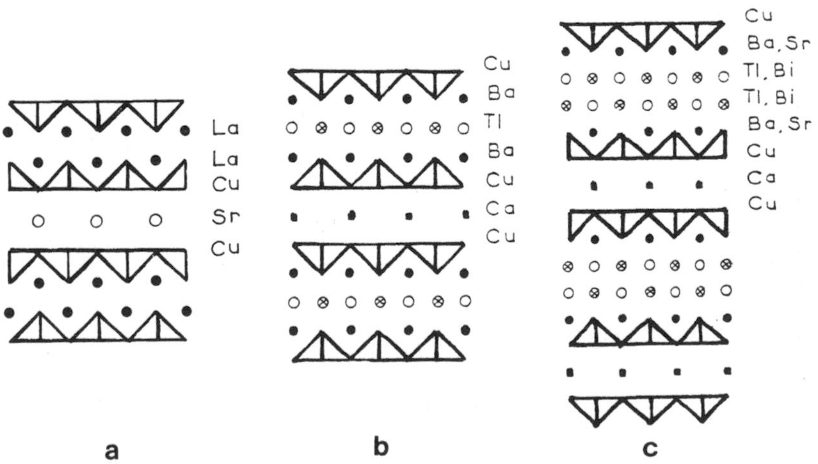

Fig. 2. Structures of the [2,n] $A_{n+1}A'Cu_2O_{5+n}$, oxides. (a) $La_2SrCu_2O_6$ (n = 1); (b) $TlBa_2CaCu_2O_7$ (n = 2); (c) $Tl_2Ba_2CaCu_2O_8$ and $Bi_2Sr_2CaCu_2O_8$ (n = 3).

Table 1. The Different Layered Copper Oxides $(AO)_n(A'CuO_{3-y})_m$ Symbolized by [m,n]. N.S. are not oxide superconductors

m \ n	1	2	3
1	[1,1] $La_{2-x}A_xCuO_4$ La_2CuO_4 (A = Ca,Sr,Ba) Tc≈20K-40K	[1,2] $TlBa_2CuO_{5-\delta}$ N.S. $TlSr_2CuO_{5-\delta}$ Traces $Tl_{1-x}Pr_xSr_{2-y}Pr_yCuO_{5-\delta}$ Tc≈40K (x = 0.2, y = 0.4) $Tl_{0.5}Pb_{0.5}Sr_2CuO_{5-\delta}$ N.S. $TlBa_{1+x}La_{1-x}CuO_{5-\delta}$ T_c≈50K (x = 0.2)	[1,3] $Tl_2Ba_2CuO_6$ Tc≈30K $Bi_2Sr_2CuO_6$ Tc≈10-22K
1.5	[1.5,1] $Pb_2Sr_2Y_{1-x}Ca_xCu_3O_8$ T_c = 50K (x = 0.5) $Pb_{2-x}Bi_xSr_2Y_{1-y}Ca_yCu_3O_8$ T_c = 79K (x = 0.6, y = 0.5, 1)		
2	[2,1] $La_{2-x}A_{1+x}Cu_2O_6$ A = Ca,Sr N.S.	[2,2] $TlBa_2CaCu_2O_7$ Tc≈60K $TlSr_2CaCu_2O_7$ Tc≈50K $Tl_{0.5}Pb_{0.5}Sr_2CaCu_2O_7$ Tc≈85K $TlBa_2LnCu_2O_7$ Ln = Pr, Y, Nd N.S. $Pb_{0.5}Sr_{2.5}Y_{0.5}Ca_{0.5}Cu_2O_{7-\delta}$ T_c = 59K	[2,3] $Tl_2Ba_2CaCu_2O_8$ Tc≈105K $Bi_2Sr_2CaCu_2O_8$ Tc≈85K $Bi_{2-x}Pb_xSr_2Ca_{1-x}Y_xCu_2O_8$ Tc≈85K to N.S. $Tl_{3-4x/3}Ba_{1+x}LnCu_2O_8$ Ln = Pr,Nd,Sm (x = 0.25) N.S.
3	[3,1] $PbBaYSrCu_3O_8$ N.S.	[3,2] $TlBa_2Ca_2Cu_3O_9$ Tc≈120K $Tl_{0.5}Pb_{0.5}Sr_2Ca_2Cu_3O_9$ Tc≈120K	[3,3] $Tl_2Ba_2Ca_2Cu_3O_{10}$ Tc≈125K $Bi_{2-x}Pb_xSr_2Ca_2Cu_3O_{10}$ Tc≈110K
4	[4,1]	[4,2] $TlBa_2Ca_3Cu_4O_{11}$ Tc≈108K	[4,3] $Tl_2Ba_2Ca_3Cu_4O_{12}$ Tc≈115K

Cu_3O_{8+n} as shown for instance for the hypothetical oxide "$Pb_2Sr_2YCu_3O_{10}$" (Fig. 5a). The $Pb_2Sr_2YCu_3O_8$-type superconductors (Fig. 5b) derives from this structure by elimination of additional oxygen atoms in the single perovskite layers.

The consideration of the horizontal periods allows three structural series to be formulated : (i) the [m,1] compounds, which are formulated $A_2A'Cu_{m-1}Cu_mO_{2m+2}$ all exhibit single rock salt type layers. They are obtained for A = La, Sr, Ba, Pb and A' = Ca, Sr, Y; (ii) the [m,2] oxides, whose general formulation is $A_3A'_{m-1}Cu_mO_{2m+3}$ are characterized by double rock salt type layers. They have only been synthesized in the case of thallium, i.e. the rock salt layers are always built up from thallium monolayers associated with barium or strontium monolayers (A = Tl + Ba or Sr). In those phases the A' cation interleaved between the copper is most of the time Ca but can also be a rare earth element and especially yttrium; (iii) the [m,3] compounds, $A_4A'_{m-1}Cu_mO_{2m+4}$ whose common features is the triple rock salt type layers, always involve either thallium or bismuth bilayers associated with barium or strontium monolayers (A = Bi or Tl + Sr or Ba). They are mainly obtained for A'= Ca and more rarely for A'= Ln, Y.

The 92 K-orthorhombic superconductor $YBa_2Cu_3O_7$ and its relatives in which yttrium is replaced by a lanthanide belong also to this large family. Their structure (Fig. 6a) formed of

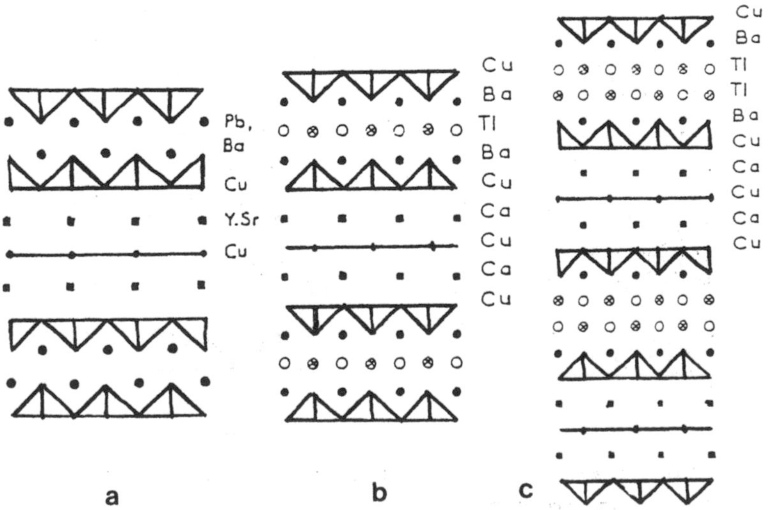

Fig. 3. Structures of the [3,n], $A_{n+1}A'_2Cu_3O_{7+n}$, oxides. (a) PbBaYSr Cu_3O_8 (n = 1); (b) $TlBa_2Ca_2Cu_3O_9$ (n = 2); (c) $Tl_2Ba_2Ca_2Cu_3O_{10}$ (n = 3)

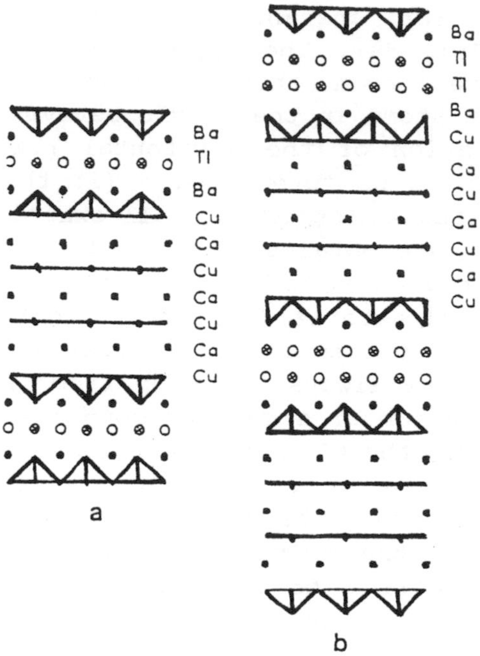

Fig. 4. Structures of the [4,n], $A_{n+1}A'_3Cu_4O_{9+n}$, oxides. (a) $TlBa_2Ca_3Cu_4O_{11}$ (n= 2); (b) $Tl_2Ba_2Ca_3Cu_4O_{12}$ (n = 3).

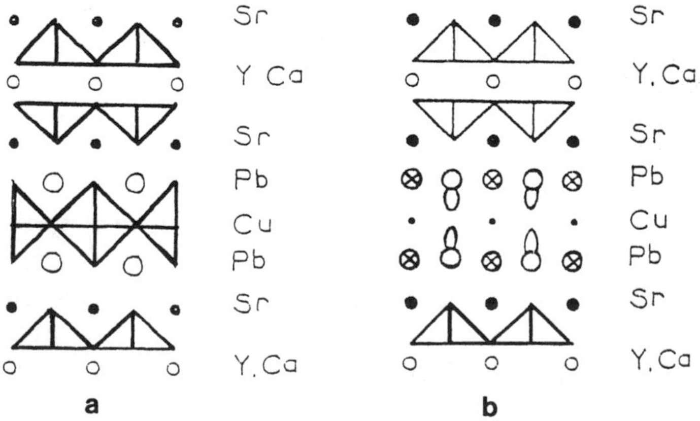

Fig. 5. Structures of the [1.5,n] oxides: (a) hypothetical "$Pb_2Sr_2YCu_3O_{10}$" intergrowth of [1,n] and [2,n], (b) $Pb_2Sr_2Y_{0.5}Ca_{0.5}Cu_3O_8$-type superconductor.

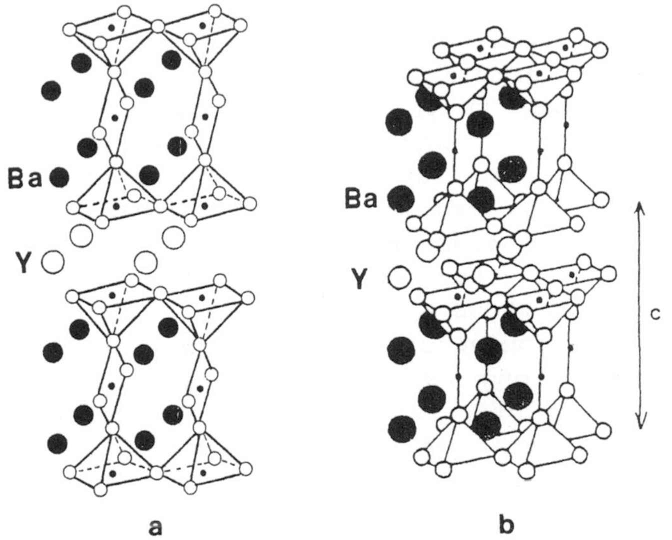

Fig. 6. Structures of orthorhombic superconductive $YBa_2Cu_3O_7$ (a) and insulating tetragonal $YBa_2Cu_3O_6$ (b).

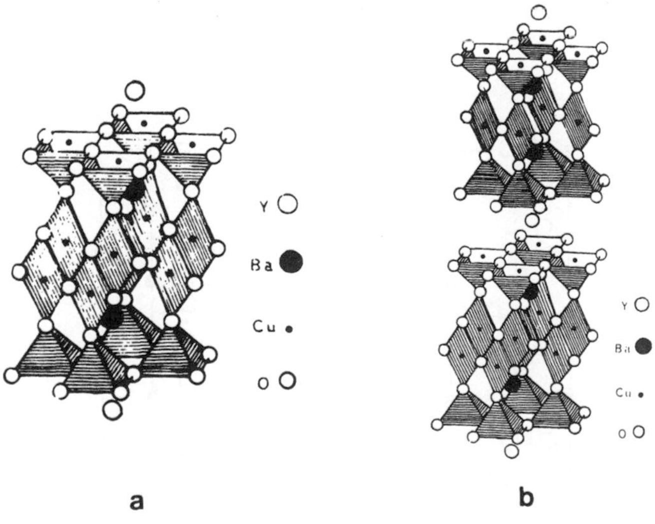

Fig. 7. Structures of (a) $YBa_2Cu_4O_8$ and (b) $Y_2Ba_4Cu_7O_{15}$.

triple copper layers, involving CuO_5 pyramids and CuO_4 square planar groups, interleaved with yttrium (or lanthanide) ions correspond to the formulation $[ACuO_{3-x}]_\infty$ with A = Ba, Y and x=2/3. Thus, it corresponds to the symbol [∞,0]. In the same way the insulating tetragonal oxide $YBa_2Cu_3O_6$ (56-58) whose structure (Fig. 6b) is formed of layers of corner-sharing CuO_5 pyramids linked through CuO_2 sticks, correspond also to the symbol [∞,0]. It is indeed derived from $YBa_2Cu_3O_7$ by removing the oxygen atoms of the square planar groups between the pyramidal layers, leading to the formulation $[ACuO_2]_\infty$ i.e. x = 1.

The hole superconductors $YBa_2Cu_4O_8$ (55, 59-61) and $Y_2Ba_4Cu_7O_{15}$ (60,62) do not belong to this very large structural family. Nevertheless their structures are closely related to that of $YBa_2Cu_3O_7$. The structure of $YBa_2Cu_4O_8$ (Fig. 7a) is indeed deduced from that of $YBa_2Cu_3O_7$, by replacing the single rows of corner-sharing CuO_4 square planar groups by double rows of edge-sharing CuO_4 square-planar groups. This shearing phenomenon does not kill the superconductivity since this compound exhibits a T_c of 80 K. The great similarity of the two structures $YBa_2Cu_3O_7$ and $YBa_2Cu_4O_8$, allows intergrowths to be predicted along c owing to their bidimensional accord in the (001) plane. Such intergrowths formed of the two sorts of slabs correspond to the general formulation $(YBa_2Cu_3O_7)_n(YBa_2Cu_4O_8)_{n'}$. The structure of $Y_2Ba_4Cu_7O_{15}$ (Fig. 7b) shows that this phase represents the first member (n = n' = 1) of this series.

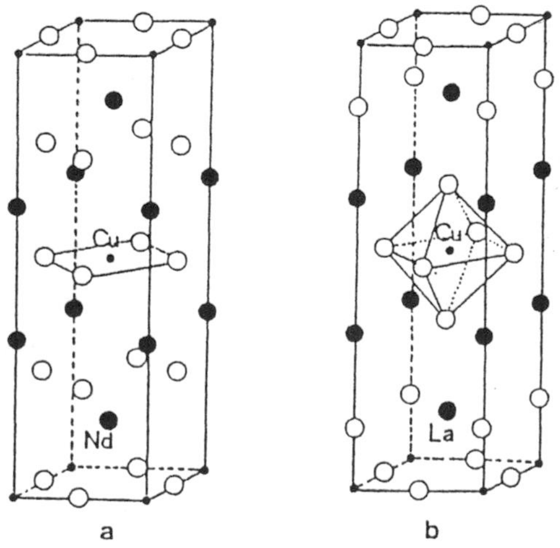

Fig. 8. Structure of Nd_2CuO_4 (a) compared to that of La_2CuO_4 (b).

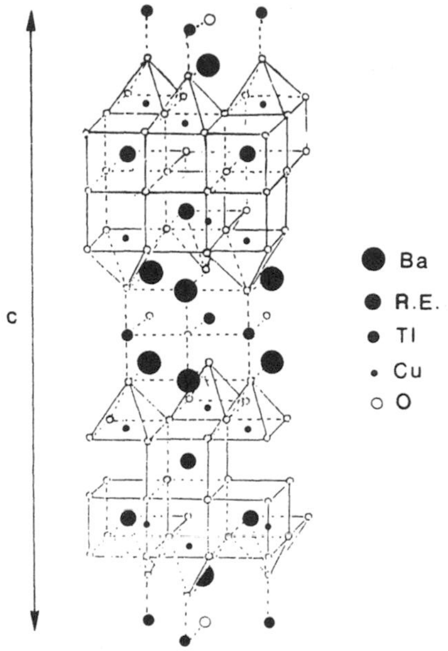

Fig. 9. Structure of the oxides $Tl_{1+x}A_{2-y}Ln_2Cu_2O_9$ intergrowths of double fluorite, double rock salt and pyramidal layers.

Although it does not belong to the same structural family, the electron superconductor $Nd_{2-x}Ce_xCuO_4$ is closely related to the other layered cuprates. Its tetragonal structure (Fig. 8a) exhibits the same cationic positions as those observed for the La_2Cu_4-type (Fig. 8b). Both structures exhibit identical $[CuO_2]_\infty$ layers of corner-sharing CuO_4 square planar groups interleaved with lanthanide ions. They only differ by the positions of two oxygen atoms out of four per Ln_2CuO_4 formula. It results that the Nd_2CuO_4-type structure can also be described as built up from double fluorite-type layers sharing the faces of their NdO_8 pseudo-cubic groups (Fig. 8a). Consequently, this great similarity allows inter-growths between rock salt-type, perovskite and fluorite-type layers to be expected. Although they do not superconduct, the oxides $Tl_{1+x}A_{2-y}Ln_2Cu_2O_9$ (65) are of interest since they confirm this point of view. Their structure (Fig. 9) can indeed be described as an intergrowths of double rock salt-type layers $[(AO)_2]_\infty$ (A = Tl/I) + Ba or Sr), double fluorite type layers $2[Ln_2O_4]_\infty$ and of single layers of corner-sharing CuO_5 pyramids.

III. Nonstoichiometry in Layered Cuprates

The nonstoichiometry phenomena in these materials can be classified into three categories: oxygen nonstoichiometry, intergrowth defects, and cationic disordering. These features are of importance, since they influence the Cu(III)/Cu(II) ratio, and consequently affect the superconducting properties of the materials.

A. Oxygen Nonstoichiometry

The oxygen nonstoichiometry results from the great flexibility of the perovskite framework (64) and from the ability of copper to take coordinations smaller than six, i.e. pyramidal and square planar coordinations.

The most spectacular oxygen nonstoichiometry concerns the "123" oxide $YBa_2Cu_3O_{7-\delta}$, whose homogeneity range, $0<\delta<1$, is characterized by a transition from the orthorhombic to the tetragonal symmetry (65-67) as δ increases leading to a dramatic decrease of the critical temperature (Fig. 10). The X-ray and neutron diffraction studies allow only an average structure to be established (Fig. 11) for the intermediate δ values. This

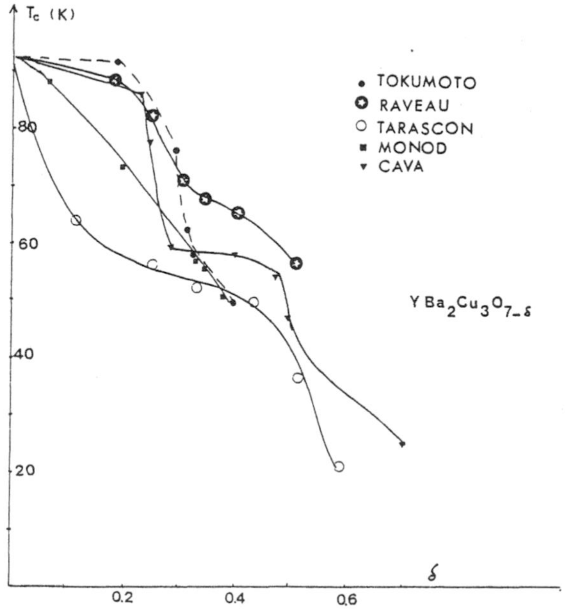

Fig. 10. Evolution of the critical temperature of $YBa_2Cu_3O_{7-\delta}$ versus δ.

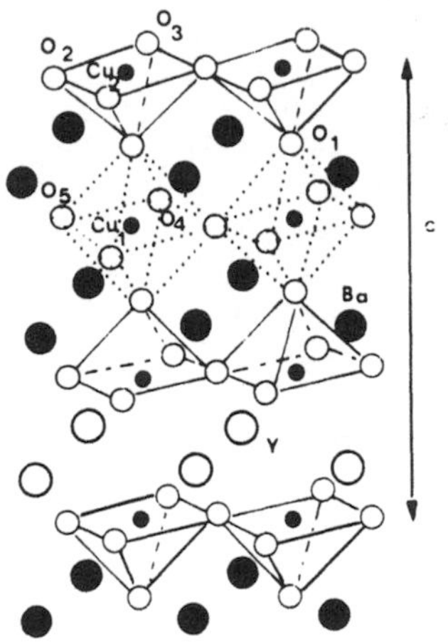

Fig. 11. Average structure deduced from X-ray and neutron diffraction for tetragonal intermediate phases $YBa_2Cu_3O_{7-\delta}$.

latter structure which corresponds to a statistical distribution of the oxygen atoms and vacancies in the intermediate plane (z = 0) between the pyramidal layers would involve for the Cu(1) atoms located at this level (Fig. 11) a threefold coordination not likely for those oxides.

The high resolution electron microscopy studies (68-79) shed light on this problem. One indeed observes for those intermediate compositions, modulations of the structure involving various superstructures limited to small areas in the same crystal (Fig. 12). These microdomains, in which one observes doubling of the a parameter of $YBa_2Cu_3O_7$ (Fig. 12a) or a doubling of a and c parameters (Fig. 12b) or more complex superstructures which set up along [110] (Fig. 12c) or [210], [110], [310] (Fig. 12d), are explained by structural models in which Cu(1) takes only three sorts of coordinations, twofold and square planar and pyramidal coordinations (Fig. 13). In fact such "local superstructures" explain the superconducting properties of the intermediate compositions. For instance $YBa_2Cu_3O_{6.50}$ (δ = 0.5), has been observed to be superconductive in spite of the charge balance which would only involve Cu(II). For all these intermediate δ values, a partial disproportionation of Cu(II) into Cu(III) and Cu(I) can be proposed (80) according to the equation : 2Cu(II)—> Cu(III) + Cu(I). Thus, it is easy to understand that these superstructures (Fig. 13) correspond to the association of

(a)

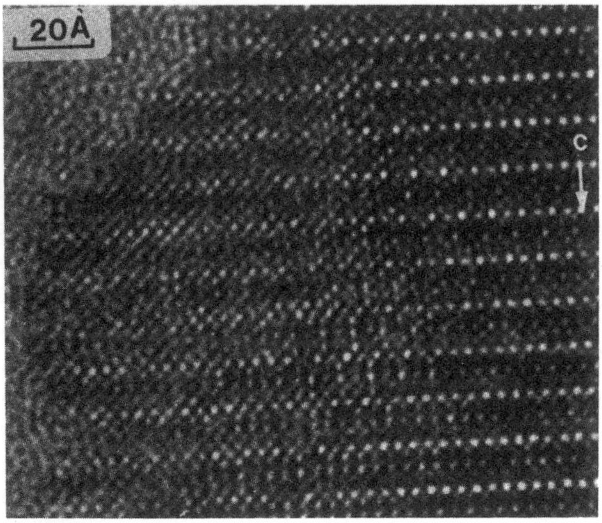

(b)

Fig. 12. High resolution electron microscopy micrographs of superconducting $YBa_2Cu_3O_{7-\delta}$ (δ = 0.4) showing (a) 2 x a superstructure (b) 2a x 2c superstructure and superstructures setting up along (c) [110], (d) [210], [110] and [310] in the same crystal.

insulating $YBa_2Cu_3O_6$-type regions only characterized by the couple Cu(II)-Cu(I) in which Cu(I) exhibits the classical twofold coordination with superconductive $YBa_2Cu_3O_7$-type regions in which the mixed valence Cu(II)-Cu(III) allows both square planar and

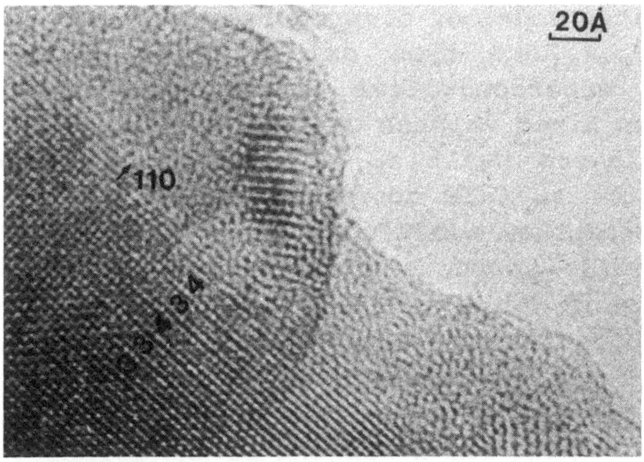

(c)

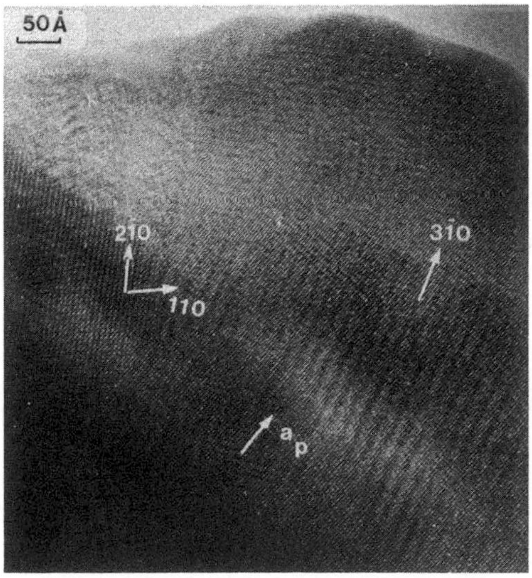

(d)

pyramidal coordinations to be satisfied. Consequently, this model of disproportionation can be represented for all δ values by the formulation $(YBa_2Cu^{II}{}_2Cu^{III}O_7)_{1-\delta}(YBa_2Cu^{II}{}_2Cu^{I}O_6)_\delta$. It must also be pointed out that such a model is strongly supported by the X-ray absorption studies of that phase (81) which show without any ambiguity the presence of univalent copper whatever δ may be ($\delta \neq 0$) and that the Cu(I) content increases progressively as δ increases.

This static model of disproportionation is also valuable for the $Pb_2Sr_2Ca_{0.5}Y_{0.5}Cu_3O_8$-type oxides whose structure (Fig. 5b) consists of superconductive layers (double pyramidal layers) involving the mixed valence Cu(II)-Cu(III), which alternate with insulating layers built up of CuO_2 sticks in which univalent copper has the twofold coordination. This latter structure is also able to capture additional oxygen, leading to the formation of CuO_5 pyramids in the insulating layer. Moreover this oxygen substoichiometry tends to kill the superconductivity.

On the opposite the La_2CuO_4-type oxides do not exhibit any copper disproportionation from the X-ray adsorption measurements (82). The formation of oxygen vacancies in the basal plane of the CuO_6 octahedra (Fig. 1a) according to the formulation $La_{2-x}A_xCuO_{4-\delta}$,

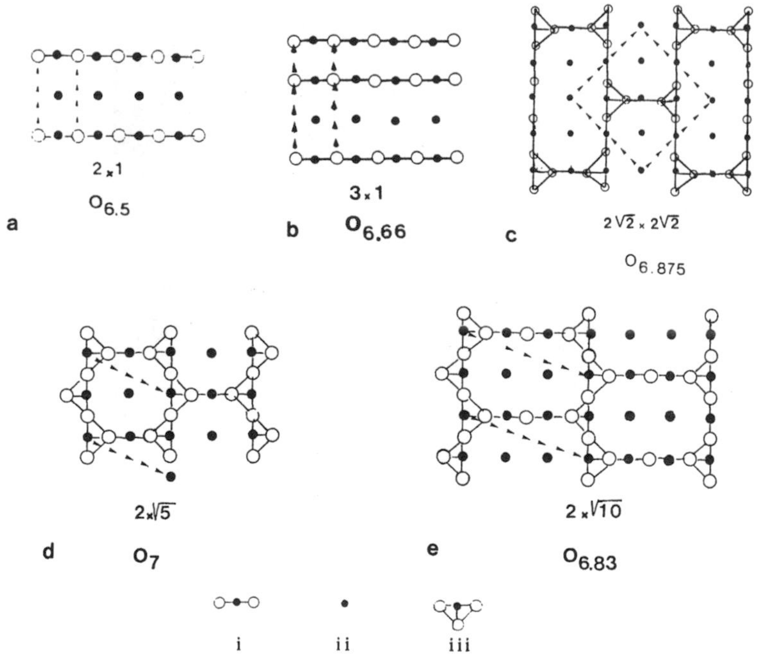

Fig. 13. Possible models resulting from ordering of oxygen vacancies in the $[CuO_2]_\infty$ layers at z = 0. Just the intermediate layer of polyhedra is drawn with Cu at z = 0. The new supercells are : (a) 2a x b, (b) 3a x b. (c) 2a $\sqrt{2}$ x 2a $\sqrt{2}$, (d) 2a x a $\sqrt{5}$ and (e) 2a x a $\sqrt{10}$. The symbols used are: i) CuO_4 square planar group perpendicular to the projection, ii) CuO_2 group (corresponding to Cu(I) only) perpendicular to the projection, iii) CuO_5 pyramid whose basis is perpendicular to the projection.

leads indeed either to a pyramidal or to a square planar coordination which is always compatible with the mixed valence Cu(II)-Cu(III).

The possibility of introduction of an oxygen excess in the $YBa_2Cu_3O_7$-type structure has been observed by high resolution electron microscopy leading to the formation of small regions in the crystal (some nanometers) of the $YBa_2Cu_3O_8$-type in which Cu(1) can present the pyramidal or the octahedral coordination due to the introduction of additional oxygen between the rows of CuO_4 groups at the Cu(1) level. On the opposite the possibility of an excess of oxygen in the $La_2CuO_{4+\delta}$ has not yet been proved. Nevertheless, the isotypic oxide $La_2NiO_{4+\delta}$ was found to present an excess of oxygen located in the rock salt-type layers (83).

The thallium cuprates do not seem to present any oxygen vacancies if one excepts the oxide $Tl_2Ba_2CuO_{6-\delta}$ and $Tl_{1-x}Pr_xa_2CuO_{5-\delta}$ for which oxygen vacancies may appear in the basal planes of the CuO_6 octahedra of the single perovskite layers (see Fig. 1b-c). Such a phenomenon is also susceptible to appear in the oxide $Bi_2Sr_2CuO_6$ (Fig. 1c) but has not been proved. The question of oxygen non-stoichiometry in bismuth and lead cuprates, especially in the rock salt-type layers, is so far not clear owing to the problems of incommensurability of the structure which are not resolved.

B. Intergrowth Nonstoichiometry

The ability of all these phases to form numerous intergrowths of the perovskite and rock salt-type layers suggests the possibility of existence of numerous stacking faults, called intergrowth defects. It has been shown (84) that high resolution electron microscopy is a powerful tool to study this non-stoichiometry phenomenon. Thus the HREM observations will not be presented here.

It must be pointed out that the La_2CuO_4-type oxides as well as the bismuth cuprates rarely exhibit intergrowth defects contrary to the thallium cuprates. It is too early to draw conclusions for the lead compounds whose investigation started only recently.

The thallium cuprates exhibit two sorts of defects concerning the variation of the thickness of the perovskite slabs and of the rock salt-type respectively.

In the perovskite slabs, the defects can be called m', with

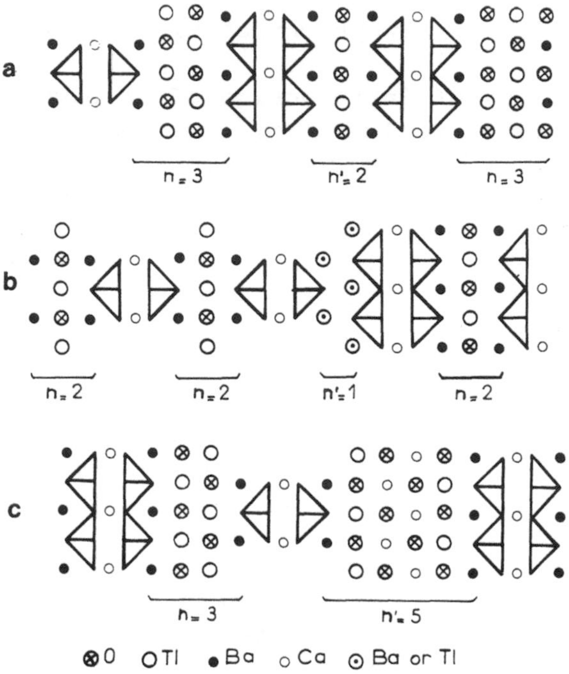

Fig. 14. Structural models showing (a) a n'= 2 "$TlBa_2CaCu_2O_7$" defect in the regular matrix of $Tl_2Ba_2CaCu_2O_8$ (n = 3), (b) a n' = 1 defect in a regular matrix of $TlBa_2CaCu_2O_7$ (n = 2), (c) a n'= 5 defect corresponding to the introduction of two additional CaO layers in the n = 3 matrix of $Tl_2Ba_2CaCu_2O_8$.

respect to the matrix characterized by the number m, where m and m' correspond to the number of copper layers forming the perovskite slabs in the matrix and in the defect respectively. For low m values, especially for m = 1 or 2, the defects are rather rare and m' differs generally from m by one unit (m' = m+1). For higher m values, the perovskite slabs can present thicker defects and the defects are more numerous ; this is the case for instance of the m' = 7 defects which appear in a m = 4 matrix of $Tl_2Ba_2Ca_3Cu_4O_{12}$ (84). Thus, a first general rule is that the frequency of intergrowth defects increases as m increases.

Variations of the thickness of the rock salt-type layers are currently observed in the thallium cuprates. This is due to the great mobility of thallium at high temperature and also to the fact that the thallium phases exhibit two sorts of rock salt-type slabs, double slabs (n = 2) involving thallium monolayers, and triple slabs (n = 3) involving thallium bilayers. Thus they are quite numerous for n = 3 or n = 2, as shown for instance from the existence of n' = 2 defects ("$TlBaCaCu_2O_7$") in the regular matrix of $Tl_2Ba_2CaCu_2O_8$ (Fig. 14a) which is characterized by triple rock salt-type layers (n = 3). On the opposite one observes rarely n' = 1 defects. Such a defect corresponding to the formation of simple rock salt-type layers has been observed for the oxide $TlBa_2CaCu_2O_7$ (n = 2) (Fig. 14b). The thickness of the rock salt layers can be increased by introduction of additional $[CaO]_\infty$ layers between the thallium layers as shown for instance for the n' = 5 defects observed in the n = 3 matrix of $Tl_2Ba_2CaCu_2O_8$ (Fig. 14c).

The possible intergrowth of fluorite-type layers with rock-salt type and perovskite layers leads also to the formation of intergrowths defects which can be detected by high resolution electron microscopy as shown for instance for the double fluorite-type defects observed in $TlBa_2NdCu_2O_7$ (85).

C. Cationic Disordering

The great number of cations often present in the structures is susceptible to order-disorder phenomena, depending on the experimental conditions of synthesis.

Such phenomena are sometimes observed in the $YBa_2Cu_3O_7$ oxides.

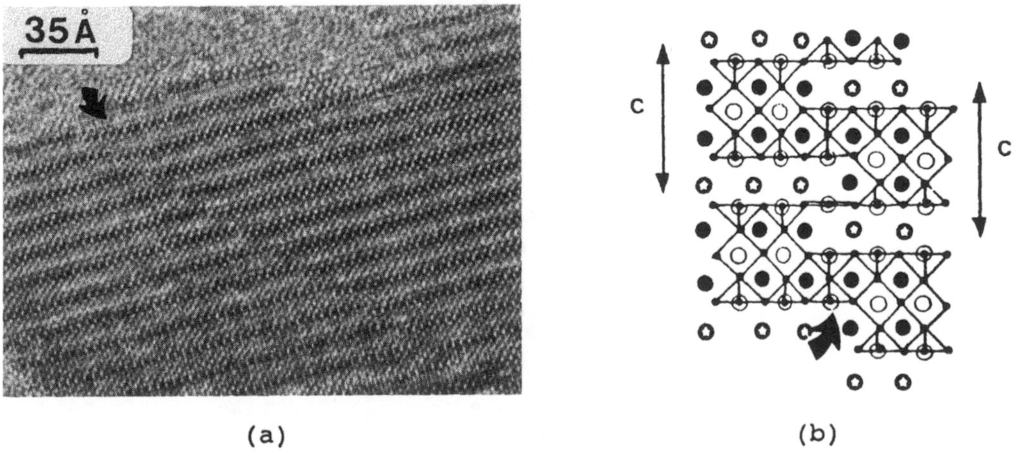

(a) (b)

Fig. 15. (a) HREM image and (b) idealized model of a reversal between barium layers and yttrium layers in $YBa_2Cu_3O_{7-\delta}$. The additional CuO_5 pyramids are labelled by a curved arrow.

High resolution electron microscopy studies (55) have indeed shown that a reversal between barium and yttrium layers sometimes appears (Fig. 15) in which barium layers are suddenly stopped and replaced by yttrium layers and reciprocally. Such extended defects which are rather rare, involve a variation of the oxygen stoichiometry since they imply the formation of additional CuO_5 pyramids at the junction of the two parts of the crystals (see arrow Fig. 15b).

Another example of cation disordering is more often observed in the thallium cuprates. It concerns the accidental replacement of a calcium layer by a thallium layer. Such a phenomenon often induces complex defects involving both, intergrowth defects and thallium-calcium substitution. This has been observed by HREM in $Tl_2Ba_2CaCu_2O_8$. Such a defect (Fig. 16) corresponds to the transfer of one $[TlO]_\infty$ layer from the triple rock salt type slab (n = 3) to the normal calcium layer leading to a double defect involving a double rock salt type layer (n' = 2) associated with a thallium layer.

The synthesis of the oxides $Tl_{3-4x/3}Ba_{1-x}LnCu_2O_8$ with x=0.25 (86) is interesting although these materials do not show superconductivity. It shows that thallium can be distributed between two

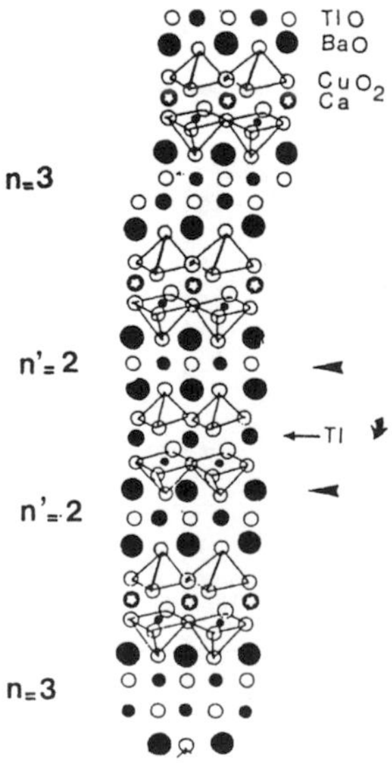

Fig. 16. Structural model showing a complex defect corresponding to the replacement of Ca by Tl, and the formation of a n' = 2 defect ("$TlBa_2CaCu_2O_7$") in a n = 3 matrix of $Tl_2Ba_2CaCu_2O_8$.

sorts of layers in the rock-salt type slabs, thallium bilayers and the nominal barium layers. Such a phenomenon could result from the existence for thallium of a mixed valence Tl(I)/Tl(III). The possibility of thallium nonstoichiometry in the superconductors of the system Tl-Ba-Ca-Cu-O by introducing the mixed valence Tl(I)-Tl(III) should also be considered.

IV. Concluding Remarks

The classification of the layered copper oxides shows that the two main factors which govern superconductivity in those oxides are the low dimensionality of the structure and the mixed valence of copper. The analysis of the non-stoichiometry phenomena in those structures shows that their mechanism is complex and that they may dramatically influence the superconducting properties of the different phases. Consequently, it is worth pointing out that the conditions of synthesis must be carefully controlled in order to optimize these materials. Finally attention must be drawn to a structural feature, which has not been exposed here and which deals with the problem of structure incommensurability. This latter phenomenon, which corresponds to the observation of extra spots and especially satellites in electron diffraction patterns in bismuth and lead cuprates (87), is in fact related to the presence for Bi(III) and Pb(II) of the $6s^2$ electronic pair which is known to be stereochemically active. Detailed work will have to be performed in order to establish relationships between nonstoichiometry and incommensurability.

References

1. Bednorz, J.G. and K.A. Müller, Possible high T_c superconductivity in the barium-lanthanum copper oxygen system, *Z. Phys. B*, **64:** 189 (1986).
2. Michel, C., B. Raveau, Copper-doped barium yttrium zinc oxide Y_2BaZnO_5: ESR and optical investigation, *Rev. Chim. Min.*, **21:** 407 (1984).
3. Fujita, T., Y. Aoki, Y. Maeno, J. Sakurai, H. Fukuba, H. Fujil, Crystallographic, magnetic and superconductive transition in lanthanum barium copper oxide, $((La_{1-x}Ba_x)_2 CuO_{4-y})$, *Jpn. J. Appl. Phys.*, **26:** 202 (1987).

4. Nguyen, N., J. Choisnet, M. Hervieu and B. Raveau, Oxygen defect potassium nickel fluoride (K_2NiF_2)-type oxides: the compounds lanthanum strontium copper oxide ($La_{2-x}Sr_xCu_{4x/2+\delta}$), *J. Solid State Chem.*, **39:** 20 (1981).
5. Goodenough, J.B., N.F. Mott, G. Demazeau, M. Pouchard and P. Hagenmuller, Comparative study of the magnetic behavior of the lanthanum nickel trioxide and lanthanum copper trioxide phases, *J. Solid State Chem.*, **8:** 325 (1973).
6. Wu, M.K., J.R. Ashburn, C.J. Torng, P.H. Hor, R.L. Meng, L. Gao, Z.J. Huang, Y.Z. Wang and C.W. Chu, Superconductivity at 93 K of a new mixed-phase yttrium barium-copper-oxygen compound system at ambient pressure, *Phys. Rev. Lett.*, **58:** 908 (1987).
7. Cava, R.J., B. Battlog, R.B. Vandover, D.N. Murphy, S. Sunshine, T. Siegrist, J.P. Remeika, E.A. Rietman, S. Zahurak and G.P. Espinosa, Bulk superconductivity at 91 K in single-phase oxygen deficient perovskite barium yttrium copper oxide ($Ba_2YCu_3O_{9-\delta}$, *Phys. Rev. Lett.*, **58:** 1676 (1987).
8. Michel, C., F. Deslandes, J. Provost, P. Lejay, R. Tournier, M. Hervieu and B. Raveau, The oxide yttium barium cuprate ($YBa_2Cu_3O_{8-y}$) a novel mixed valence copper perovskite, with oxygen defect, superconductive below 91 K, *C.R. Acad. Sci.*, **304:** 1059 (1987).
9. Lepage, Y., W.R. McKinnon, J.M. Tarascon, L.H. Greene, G.W. Hull, D.M. Hwang, Room temperature structure of the 90 K bulk superconductor yttrium barium copper oxide ($YBa_2Cu_3O_{8-x}$) *Phys. Rev. B*, **35:** 7245 (1987).
10. Roth, G., D. Ewert, G. Heger, C. Michel, M. Hervieu, B. Raveau, F. d'Yvoire and A. Revcolevschi, Phase transformations and microtwinning in crystals of the high T_c superconductor yttrium barium copper oxide ($YBa_2Cu_3O_{8-x}$, x ≈ 1.0), *Z. Phys. B*, **69:** 21 (1987).
11. Beno, M.A., L. Soderholm, D.W. Capone, D. Hinks, J.D. Jorgensen, I.K. Schuller, C.U. Segre, K. Zhang and J.D. Grace, Structure of the single phase high temperature superconductor yttrium barium copper oxide ($YBa_2Cu_3O_{7-\delta}$), *Appl. Phys. Lett.*, **51:** 57 (1987).
12. Capponi, J.J., C. Chaillout, A.W. Hewat, P. Lejay, M. Marezio, N. Nguyen, B. Raveau, J.L. Soubeyroux, J.L. Tholence and R. Tournier, Structure of the 100 K superconductor barium yttrium copper oxide ($Ba_2YCu_3O_7$) at 5-300 K by neutron powder difraction, *Europhys. Lett.*, **12:** 1301 (1987).

13. Michel, C., M. Hervieu, M.M. Borel, A. Grandin, F. Deslandes, J. Provost and B. Raveau, Superconductivity in the bismuth strontium copper oxide system, *Z. Phys.B*, **68:** 421 (1987).
14. Maeda, M., Y. Tanaka, M. Fukutoni and T. Asano, Superconductivity in bismuth-strontium-copper oxides, *Jpn. J. Appl. Phys. Lett.*, **27(2):** L209 and L548 (1988).
15. Hervieu, M., C. Michel, B. Domenges, Y. Lalignant, A. Lebail, G. Ferey and B. Raveau, Electron microscopy study of the superconductor bismuth strontium calcium copper oxide ($Bi_2Sr_2CaCu_2O_{8+\delta}$), *Modern Phys. Lett. B*, **2:** 491 (1988).
16. Torardi, C.C., M.A. Subramanian, J.C. Calabrese, J. Gopalakrishnan, E.M.C. McCarron, K.J. Morrissey, T.R. Askew, R.B. Flippen, U. Chowdhry and A.W. Sleight, Structures of the superconducting oxides thallium barium cuprate ($Tl_2Ba_2CuO_6$) and bismuth strontium cuprate ($Bi_2Sr_2CuO_6$), *Phys. Rev. B*, **38:** 225 (1988).
17. Tarascon, J.M., Y. Lepage, P. Barboux, B.G. Bagley, L.H. Greene, W.R. McKinnon, G.W. Hull, M. Giroud and D.M. Hwang, Crystal substructure and physical properties of the superconducting phase bismuth strontium calcium copper oxide ($Bi_4(Ca,Sr)_6Cu_4O_{16+x}$), *Phys. Rev. B*, **37:** 9382 (1988).
18. Subramanian, M.A., C.C. Torardi, J.C. Calabrese, J. Gopalakrishnan, K.J. Morrissey, T.R. Askew, R.B. Flippen, U. Chowdry and A.W. Sleight, A new high temperature superconductor $Bi_2Sr_{3-x}Ca_xCu_2O_{8+y}$, *Science*, **239:** 1015 (1988).
19. Hervieu, M. B. Domenges, C. Michel and B. Raveau, High resolution electron microscopy of the high T_c superconductor bismuth strontium calcium copper oxide ($Bi_2Sr_2CaCu_2O_{8+\delta}$), *Modern Phys. Lett. B*, **2:** 835 (1988).
20. Bordet, P., J.J. Capponi, C. Chaillout, J. Chenavas, A.W. Hewat, E.A. Hewat, J.L. Hodeau, M. Marezio, J.L. Tholence and D. Tranqui, Powder X-ray and neutron diffraction study of the superconductor bismuth strontium calcium copper oxide, *Physica C*, **153:** 623 (1988).
21. Von Schnering, H.G., L. Walz, M. Schwartz, W. Beker, M. Hartweg, T. Popp, B. Hettich, P. Muller and G. Kampt, Structure of the superconducting bismuth strontium calcium copper oxides ($Bi_2(Sr_{1-x}Ca_x)_2Cu_{8-\delta}$ and $Bi_2(Sr_{1-y}Ca_y)_3Cu_2O_{10-\delta}$, $0 \leq x \leq 0.3$ and $0.16 \leq y \leq 0.33$), *Angew. Chem.*, **27:** 574 (1988).
22. Kasitani, T., K. Kusaba, M. Kikuchi, N. Kobayashi, Y. Syono, T.B. Williams and M. Hirabayashi, *Jpn, J. Appl. Phys.*,

27: L587 (1988).

23. Sunshine, S.A., T. Siegrist, L.F. Schneemeyer, D.W. Murphy, R.J. Cava, B. Batlogg, R.B. van Dover, R.M. Fleming, S.H. Glarum, S. Nakahara, R. Farrow, J.J. Krajewski, S.M. Zahurak, J.V. Waszczak, J.H Marshall, P. Marsh, L.W. Rupp and W.F. Peck, Structure and physical properties of single crystals of the 84 K superconductor bismuth strontium calcium copper oxide ($Bi_{2.2}Sr_2Ca_{0.8}Cu_2O_{8+\delta}$), *Phys. Rev. B*, **38:** 893 (1988).
24. Tarascon, J.M., W.R. McKinnon, P. Barboux, D.M. Hwang, B.G. Bagley, L.H. Greene, G.W. Hull, Y. Le Page, N. Stoffel and M. Giroud, High-temperature superconducting oxide synthesis and the chemical doping of the copper-oxygen planes, *Phys. Rev. B*, **38:** 8885 (1988).
25. Kijima, N., H. Endo, J. Tsuchiya, A. Sumiyama, M. Mizumo and Y. Oguri, Structural properties of two superconducting phases in the bismuth-strontium-calcium-copper-oxygen system, *Jpn. J. Appl. Phys.*, **27:** L821 (1988).
26. Torardi, C.C., M.A. Subramanian, J.C. Calabrese, J. Gopalakrishnan, K.J. Morrissey, T.R. Askew, R.B. Flippen, U. Chowdhry and A.W. Sleight, Structures of the superconducting oxides thallium barium cuprate ($Tl_2Ba_2CuO_6$) and bismuth strontium cuprate ($Bi_2Sr_2CuO_6$), *Science*, **240:** 631 (1988).
27. Zandbergen, H.W., Y.K. Huang, M.J.V. Menken, J.N. Li, K. Kadouaki, A.A. Menovsky, G. van Tendeloo and S. Amelinckx, Electron microscopy on the T_C = 110 K (mid point) phase in the system bismuth trioxide-strontium-oxide-calcium-cupric oxide, *Nature*, **332:** 620 (1988).
28. Hazen, R.M., C.T. Prewitt, R.J. Angel, N.L. Ross, L.W. Fingen, C.G. Hadi-Diacos, D.R. Veblen, P.J. Heaney, P.H. Mor, R.L. Meng, Y.Y. Sun, K. Wang, J. Huang, L. Gao, J. Bechtold and C.W. Chu, Superconductivity in the high-T_C bismuth calcium strontium copper oxide system: phase identification, *Phys. Rev. Lett.*, **60:** 1174 (1988).
29. Politis, C., Synthesis and characterization of amorphous and microcrystalline materials by mechanical alloying, *Appl. Phys. A*, **45:** 261 (1988).
30. Sheng, Z.Z. and A.M. Hermann, Bulk superconductivity at 120 K in the thallium-calcium, barium-copper-oxygen (Tl-Ca/Ba-Cu-O) system, *Nature*, **332:** 55 and 138 (1988).
31. Sheng, Z.Z., A.M. Hermann, A. El Ali, C. Almason, J. Estrada, T. Datta and R.J. Matson, Superconductivity at 90

K in the thallium barium copper oxide system, *Phys. Rev. Lett.*, **60**: 937 (1988).

32. Hazen, R.M., D.W. Finger, R.J. Angel, C.T. Previtt, N.L. Ross, C.G. Hadi-Diacos, P.J. Heaney, D.R. Veblen, Z.Z. Sheng, A. El Ali and A.M. Hermann, 100 K superconducting phase in the thallium calcium barium copper oxide system, *Phys. Rev., Lett.*, **60**: 1657 (1988).
33. Subramanian, M.A., J.C. Calabrese, C.C. Torardi, J. Gopalakrishnan, T.R. Askew, R.B. Flippen, K.J. Morrissey, U. Chowdry and A.W. Sleight, Crystal structure of the high temperature superconductor thallium barium calcium copper oxide ($Tl_2Ba_2CaCuO_8$), *Nature*, **332**: 420 (1988).
34. Politis, C., H. Luo, Thin film studies of superconductors - a review, *Modern Phys. Lett. B*, **2**: 793 (1988).
35. Maignan, A., C. Michel, M. Hervieu, C. Martin, D. Groult and B. Raveau, Thallium barium calcium copper oxide ($Tl_2Ba_2CaCu_2O_8$): structure and superconductivity, *Modern Phys. Lett. B*, **2**: 681 (1988).
36. Hervieu, M., C. Michel, A. Maignan, C. Martin and B. Raveau, The 125 K superconductor thallium barium calcium copper oxide ($Tl_{1-x}Ba_2Ca_2Cu_3O_{10+\delta}$): a tentative structural model, *J. Solid. State Chem.*, **74**: 428 (1988).
37. Parkin, S.S.P., V.Y. Lee, E.M. Engler, A.I. Nazzal, T.C. Huang, G. Gorman, R. Savoy and R. Beyers, Bulk superconductivity at 125 K in thallium calcium barium copper oxide ($Tl_2Ca_2Ba_2Cu_3O_x$), *Phys. Rev. Lett.*, **60**: 2539 (1988).
38. Martin, C., C. Michel, A. Maignan, M. Hervieu and B. Raveau, Thallium barium calcium copper oxide ($TlBa_{2-x}Ca_{2+x}Cu_3O_{10-y}$): a 120 K superconductor, new member of a large family of intergrowths of multiple sodium chloride-type layers with multiple oxygen deficient perovskite layers, *C.R. Acad. Sci.*, **307**, II, 27 (1988).
39. Morosin, B., D.S. Ginley, P.F. Hlava, M.J. Carr, R.J. Baughman, J.E. Schirber, E.L. Venturini and J.K. Kwak, Structural and compositional characterization of polycrystals and single crystals in the bismuth - and thallium - superconductor systems: crystal structure of thallium calcium barium copper oxide ($TlCaBa_2Cu_2O_7$), *Physica C*, **152**: 413 (1988).
40. Domenges, B., M. Hervieu and B. Raveau, HREM study of the 120 K-superconductor thallium barium calcium copper oxide ($TlBa_2Ca_2Cu_3O_{10-y}$), *Solid State Comm.*, **68**: 303 (1988).

41. Parkin, S.S.P., V.Y. Lee, A.I. Nazzal, R. Savoy, R. Beyers and S. La Placa, $Tl_1Ca_{n-1}Ba_2Cu_nO_{2n+3}$ (n = 1, 2, 3): A new class of crystal structures exhibiting volume superconductivity up to ~ 110 K, *Phys. Rev. Lett.*, **61**: 750 (1988).
42. Hervieu, M., A. Maignan, C. Martin, C. Michel, J. Provost and B. Raveau, A new member of the thallium superconductive series, the "1212" thallium barium calcium copper oxide ($TlBa_2CaCu_2O_{8-y}$): importance of oxygen content, *J. Solid State Chem.*, **75**: 212 (1988).
43. Hervieu, M., A. Maignan, C. Martin, C. Michel, J. Provost and B. Raveau, Thallium barium copper oxide ($Tl_2Ba_2Ca_3Cu_4O_{12}$), a new 104 K superconductor of the family $(AO)_3(A'CuO_{3-y})_m$, *Modern Phys. Lett. B*, **2**: 1103 (1988).
44. Bourgault, D., C. Martin, C. Michel, M. Hervieu, J. Provost and B. Raveau, Thallium praseodymium strontium copper oxide $Tl_{1-x}Pr_xSr_{2-y}Pr_yCu_{5-\delta}$: first member of the family $TlA_2Ca_{m-1}Cu_mO_{2m+3}$ (A = barium, strontium), *J. Solid State Chem.*, **78**: 326 (1989).
45. Subramanian, M.A., C.C. Torardi, J. Gopalakrishnan, P.L. Gai, J.C. Calabrese, T.R. Askew, R.B. Flippen and A.W. Sleight, New oxide superconductors, *Science*, **242**: 249 (1988).
46. Martin, C., J. Provost, D. Bourgault, B. Domenges, C. Michel, M. Hervieu and B. Raveau, Structural peculiarities of the "1212" superconductor thallium lead strontium calcium copper oxide ($Tl_{0.5}Pb_{0.5}Sr_2CaCu_2O_7$), *Physica C*, **157**: 460 (1989).
47. Cava, R.J., B. Batlogg, J.J. Krajewski, L.W. Rupp, L.F. Schneemeyer, T. Siegrist, R.B. van Dover, P. Marsh, W.F. Peck, P.K. Gallagher, S.H. Glarum, J.H. Marshall, R.C. Farrow, J.V. Waszczak, R. Hull and P. Trevor, Superconductivity near 70 K in a new family of layered copper oxides, *Nature*, **336**: 211 (1988).
48. Martin, C., D. Bourgault, C. Michel, J. Provost, M. Hervieu and B. Raveau, Thallium lead strontium copper oxide ($Tl_{0.5}Pb_{0.5}Sr_2CuO_{5-\delta}$), a new member of the intergrowth family $TlA_2Ca_{m-1}Cu_mO_{2m+3}$) (A = barium, strontium), *Eur. J. Inorg. Solid State Chem.*, **12**: 157 (1989).
49. Rouillon, T., R. Retoux, D. Groult, C. Michel, M. Hervieu, J. Provost and B. Raveau, Lead barium yttrium strontium copper oxide ($PbBaYSrCu_3O_8$): a new member of the intergrowth family $(ACuO_{3-x})_m(A'O)_n$, *J. Solid State Chem.*, **78**: 322 (1989).
50. Retoux, R., C. Michel, M. Hervieu and B. Raveau, Zero resistance near 80 K in the bismuth lead cuprates $Pb_{2-x}Bi_xSr_2$

$(YCa)_1Cu_3O_8$, *Modern Phys. Lett.* B, **3**: 591 (1989).

51. Grande, B., H.K., Muller-Buschbaum and W. Wollschlager, Zur Kristallstruktur von Seltenerdmetalloxocupraten: La_2CuO_4, Gd_2CuO_4, *Z. Anorg. Allg. Chem.*, **414**: 76 (1975) and **428**: 120 (1977).
52. Akimitsu, J., S. Suzuki, M. Watanabe and H. Sawa, Superconductivity in the bismuth strontium copper oxide system, *Jpn. J. Appl. Phys.*, **27**: L1859 (1988).
53. Labbe, J. and J. Bok, Superconductivity in alkalyne-earth-substituted lanthanum cuprate (La_2CuO_4): a theoretical model, *Europhys. Lett.*, **3**: 1225 (1987).
54. Friedel, J., Broadening of low dimension van Howe singularities in oxide superconductors, *J. Phys. Fr.*, **48**: 1787 (1987).
55. Raveau, B. and C. Michel, Novel superconductivity, in *Proceedings of the International Workshop on Novel Mechanisms of Superconductivity, Berkeley, June 1987, P. Wolff and V. Kresin (Eds.)*, Plenum Press, 1987, p. 599.
56. Roth, G., B. Renker, G. Heger, M. Hervieu, B. Domenges and B. Raveau, On the structure of nonsuperconducting yttrium bariumcopper oxide ($YBa_2Cu_3O_{6+\epsilon}$), *Z. Phys. B*, **69**: 53 (1987).
57. Santoro, A.,S. Miraglia, F. Beech, S.A. Sunshine, D.W. Murphy, L.F. Schneemeyer and J.Y. Waszczak, The structure and properties of barium yttrium barium copper oxide ($Ba_2YCu_3O_6$), *Mat. Res. Bull.*, **22**: 1007 (1987).
58. Bordet, P., C. Chaillout, J.J. Capponi, J. Chenavas and M. Marezio, Crystal structure of yttrium barium copper oxide ($Y_{0.9}Ba_{2.1}Cu_3O_6$), a coumpound related to the high-T_c superconductor $YBa_2Cu_3O_7$, *Nature*, **327**: 687 (1987).
59. Domenges, B., M. Hervieu, C. Michel and B. Raveau, HREM study of the defects in the orthorhombic superconductor yttrium barium copper oxide ($YBa_2Cu_3O_{7\pm\epsilon}$). II. Oxygen substoichiometry and cationic disorder, *Europhys. Lett.*, **4**: 211 (1987).
60. Bordet, P., C. Chaillout, J. Chenavas, J.L. Hodeau, M. Marezio, J. Karpinski and E. Kaldis, Structure determination of the new high-temperature superconductor yttrium barium copper oxide ($Y_2Ba_4Cu_7O_{14+x}$), *Nature*, **334**: 596 (1988).
61. Cava, R.J., J.J. Krajewski, W.F. Peck, Jr., B. Batlogg, L.W. Rupp, Jr., R.M. Fleming, A.C.W.P. James and P. Marsh, Synthesis of bulk superconducting yttrium-barium-copper

oxide ($YBaCu_4O_8$) at one atmosphere oxygen pressure, *Nature*, **338**: 328 (1989).
62. Chaillout, C., P. Bordet, J. Chenavas, J.L. Hodeau and M. Marezio, Present status of high T_c superconductivity, *Solid State Comm.*, **79**: 275 (1989).
63. Martin, C., D. Bourgault, M. Hervieu, C. Michel, J. Provost and B. Raveau, The layered thallium cuprates, $Tl_{1+x}A_{2-y}Ln_2Cu_2O_9$: a triple intergrowth of the perovskite, rock salt, and fluorite structure, *Modern Phys. Lett. B*, **3**: 93, (1989).
64. Raveau, B., Oxides with a tunnel structure characterized by a mixed framework of octahedra and tetrahedra, *Proc. Indian Acad. Sci., Chem. Sci.*, **96**: 419 (1986).
65. Izumi, F., H. Asano, T. Ishigaki, E. Takayama, Y. Uchida, N. Watanabe and T. Nishikawa, Crystal structure of orthorhombic form of barium yttrium copper oxide ($Ba_2YCu_3O_{7-x}$) at 42 K, *Jpn. J. Appl. Phys.*, **26**: 649 (1987).
66. Izumi, F., H. Asano, T. Ishigaki, E. Takayama, Y. Uchida and N. Watanabe, Crystal structure of the tetragonal form of barium yttrium copper oxide ($Ba_2YCu_3O_{7-x}$), *Jpn. J. Appl. Phys.*, **26**: 1214 (1987).
67. Jorgensen, J.D., M. Beno, D.G. Hinks, L. Soderholm, K.J. Volin, R.L. Hitterman, J.D. Grace, I.K. Schuller, C.V. Segre, K. Zhang and M.S. Kleefisch, Oxygen ordering and the orthorhombic-to-tetragonal phase transition in yttrium barium copper oxide ($YBa_2Cu_3O_{7-x}$), *Phys. Rev. B*, **36**: 3608 (1987).
68. Zandbergen, H.W., C.J.D. Hetherington and R. Gronsky, High resolution electron microscopy of the superconducting $YBa_2Cu_3O_{7-\delta}$ system, *J. Supercond.*, **3**: 117 (1988).
69. van Tendeloo, G.,H.W. Zandbergen and S. Amelinckx, The vacancy order-disorder transition in barium yttrium copper oxide ($Ba_2YCu_3O_{7-\delta}$) observed by means of electron diffraction and electron microscopy, *Solid State Comm.*, **63**: 603 (1987).
70. Hervieu, M., B. Domenges, B. Raveau, M. Post, W.R. McKinnon and J.M. Tarascon, Order-disorder phenomena in the "60 K superconductor" yttrium barium copper oxide $YBa_2Cu_3O_{7-\delta}$; $0.37 \leq \delta \leq 0.45$) by means of high resolution electron microscopy studies, *Mat. Lett.*, **8**: 73 (1989).
71. Domenges, B., M. Hervieu, V. Caignaert, B. Raveau, J.L. Tholence and R. Tournier, Superstructures in $YBa_2Cu_3O_{7-\delta}$ oxides, *J. Microsc. Spectrosc. Electron.*, **13**: 75 (1988).
72. Zandbergen, M.W., R. Gronsky, M.Y. Chy, L.C. Dejonghe, G.F.

Holland and A. Stacy, *MRS Fall Meeting*, Jan. 1988.

73. Caps, R.A., J.E. Evelts, B.A. Glowacki, S.B. Newcomb, R.E. Somekh and W.M. Stobbs, in *Proc. High T_c superconductors and potential applications*, Genoa, 1-3 July, 1987, p. 325.
74. Iichihashi, T., S. Iijima, Y. Kubo and J. Tabuchi, Oxygen nonstoichiometry in the $YBa_2Cu_3O_{7-\delta}$ superconductor revealed by electron microscopy, *Jpn. J. Appl. Phys.*, **27**: 385 (1990).
75. Chaillout, C., M.A. Alario Franco, J.J. Capponi, J. Chenavas, J.L. Hodeau and M. Marezio, Electron microscopy studies of the "123" superconductor, *Phys. Rev. B*, **36**: 7118 (1987).
76. Alario Franco, M.A., C. Chaillout, J.J. Capponi and J. Chenavas, A family of non-stoichiometric phases based on barium yttrium copper oxide $Ba_2YCu_3O_{7-\delta}$ ($0\leq\delta\leq1.0$), *Mat. Res. Bull.*, **22**: 1685 (1988).
77. Chaillout, C., M.A. Alario Franco, J.J. Capponi, J. Chenavas, P. Strobel and M. Marezio, Oxygen vacancy ordering in $Ba_2YCu_3O_{7-x}$ around x = 0.5, *Solid State Comm.*, **65**: 283 (1988).
78. van Tendeloo, G., H.W. Zandbergen, T. Okabe and S. Amelinckx, Heavy atom disorder in the high T_c superconductor $Ba_2YCu_3O_{7-\delta}$ studied by means of electron microscopy and electron diffraction, *Solid State Comm.*, **63**: 969 (1987).
79. Zandbergen, H.W., R. Gronsky and G. Thomas, High resolution electron microscopy study of grain boundaries in sintered yttrium barium copper oxide ($YBa_2Cu_3O_{7-x}$), *Phys. Stat. Solidi (a)*, **105**: 207(1988).
80. Raveau, B., C. Michel, M. Hervieu and J. Provost, Structure and nonstoichiometry, two important factors for superconductivity in mixed valence copper oxides, *Physica C*, 153 (1988).
81. Baudelet, F., G. Collin, E. Dartyge, A. Fontaine, J.P. Kappler, G. Krill, J.P. Itie, J. Jegoudez, M. Maurer, Ph. Monod, A. Revcolevski, H.T. Tolentino, G. Tourillon and M. Verdaguer, X-ray absorption spectroscopy of the 90 K superconductor $YBa_2Cu_3O_{7-\delta}$. Possible Cu ions mixed valency and its change with oxygen content and high pressure, *Z. Phys. B*, **69**: 141 (1987).
82. Bianconi, A., J. Budnick, A.M. Flanck, A. Fontaine, P. Lagarde, A. Marcelli, H. Tolentino, B. Chamberland, G. Demazeau, C. Michel and B. Raveau, Evidence of $3d^9$-ligand hole states in the superconductor lanthanum strontium

copper oxide ($La_{1.85}Sr_{0.15}CuO_4$) from L_3 X-ray absorption spectroscopy, *Phys. Lett. A*, **127:** 285 (1987).
83. Jorgensen, J.D., B. Dabrowski, S. Pei, D.R. Richards and D.G. Hinks, Structure of the interstitial oxygen defect in lanthanum nickel oxide ($La_2NiO_{4+\delta}$), *Phys. Rev.*, **40:** 2187 (1989).
84. Raveau, B., C. Michel, C. Martin and M. Hervieu, Recent trends in thallium and bismuth superconductive cuprates, *Mater. Sci. Eng. B*, 3 (1989).
85. Domenges, B., M. Hervieu, C. Martin, D. Bourgault, C. Michel and B. Raveau, Fluorite-type defects in the "1212"-type cuprate $TlBa_2NdCu_2O_7$, *Phase Trans.*, **17:** 231 (1989).
86. Bourgault, D., C. Martin, C. Michel, M. Hervieu and B. Raveau, Thallium mixed valence-thallium(I)/thallium(III) in "2212" thallium cuprates: the oxides thallium barium lanthanide copper oxide ($Tl^{III}_{2-(x/3)}Tl^{I}_{1-x}Ba_{1+x}LnCu_2O_8$), *Physica C*, **158:** 511 (1989).
87. Raveau, B., C. Michel, M. Hervieu, D. Groult and J. Provost, Structural problems and non-stoichiometry in thallium bismuth and lead cuprates, in *Proc. MRS Meeting*, San Diego, April 26, 1989.

Chapter 5

Thermoelectrical Properties of Some Copper Perovskite-Type Materials

O. Cabeza[1], J. Maza[1], Y. Yadava[1], J.A. Veira[1], F. Vidal[1], M.T. Casais[2], C. Cascales[2] and I. Rasines[2]

[1] LAFIMAS, Universidad de Santiago de Compostela, 15706 Santiago de Compostela, Spain

[2] Instituto de Ciencia de Materiales, CSIC, Serrano 113, 28006 Madrid, Spain

I. Introduction

Thermoelectric power (TEP) is recognized as a very sensitive probe for electronic properties in high-temperature superconductors (HTSC), and many works have addressed the study of the normal-state temperature behavior of TEP (1-5). However, much less effort has been devoted to its critical properties near the superconducting transition. In fact, from the few works known to us on critical behavior of TEP the emerging picture is rather confusing. Proposals include, for instance, the attribution of the precursor peak near T_c to thermodynamic fluctuation of the superconducting order parameter (OPF) (6) including a very singular peak not later reproduced (7), phonon-drag enhancement (8), or electro-phonon enhancement (9). We may also quote two and four dimensional OPF for $Y_1Ba_2Cu_3O_{7-\delta}$ compounds (10,11) or fractal dimensionality for Bi-based materials (6,11,12). These latter results are in contrast with extensive work made in HTSC's on electrical resistivity (13-15) or magnetic susceptibility (16) which, in particular, clearly support the 3D and 2D nature of OPF in the mean field region (MFR), i.e., from approximately 1 K to 10 K from the transition for, respectively, Y-based and Bi-based HTSC's. We also note, on the theoretical side, the scarcity of results applicable to critical behavior of the thermoelectric power coefficient, and, in fact, we can only quote an early work by Maki (17).

This chapter compiles simultaneous measurements of electrical resistivity, $\rho(T)$, and thermoelectric power, $S(T)$, for various $Y_1Ba_2Cu_3O_{7-\delta}$ samples. For completeness, some results for $Bi_{1.5}Pb_{0.5}Sr_2Ca_2Cu_3O_y$ samples are also included. The analysis is focused on the excesses of ρ and S in the MFR as a function of temperature. Critical exponents and rounding amplitudes for both observables are compared and correlated (18).

II. Characterization of Samples and TEP Measurements

The $Y_1Ba_2Cu_3O_{7-\delta}$ superconducting polycrystalline samples were prepared from analytical grade Y_2O_3, BaO_2 and CuO which were mixed, ground, pelletized, heated in air at 975 K for 72 hours, and slowly cooled in an oxygen-enriched air atmosphere. The samples were characterized by X-ray powder diffraction, weight loss in hydrogen flow and scanning electron microscopy. Each sample was single phase within 4%, as shown by X-ray analysis, although the oxygen-content differences may be more important from sample to sample. These small stoichiometric differences between different samples lead to relatively small differences in T_{CI}, the temperature where $\rho(T)$ around the transition shows its inflexion point (see Table 1). In contrast, the good chemical homogeneity of each sample is confirmed by the fact that the resistivity transition half-width is very small, of the order of 0.3 K or less. The three samples have very distinct long-scale structural inhomogeneity characteristics, as observed by SEM and optical microscopy. As it is now well understood (13) these long scale structural differences do not change T_{CI}, but cause important differences in the normal resistivity behavior. For instance, $\rho(300\ K)$ is more than ten times larger for the Y1 sample than for the other two, and also $d\rho/dT$ in the normal region varies one order of magnitude from Y1 to sample Y3 (see

Table 1). However, as noted before, these differences can easely be accounted for in terms of an empirical model described elsewere (13). Other details about sample preparation and structural characterization are given elsewhere (19).

TEP of the samples was measured by employing a standard dc differential method. Samples (with typical size 10 x 2 x 1 mm^3) were mounted on a specially designed sample-holder assembly made up of copper. One end of the sample was fixed onto the copper block, with help of GE varnish, the copper block being in contact with the cold tip of the cryostat and serving as the heat sink. On the other end of the sample, a microheater was fitted, for creating the temperature gradient across the sample. The microheater was made up of aluminum with insulated constantan heater wire (0.1 mm diameter) winding giving several mW power. GE varnish was applied at both ends of the sample for better thermal contact of sample with heater and heat sink.

Two copper-constantan thermocouples calibrated with RhFe thermometer were used to monitor the temperature gradient across the sample. Two copper leads (0.1 mm diameter), attached to the sample with a minute amount of silver epoxy, were used to record the thermal voltage. The temperature difference was kept between 1 and 1.5 K/cm, and the temperature sweeping rate was about 5 K/h. An automated data acquisition system comprised of an "8 1/2-

Table 1. General Characteristics of the Samples as Deduced from Electrical Resistivity

Sample	ρ	$d\rho/dT$	T_c	T_{CI}	ΔT_{CI}	$\rho(T_{CI})$	p	ρ_{ct}
	mΩcm	μΩcm/K	K	K	K	mΩcm	10^2	mΩcm
Y1	10.5	21.5	89.6	91.32	0.39	2.27	1.99	2.63
Y2	1.56	4.2	90.6	91.09	0.21	0.28	11.89	0.25
Y3	0.81	2.1	88.5	89.50	0.17	0.17	22.74	0.15

The notations are explained in the main text.

digit" voltmeter with scanner and controller was used for data acquisition. Electrical resistivity of the samples was measured with a four-probe technique using the same data acquisition system as for TEP. Electrical resistivity and TEP resolution were, respectively, 1 mWcm and 0.1 mV/K. Temperature resolution was 10 mK. Measured TEP data were corrected from the copper leads contribution according to the standard tables (20).

III. Extraction of the Fluctuation-Induced TEP

The measurements of TEP and electrical conductivity for the Y2 sample are shown in Fig. 1. In the TEP coefficient a positive value is clearly observed, which indicates that the charge carriers are positive. This section summarizes the main steps in obtaining the excess TEP from these data.

A. Intragrain TEP

A very preliminary question is how measured TEP in HTSC polycrystals correlates with intragrain TEP, i.e., with single crystal TEP data. Because of the random orientations of the grains within a polycrystalline sample, one can imagine a multiplicity of conductive paths, the extreme cases being those conveying along ab planes and those perpendicular to those planes. These paths must be considered as contributing in parallel to the total TEP, similar to stratified multimedia materials or multiband processes. In that case, the known result for the total measured S_M is (20)

$$S_M = \frac{\Sigma\ \sigma_n . S_n}{\Sigma\ \sigma_n} \tag{1}$$

where the subscript n index₋s the distinct paths contributions. Two important facts for the use of this result are that whereas anisotropy of TEP in HTSC's is small, both S_{ab} and S_c being the same order of magnitude (21), conductivity is highly anisotropic verifying $\sigma_{ab} >> \sigma_c$. Hence, a sensible estimate from Eq. (1) for the measured TEP in granular HTSC's samples is $S_M \sim S_{ab}$. This approach is consistent with in-plane TEP data for single crystals being within the order of magnitude to those for granular samples (2-4,21).

Note that the same line of reasoning applied to electrical conductivity does not lead to $\sigma_M \sim \sigma_{ab}$ for granular samples. The basic reason is that the measured quantity is resistance, and the latter is related to resistivity through geometrical parameters of the samples. Stated differently, for the conductivity's extraction from resistance it matters how many paths inside a sample are well connected (i.e., the effective cross-section area), while it does not for TEP.

B. Non-Intrinsic TEP Rounding

The use of a finite temperature difference for measuring TEP causes necessarily a spurious rounding near the transition temperature which may distort the intrinsic rounding effects due to critical fluctuations. If DT is the temperature difference between both ends of the sample, the measured TEP is simply

$$S_M(T) = \int_{T-\Delta T}^{T+\Delta T} S(T')dT' \qquad (2)$$

where T is the mean of the sample-end temperatures. Of course, the difference between S(T) and $S_M(T)$ will increase as the transition is approached. In fact on very general basis, one should expect relevant ΔT-induced rounding effects within $\sim\Delta T/2$ from the transition temperature, T_{CI}, of S(T). When analyzing TEP rounding in the MFR, i.e., from ~ 1 K to ~ 10 K from T_{CI}, this spurious effect must be dealt with if the ΔT utilized is above ~ 1 K. In our case, instead of using a very small ΔT (which usually gives a bad signal-to-noise ratio) we have taken the alternative of removing the ΔT-induced rounding numerically. In concrete terms, Eq.(2) can be recast into the form

$$S(T-n\Delta T) = S(T)-\Delta T \sum_{m=0}^{n-1} \frac{dS_M}{dT} (T-\Delta T/2-m\Delta T) \qquad (3)$$

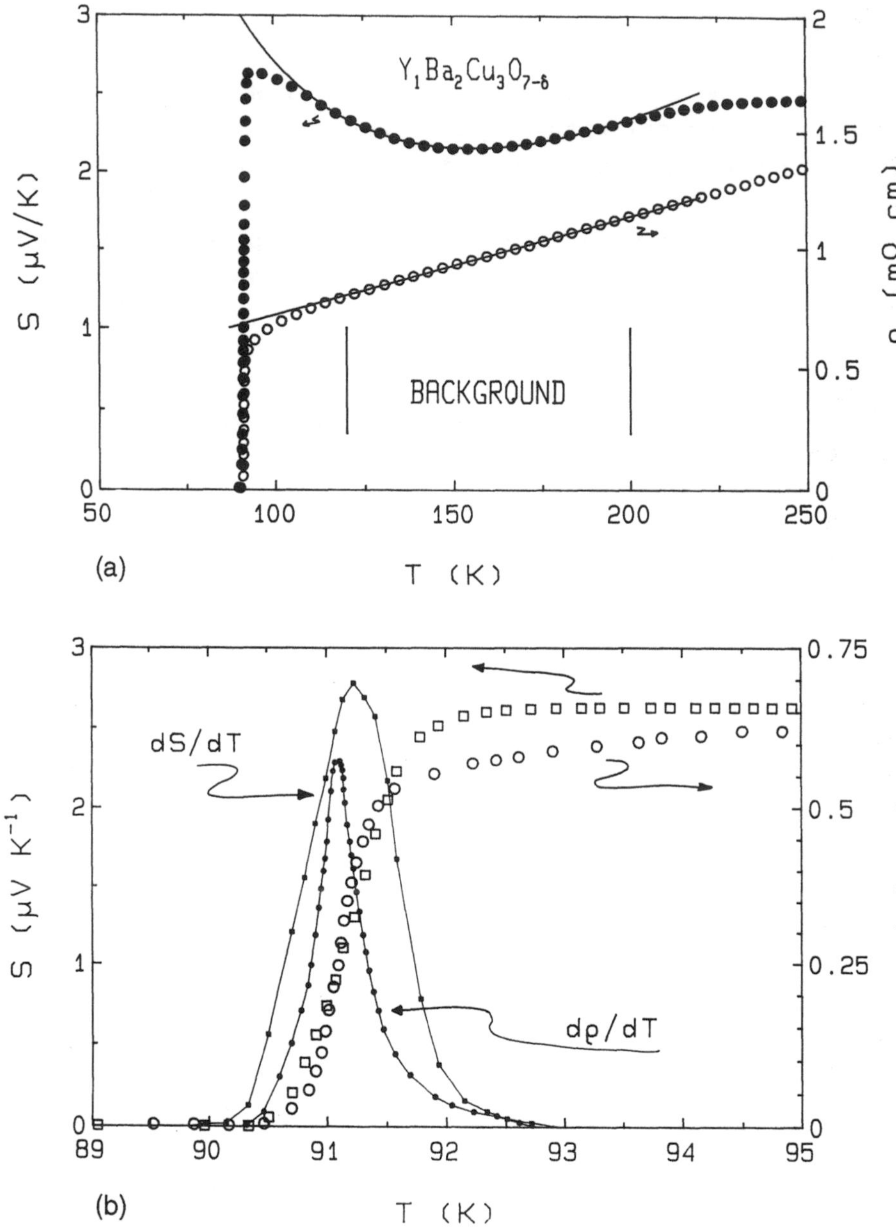

Fig. 1. (a) Thermoelectric power (solid symbols) and resistivity (open symbols) of a typical $Y_1Ba_2Cu_3O_{7-\delta}$ granular sample (Y2). The solid lines show the background dependence: linear for ρ and exponential for S (for details see the main text). (b) Thermoelectric power (squares) and resistivity (circles) in the superconducting transition region. The solid lines are the corresponding derivatives of both magnitudes. Note the closeness of both T_{CI}, which will be taken as the mean field critical temperature to analyze excesses.

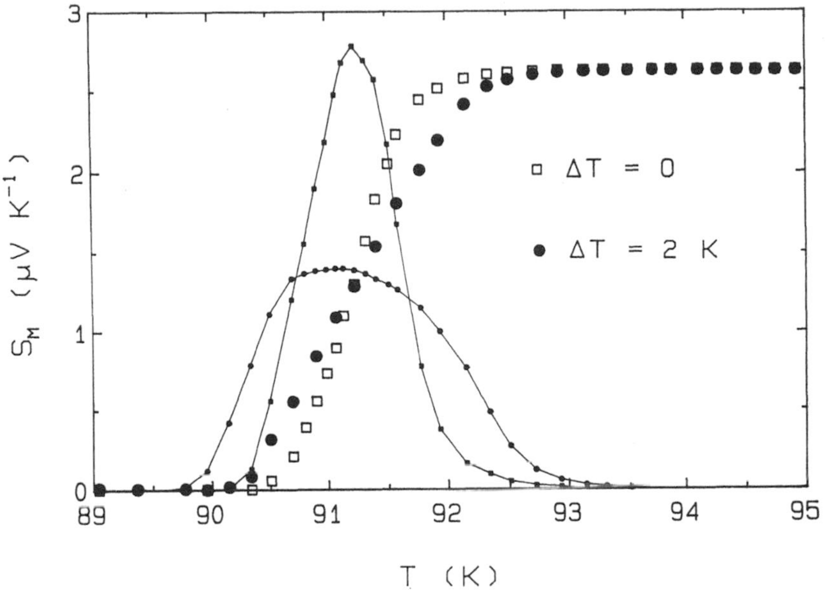

Fig. 2. Thermoelectric power data affected (circles) and free (open squares) from the spurious rounding effect due to a finite thermal gradient. The solid lines are the corresponding derivatives of both quantities. Note the closeness of both inflexion-point-temperatures. For details see the main text.

which, together with the fact that far away from the transition $S \sim S_M$ enables us to construct S(T) from the knowledge of the temperature derivative of the measured $S_M(T)$. Fig. 2 illustrates the outcome of such a procedure when applied to a $Y_1Ba_2Cu_3O_{7-\delta}$ sample. Note in particular the closeness of both inflexion-point temperatures. In fact, as it can be easily seen by Eq. (2), no shift in T_{cI} should occur as long as S(T) is antisymmetrical around T_{cI} over the interval ΔT.

C. Background TEP and Critical Temperature

The background or noncritical part of any critical quantity is associated with short-wavelength fluctuations. The necessity of including it as a main ingredient when analyzing critical phenomena stems from theoretical formalisms being unable to allow for these short-wavelength fluctuations, and thus a meaningful comparison with theories requires an independent estimation of the background (22). In practice, one looks for any functional form, provided or not by the theory of the normal state, that fits closely the normal-state temperature trend over a wide temperature region away from the transition, and that extrapolates smoothly through the transition. Usually, the background choice is a source of uncertainty unless a functional form for the normal-state temperature dependence is firmly established.

Among the various functional forms for the background TEP, $S_B(T)$, for Y-based compounds, that proposed by Mukherjee and coworkers (23)

$$S = aT(1 + be^{-T/T_o}) \tag{4}$$

has been chosen where a, b and T_0 are free parameters. According to the above reasoning, we do not claim this functional form to represent better than others the physics of transport mechanism in the normal state. Eq. (4) has been fitted to TEP data for the Y samples over the interval 120 - 200 K, and, as can be seen in the example in Fig. 1a, the fit quality is excellent.

The background resistivity for all the samples are straight lines like the one shown in the same figure, fitted to the same temperature intervals as their corresponding TEP data. It is worth noting that the background TEP follows up the peaklike enhancement relatively near T_c which is systematically seen in these compounds. Correspondingly, it is assumed that the precursor peak in thermopower is not associated with critical behavior, as proposed by other authors (6,8). In that interpretation it is argued that superconducting fluctuations enhance the phonon-drag mechanism, thus giving an extra contribution to S(T). Phonon-drag relevance is, however, controversial (24) and, moreover, there are alternative explanations based on existing theory of electrophonon enhancement for the TEP enhancement near T_c (9). But the main reason for discarding in these data the critical nature of the peak enhancement in S(T) in $Y_1Ba_2Cu_3O_{7-\delta}$ compounds near T_c is the very different onset temperature of the resistive transition, i.e., the departure from its linear T dependence (around 115 K, as seen in Fig. 1a) and that of the TEP peak (around 140 K, as also seen in Fig. 1 a). This fact will indicate certainly distinct origins for both effects.

A final main ingredient is the critical temperature. One should remember that the critical temperatures predicted by theories are not accessible directly, and thus resort to some phenomenological temperatures must be made. The inflexion-point temperature around the transition has proved its significance in the analysis of the critical behavior of electrical conductivity (13,15) and will also be used for TEP. The closeness in the inflexion-point temperatures of ρ(T) and S(T), as seen in Fig. 1b, indicates in fact the consistency of that choice, the difference being of the order of the uncertainty in absolute temperature.

IV. Comparative Analysis of the Critical Behaviors of Thermopower and Electrical Conductivity in the Mean Field Region

Following the precedent discussion and choices the determination of the excess thermopower, ΔS, defined as

$$\Delta S = \Delta_B - S \qquad (5)$$

is now straightforward. The corresponding values are shown in Fig. 3 as a function of the reduced temperature $\epsilon = (T-T_{CI})/T_{CI}$ for two of the $Y_1Ba_2Cu_3O_{7-\delta}$ samples studied here. The intragrain excess conductivity $\Delta\sigma_{ab}$ (which because of the high anisotropy in σ should be naturally identified with the in-plane conductivity in single crystals) defined by

$$\Delta\sigma = \sigma_{ab} - \sigma_{abB} \qquad (6)$$

is also shown in Fig. 3. The extraction of $\Delta\sigma_{ab}$ from the measured conductivity in granular samples has been reported extensively earlier (13,26). We only recall again that this extraction requires a previous determination of sample-dependent parameters associated with the effective cross-section and effective length, and intergrain resistivity (parameters p and ρ_c, respectively of

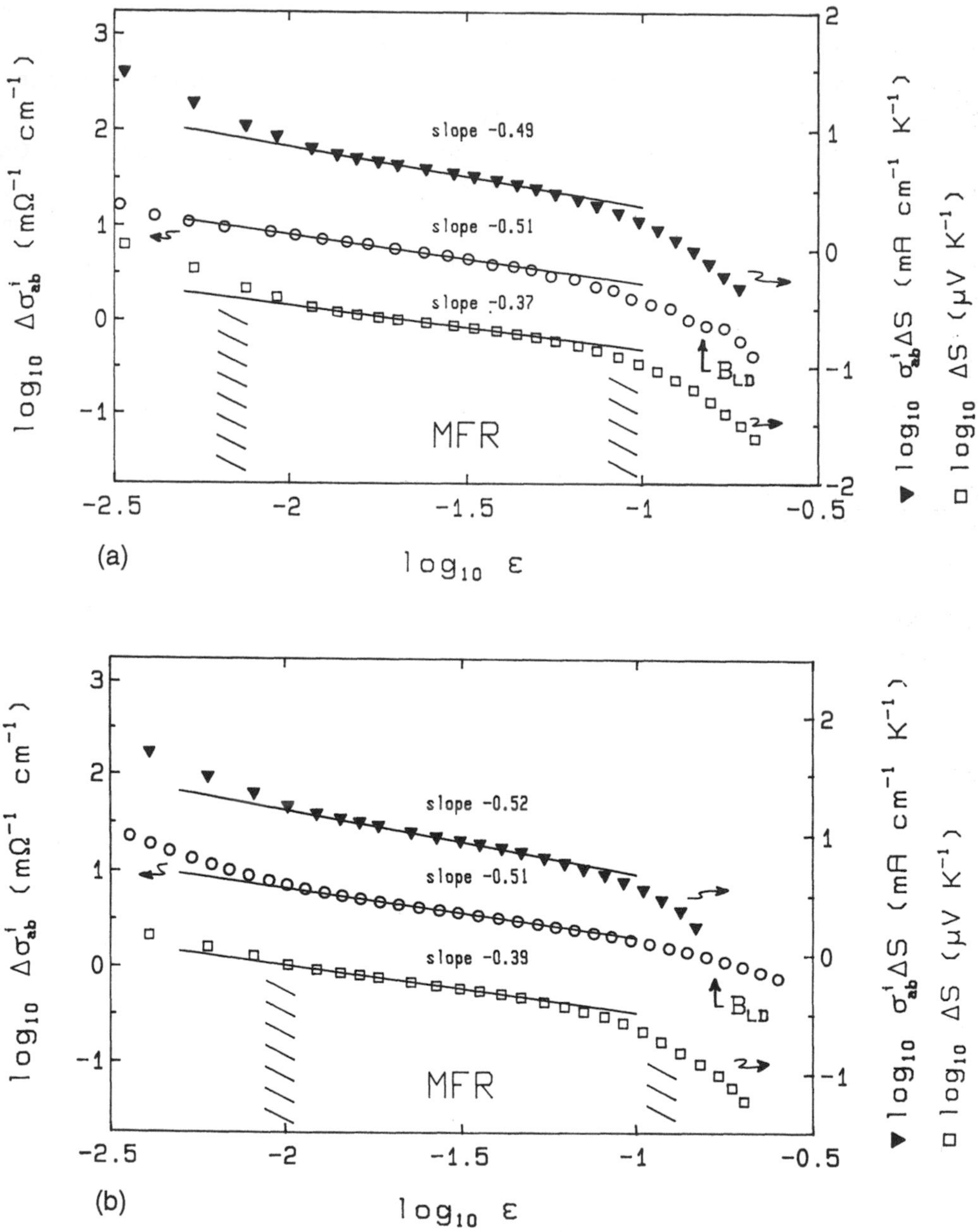

Fig. 3. Logarithmic representation of the excesses of the intrinsic paraconductivity in the ab plane (circles), thermopower (squares) and thermopower times intrinsic conductivity $\sigma_{ab}\Delta S$ (triangles) for the Y1 (a) and Y2 (b) samples as a function of the reduced temperature $\epsilon = (T-T_{CI})/T_{CI}$.

Table 1).

We summarize first the analysis of $\Delta\sigma_{ab}$. Experimental data in the mean-field region (MFR) are compared with existing Aslamazov-Larking approaches for OPF in layered superconductors (27,28). These predictions yield for $\Delta\sigma_{ab}$ in the MFR

$$\Delta\sigma_{ab} = \frac{A_{AL}}{\epsilon} \left(1 + \frac{B_{LD}}{\epsilon}\right)^{-1/2} \tag{7}$$

the limiting behaviors of Eq.(7) are $\Delta\sigma_{ab} \sim \epsilon^{-1/2}$ for $\epsilon << B_{LD}$, and $\Delta\sigma_{ab} \sim \epsilon^{-1}$ for $\epsilon >> B_{LD}$, the crossover reduced temperature being $B_{LD} = (2\xi_c(0)/d_e)^2$, where d_e is the effective distance between adjacent ab superconducting planes and $\xi_c(0)$ is the amplitude of the correlation length in the c direction. The functional form of Eq.(7) is found to account for the experimental excess conductivity of either mono- or polycrystalline samples, and no anomalous contributions of the Maki-Thompson type are compatible with experimental data (25,26).

Turning again to Fig. 3, since the Lawrence-Doniach crossover $B_{LD} \sim 0.14$ the behavior of $\Delta\sigma_{ab}$ within the MFR corresponds essentially to three dimensional (3D) OPF with the characteristic exponent of -1/2, as singled out in the figure itself. From this same figure one can see the similarity in ϵ dependence between ΔS and $\Delta\sigma_{ab}$. Its explanation constitutes one of the main objectives of this contribution. In an attempt to quantify matters, the most natural choice is to resort to the simplest mechanism, namely, the electronic diffusive contribution to TEP

$$S = L/\sigma \tag{8}$$

where L is the thermoelectric coefficient. One might think to be more sensible to apply this relationship to intragrain (i.e., approximating single crystals) quantities, but the connection between observed and intragrain values in TEP for polycrystalline samples has not yet been, unlike conductivity, worked out quantitatively. Note that Eq.(8) can include effects going beyond the simple Mott approximation, such as electron-phonon mass or energy enhancement (29).

Equation (8) predicts a close relationship of S and σ. To simplify, we might neglect the critical behavior of L, at least within the MFR, i.e., L ~ L_B. This would be consistent with the milder singularity for L, as compared with that of σ, predicted theoretically (17). However, this statement may be subjected to an experimental test since we have independent access to both S and σ and thus to L (of course, assuming the validity of Eq.(8)). Fig.4 shows the temperature dependence of L and, as can be seen, no critical behavior is observed up to the closer boundary to T_c of the MFR. This result amounts to approximating L ~ $L_B = S_B\sigma_B$, and hence from Eq.(8),

$$\Delta S = S_B \Delta\sigma/\sigma \tag{9}$$

thus, ΔS correlates to $\Delta\sigma/\sigma$, or more precisely the critical exponents of $\sigma\Delta S$ should be those of $\Delta\sigma$. Fig. 3 confirms the coincidence, within experimental uncertainties, of both critical exponents, which sign a 3D behavior. It is important to remark

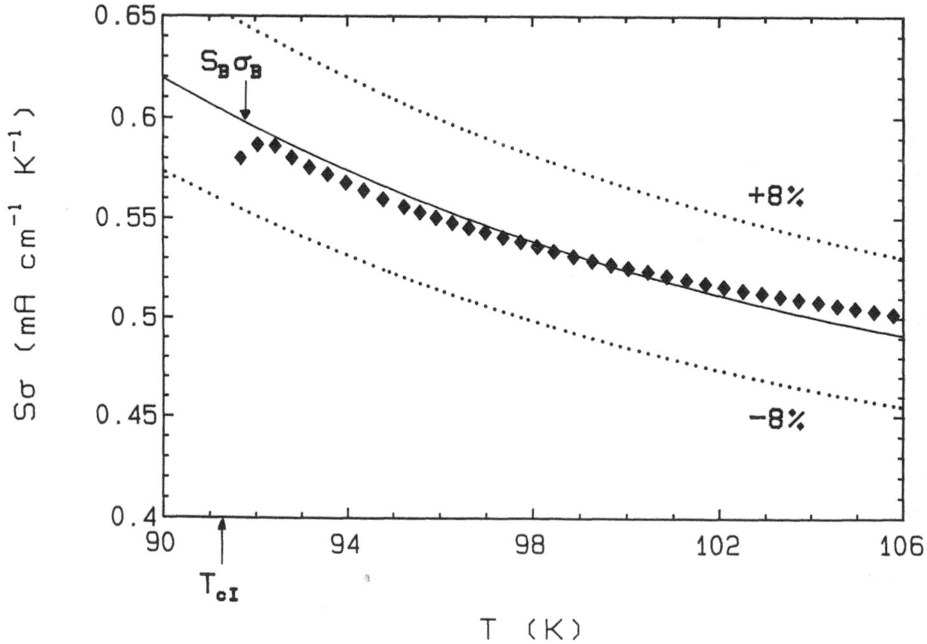

Fig. 4. Temperature dependence of the thermoelectric coefficient (L = S σ) close to T_{CI} in the mean field region (MFR) for a $Y_1Ba_2Cu_3O_{7-\delta}$ granular sample (Y1).

that Eq. (9) correlates $\Delta\sigma$ with ΔS, but Fig. 3 displays $\Delta\sigma_{ab}$ and ΔS. Since the critical exponents of $\Delta\sigma$ and $\Delta\sigma_{ab}$ in the MFR are found to be the same within a few hundredths (13), Fig. 3 provides then a first corroboration of Eq. (9). A more thorough test is later given in Table 2.

At this point we want to briefly comment on a particular feature of the data analyses of some authors (10,11). Excess data, say those of paraconductivity, may be represented in a variety of ways, for instance in terms of $d\Delta\rho/dT$ as used by those authors. However, theoretical critical exponents are predicted only for $\Delta\sigma$, and, as easily verified, they equal those for $d\Delta\rho/dT$ only as long as the temperature variation of σ in the analysis region can be neglected. Since the MFR is phenomenologically bounded by $0.1 \leq \Delta\sigma/\sigma \leq 0.6$, analyses of critical exponents in terms of $d\Delta\rho/dT$ for $\Delta T \leq 10$ K may lead to appreciable errors, precisely the difference illustrated in Fig. 3 when switching from ΔS to $\sigma\Delta S$. Therefore, one should take skeptically proposals such as identifying a critical exponent of $-5/6$ in $d\Delta\rho/dT$ for Bi-bases compounds with fractal behavior (11,12) or a OPF

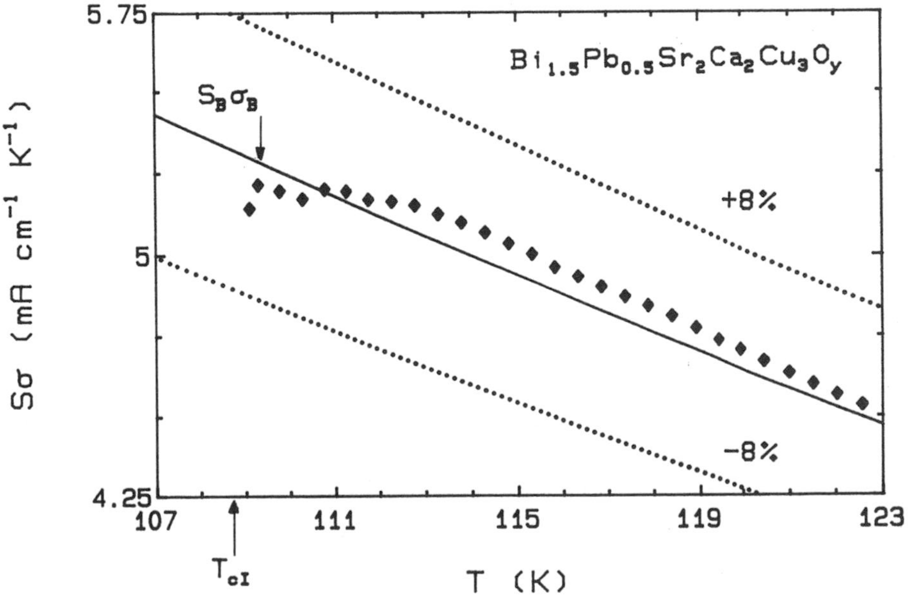

Fig. 5. Temperature dependence of the thermoelectric coefficient ($L = S\ \sigma$) close to T_{cI} in the mean field region (MFR) for a $Bi_{1.5}Pb_{0.5}Sr_2Ca_2Cu_3O_y$ granular sample.

Table 2. Excess of the Conductivity ($\Delta\sigma$) and Thermopower (ΔS) for Selected Reduced Temperatures $\epsilon = (T-T_{CI})/T_{CI}$ within the Mean Field Region for the Three Samples

Sample	T_{CI} K	ϵ 10^{-2}	$\Delta\sigma$ $m\Omega^{-1}cm^{-1}$	ΔS μVK^{-1}	$\sigma\Delta S$ mAK^{-1}	$S_B\Delta\sigma$ cm^{-1}
Y1	91.3	1.09	0.035	0.73	0.14	0.13
		8.00	0.014	0.31	0.05	0.05
Y2	91.1	1.32	0.29	0.49	0.85	0.82
		7.13	0.11	0.22	0.33	0.28
Y3	89.5	1.01	0.67	0.87	3.00	2.88
		3.91	0.35	0.40	1.25	1.43

The last two columns provide an experimental verification of Eq. (9).

dimensionality D = 4 for Y-based materials (12). The consistency of Eq. (9) for describing the ϵ dependence of the excess TEP can be exploited further by also checking numerically the amplitudes. Note that the fulfillment of Eq. (9) is very demanding in that it involves both relative and absolute values. It is also worth remarking again that since Eq.(9) involves observed quantities, granularity effects are altogether included. Table 2 shows the numerical agreement found between
Eq.(9) and the experimental data for the three $Y_1Ba_2Cu_3O_{7-\delta}$ samples and two different temperatures within the MFR. The good accord (within 8 %) gives strong support to the dominance of the diffusion electronic contribution to TEP in the MFR with a non-critical thermoelectric coefficient. Simultaneously, that agreement gives overall solidity to the extraction procedure of the excess TEP.

V. Conclusions

We have presented simultaneous measurements of the roundings of the thermopower and of the electrical conductivity for a batch of granular $Y_1Ba_2Cu_3O_{7-\delta}$ samples. Quantitative explanation of the observed excess thermopower ΔS has been achieved by correlating it with $\Delta\sigma$ through an electronic diffusion mechanism alone. It has been found that the thermoelectric coefficient $L = \Delta\sigma$ shows no (critical) rounding up to $\epsilon = 10^{-2}$. Since L directly involves the dynamics of normal electrons, this result is again suggestive of the non-interaction between superconducting and normal electrons, that is, the absence of Maki-Thompson terms in the dynamics of the superconducting transition for HTSC in agreement with our previous proposals (25,26). Fig. 5 also shows the non-critical behavior of L for a Bi-based HTSC sample, which complements and extends our results from Y-based samples to all copper perovskite-type superconducting materials.

Also consistent with the electronic diffusion mechanism, the combination $\sigma\Delta S$ has been found to follow the same ϵ dependence and amplitude as $\Delta\sigma$, showing in particular a pronounced three dimensional behavior for fluctuations within the mean field region. This result means that the rounding of TEP in HTSC is driven exclusively by that of electrical conductivity. These findings open new possibilities in the analysis of TEP near and above T_c in HTSC and offer partial but firm contributions to the knowledge of their critical behavior.

Acknowledgements

This work has been supported by the CICYT Grant MAT 88-0769 and Grant MAT 88-0250-C02-01, by the Programa MIDAS Grant 89-3800, and by the Fundación Ramón Areces, Spain.

References

1. Ginzburg, V.L., Thermoelectric effects in high-temperature superconductors, *Physica C* 162-164: 277 (1989); Thermoelec-

tric effects in superconductors, *J. Superconductivity* **2:** 323 (1989); Thermoelectric effects in superconductors, *Supercond. Sci. Technol.* **4:** S1–10 (1991).
2. Devaux, F., A. Manthiram and J.B. Goodenough, Thermoelectric power of high-T_c superconductors, *Phys. Rev. B* **41:** 8723 (1990); Ouseph, P.J. and M. Ray O'Brya, Thermoelectric power of $YBa_2Cu_3O_{7-\delta}$, *Phys. Rev. B* **41:** 4123 (1990).
3. Bhatnagar, A.K., R. Pan, D.G. Naugle, G.R. Gilbert and R.K. Pandey, Thermopower of $RBa_2Cu_3O_{7-\delta}$ (R = Y, Er), *Phys. Rev. B* **41:** 4002 (1990).
4. Fisher, B., J. Genossar, L. Patlagan and J. Ashkenazi, Resistivity and thermoelectric-power measurements of $Pr_x Y_{1-x} Ba_2Cu_3O_{7-\delta}$ up to 1200 K and an electronic structure analysis, *Phys. Rev. B* **43:** 2821 (1991).
5. Gridin, V.V., W.R. Datars and P.K. Ummat, Electrical transport in superconducting BiPbSrCaCuO, *Solid State Commun.* **74:** 187 (1990).
6. Dey, T.K., S.K. Ghatak, S. Srinivasan, D. Bhattacharya and K.L. Chopra, Thermoelectric power of undoped and doped Y-Ba-Cu-O superconductor between 77 and 300 K, *Solid State Commun.* **72:** 525 (1989); Dey, T.K., K. Radha, H.K. Barik, D. Bhattacharya and K.L. Chopra, Excess electrical conductivity and thermoelectric power of $(YBa_2Cu_3O_x)Ag_n$ pellets, *Solid State Commun.* **74:** 1315 (1990).
7. Howson, M.A., M.B. Salamon, T.A. Friedmann, J. P. Rice and D. Ginsberg, Anomalous peak in the thermopower of $YBa_2Cu_3O_{7-\delta}$ single crystals: A possible fluctuation effect, *Phys. Rev. B* **41:** 300 (1990).
8. Trodahl, H.J. and A. Mawdsley, Thermopower of $YBa_2Cu_3O_7$ and related superconducting oxides, *Phys. Rev. B* **36:** 8881 (1987).
9. Kaiser, A.B. and G. Mountjoy, Consistency with anomalous electro-phonon interactions of the thermopower of high-T_c superconductors, *Phys. Rev. B* **43:** 6266 (1991).
10. Ausloos, M., K. Durczewski, S.K. Patapis, Ch. Laurent and H.W. Vanderschueren, Thermoelectric power of granular ceramic oxide superconductors, *Solid State Commun*, **65:** 365 (1988); Laurent, Ch., S.K. Patapis, M. Laguesse, H.W. Vanderschueren, A. Rulmont, P. Tarte and M. Ausloos,

Thermoelectric power and magneto Seebeck-effect near the critical temperature of granular ceramic oxide superconductors $YBa_2Cu_3O_{7-y}$, *Solid State Commun.* **66**: 445 (1988).

11. Clippe, P., Ch. Laurent, S.K. Patapis and M. Ausloos, Superconductivity fluctuations in electrical and thermoelectrical properties of granular ceramic superconductors: Homogeneous versus fractal behavior, *Phys. Rev. B* **42**: 8611 (1990).
12. Laurent, Ch., S.K. Patapis, S.M. Green, H.L. Luo, C. Politis, K. Durczewski and M. Ausloos, Fluctuation conductivity effects on thermoelectric power of granular $Bi_{1.75}Pb_{0.25}Ca_2Sr_2Cu_3O_{10}$ superconductor, *Mod. Phys. Lett. B* **3**: 241 (1989).
13. Veira J.A. and F. Vidal, Paraconductivity of ceramic $YBa_2Cu_3O_{7-\delta}$ in the Mean-Field-Like region, *Physica C* **159**: 468 (1989).
14. Paracchini, C., Electrical transport properties of $YBa_2Cu_3O_x$ films, *Solid State Commun.* **74**: 1113 (1990); Padmanaban, V.P.N. and K. Shahi, Fluctuation induced paraconductivity in $YBa_2Cu_3O_{7-\delta}$, *Physica C* **172**: 427 (1991).
15. Vidal, F., J.A. Veira, J. Maza, J.J. Ponte, F. García-Alvarado, E. Morán, J. Amador, C. Cascales, A. Castro, M.T. Casais and I. Rasines, Excess electrical conducti-vity in polycrystalline Bi-Ca-Sr-Cu-O compounds and thermodynamic fluctuations of the amplitude of the superconducting order parameter, *Physica C* **156**: 807 (1988); Akinaga, M. and L. Rinderer, Thermodynamic fluctuation in the Bi(Pb,Sb)-Sr-Ca-Cu-O superconductors, *Physica B* **165-166**: 1373 (1990); Balestino, G., A. Nigro, R. Vaglio and M. Marinelli, Thermoelectric fluctuations in the 110 K Bi-Sr-Ca-Cu-O superconductor: Evidence for two dimensional behavior, *Phys. Rev. B* **39**: 12264 (1989).
16. Kanoda, K., T. Kawagoe, M. Hasumi, T. Takahashi, S. Kagoshima and T. Mizoguchi, Dimensionality of the Superconductivity in $YBa_2Cu_3O_{7-\delta}$, *J. Phys. Soc. Jpn* . **57**: 1554 (1988); Lee, W.G., R.A. Klemm and D.C. Johnston, Superconducting fluctuation diamagnetic above T_c in $YBa_2Cu_3O_7$, $La_{1.8}Sr_{0.2}CuO_4$ and $Bi_{2-x}Pb_xSr_2CaCu_2O_{8+\delta}$, *Phys. Rev. Lett.* **63**: 1012 (1989).

17. Maki, K., Thermoelectric power above the superconducting transition, *J. Low Temp. Phys.* **14**: 419 (1974).
18. López, A.J., J. Maza, Y.P. Yadava, F. Vidal, F. García-Alvarado, E. Morán and M.A. Señarís-Rodríguez, Thermoelectric Power in Lead-Doped Polycrystalline BiSrCaCuO Superconductors, *Supercond. Sci. Technol.* **4**: S292 (1991); Cabeza, O.,Y.P. Yadava, J. Maza, C. Torrón and F. Vidal, Thermoelectric power rounding above T_c versus paraconductivity in copper oxide superconductors, *Physica C*, (in print).
19. Amador, J., C. Cascales and I. Rasines, Synthesis and characterizacion of $Ba_2Cu_3PrO_{7-x}$, in *High-Temperature Superconductors, M.B. Brodsky, R.C. Dynes, K. Kitazawa and H.L. Tuller (Eds.)*, Materials Research Society, Pittsburgh, 1988, p. 249.
20. Blatt, F.J., P.A. Schroeder, C.L. Foiles and D. Greig, *Thermoelectric Power of Metals*, Plenum, New York, 1976.
21. Li, L., M. Bei-Hai, L. Shu-Yuan, D. Hong-Min and Z. Dian-Lin, Anisotropic thermoelectric and the possible existence of a nonsuperconducting phase in $YBa_2Cu_3O_{7-\delta}$ simgle crystals, *Europhys. Lett.* **7**: 555 (1988); Pekala, M., K. Kitazawa, A.M. Balbashov, A. Polaczek, I. Tanaka and H. Kojima, Anisotropic thermoelectric power and thermal conductivity in superconducting single crystals Bi-Sr-Ca-Cu-O, *Solid State Commun.* **76**: 419 (1990).
22. See, e.g., Vidal, F., Background contributions to sound waves at the liquid-helium l transition, *Phys, Rev, B* **26**: 3986 (1982), and referecences therein.
23. Mukherjje, C.D., R. Ranganathan, A.K. Raychandhuri and N. Chatterjee, Thermoelectric power of the Ni and Cd substituted YBCO system, *Phys. Stat. Sol. (b)* **154**: K55 (1989).
24. Radharkrishnan, V., C.K. Subramanian, V. Sankaranarayanan, G.V. Subba Rao and R. Srinivasan, Thermopower of Zinc- doped Y-Ba-Cu-O, *Phys. Rev. B* **40**: 6850 (1989); Kang, W.N., K.C. Cho, Y.M. Kim and Mu-Yong Choi, Oxygen-deficiency dependence of the thermopower of $YBa_2Cu_3O_{7-\delta}$, *Phys. Rev. B* **39**: 2763 (1989); Tsidilkovski, I.M. and V.I. Tsidilkovski, Resistivity and thermoelectric power of ceramics at $T > T_c$, *Solid State Commun.*, **66**: 51 (1988).
25. Vidal, F., C. Torrón, J.A. Veira, F. Miguelez and J. Maza,

Electrical resistivity and magnetic susceptibility roundings above the superconducting transtion in $YBa_2Cu_3O_{7-\delta}$, *J. Phys. Condensed Matter* 3: 5219 (1991).
26. Veira, J.A. and F. Vidal, Scattering of normal excitations by superconducting fluctuations in single-crystal and granular copper oxide superconductors, *Phys. Rev. B* 42: 8748 (1990).
27. Lawrence, W.E. and S. Doniach, Theory of layer structure superconductors, *Proceedings of the Twelfth International Conference on Low-Temperature Physics, E. Kanda (Ed.)*, Keigatu, Tokyo, 1971, p. 361.
28. Yip, S., Paraconductivity in an unconventional layered superconductorr, *J. Low Temp. Phys.* 81: 129 (1990).
29. Kaiser, A.B., Electron-phonon enhancement of thermopower: Application to metallic glasses, *Phys. Rev. B* 29: 7088 (1984).

Chapter 6

Optical Spectroscopy of Perovskite-Type Oxides

G. Blasse

Debye Research Institute, University of Utrecht,
3508 TA Utrecht, The Netherlands

I. Introduction

Compounds with perovskite structure have frequently served as model compounds for the study of physical properties. In this chapter we wish to review in an illustrative way how compounds of this type have been used as model compounds for spectroscopic studies in the infrared, the visible, and the (conventional) ultraviolet. There are several factors which make the use of the perovskite structure in these studies so attractive: (i) the metal ions have a site symmetry with an inversion center (in the ideal case O_h); there is a six- and a twelve-coordinated site available; (ii) by changing the chemical composition distances can be varied in a controlled way (e.g. in the series $CaTiO_3$, $SrTiO_3$ and $BaTiO_3$); (iii) interactions between metal ions can be studied; the sublattice of octahedrally coordinated ions especially offers interesting possibilities which have been used in the past in the study of magnetic interactions (1); (iv) all types of metal ions can be studied in this crystal structure, from the smaller transition metal ions to the larger lanthanide ions; (v) the octahedral sublattice shows several superstructures, especially the 1:1 ordering (elpasolite; A_2BCX_6 with the long-range order between B and C on octahedral sites) has been used extensively to create new model systems; (vi) the chemical flexibility of the structure is enormous, i.e. many compositions of a very different chemical nature crystallize in the perovskite structure, a nice example being $BaLiH_3$ and Fe_4N (both with cubic perovskite structure) (1).

In this chapter space limitations restrict the choice of topics to be dealt with. Therefore, we limit ourselves as follows: (i) only compositions with the following crystal structures will be considered: the (slightly distorted) cubic perovskite structure (composition ABX_3), its 1:1 ordered variant often called elpasolite (A_2BCX_6), and the K_2PtCl_6 structure which can be structurally derived from elapsolite by leaving C vacant (composition A_2BX_6); (ii) in addition to oxides we will also consider some halides since the spectroscopy of these latter compounds is in some cases of utmost importance; (iii) the following groups of optically active metal ions will be considered: transition metal ions with empty d-shell, transition metal ions with partly filled d-shell, metal ions with s^2 configuration, the hexavalent uranium ion, and the lanthanides; (iv) in many cases the author will follow his own preference; others would undoubtedly have made different, but fully justified, choices; (v) for the fundamentals of the spectroscopy and the energy level schemes the reader is referred to other texts (2-7).

II. Transition Metal Ions with Empty d-Shell

Well-known examples of ions to be discussed in this section are Ti^{4+}, Nb^{5+}, and W^{6+}. The complex ions consisting of a central metal ion of this type and a number (usually 4 or 6) of oxygen ions as ligands are known for their strong optical absorption in the ultraviolet (and some times in the visible) and their ability to show efficient luminiscence (4). The optical transitions involved are of the charge-transfer type. Within the perovskite family the titanates $MTiO_3$ (M = Ca, Sr, Ba) have been studied intensively at length.

Before considering these titanates, we first have a look at the spectroscopy of the isolated titanate group. This group can, for example, be studied in $SrZrO_3$:Ti (8) or $BaZrO_3$:Ti (9) (Fig. 1). The titanate absorption band has a maximum at about 280 nm. Excitation into this band results in broad band emissions with a maximum at 460 nm, a quenching temperature of about 150 K and a Stokes shift of 12,100 cm^{-1}. The excited charge-transfer state has a larger equilibrium distance than the ground state. All luminescence properties can be explained in the simple configurational coordinate diagram (2,6). The electron-lattice coupling in this titanate group is large.

The corresponding properties of $SrTiO_3$ (Fig. 1) are different. The absorption band has shifted to much lower energy. Excitation into this band yields a luminiscence which is independent of sample history and contamination (8,10-12) and shows a band maximum at about 500 nm. The quenching temperature is low (35 K) and the Stokes shift small (7,000 cm^{-1}).

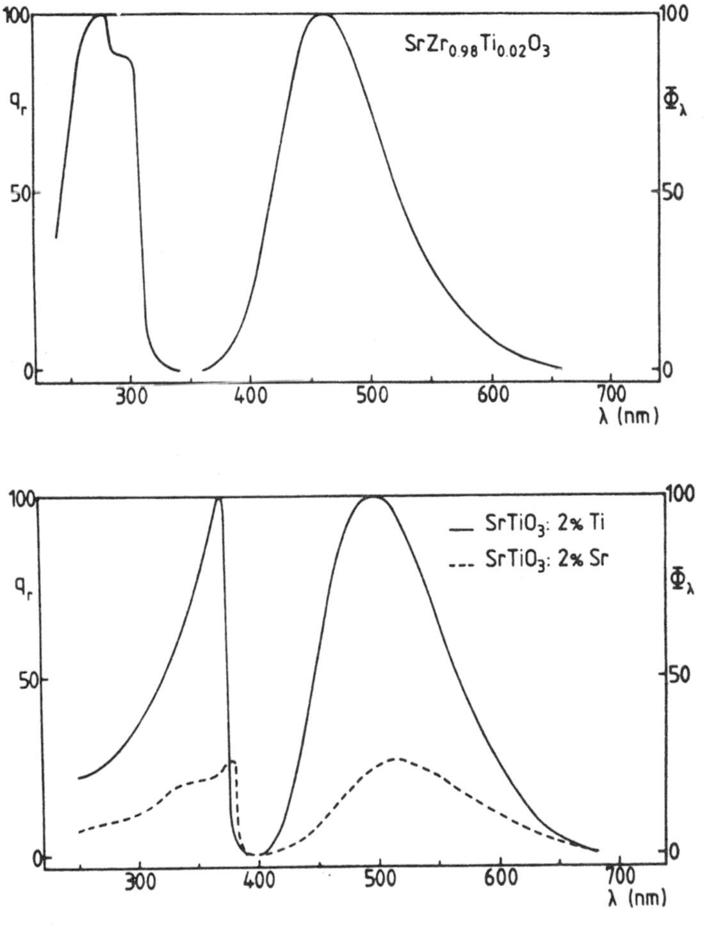

Fig. 1. (top) Emission and excitation spectra of the luminescence of $SrZrO_3$:Ti at 4.2 K. (bottom) Emission and excitation spectra of the luminescence of $SrTiO_3$ with different stoichiometries at 4.2 K. Emission spectra on the right-hand side, excitation spectra on the left-hand side. After L.G.J. de Haart, Thesis, Utrecht, 1985.

In $SrTiO_3$ the titanate octahedra share corners. This is favorable for energy band formation. The fact that the energy levels of the titanate octahedron broaden into bands results in a considerably lower energy for the first absorption transition. A more important consequence is that the excited state of the titanate octahedron becomes mobile. It has been assumed that the situation is best described by the coexistence of free and delocalized (self-trapped) excitons (6,13). This is possible if the free and self-trapped exciton states are separated by a potential barrier. In TiO_2 the free-exciton emission has actually been observed, together with self-trapped exciton emission (6). In $SrTiO_3$ only the self-trapped exciton emission has been observed. The nature of the optical transition involved is the same as in that on the isolated titanate octahedron. The mobility of the exciton state is illustrated by the temperature dependence of the luminescence intensity of $SrTiO_3$ doped with a luminescent impurity, for example Cr^{3+} (ref. 8, see below). Fig. 2 shows that the titanate emission drops rapidly and that the Cr^{3+} emission builds up simultaneously.

The absorption and emission spectra of the niobates $MNbO_3$ (M = Li, Na, K) have been explained along the same lines (14). The electronic states obtain a more delocalized character in the sequence Li, Na, K. This can be correlated to the increase of the Nb-O-Nb angle towards 180° in this sequence. Whereas $LiNbO_3$ has

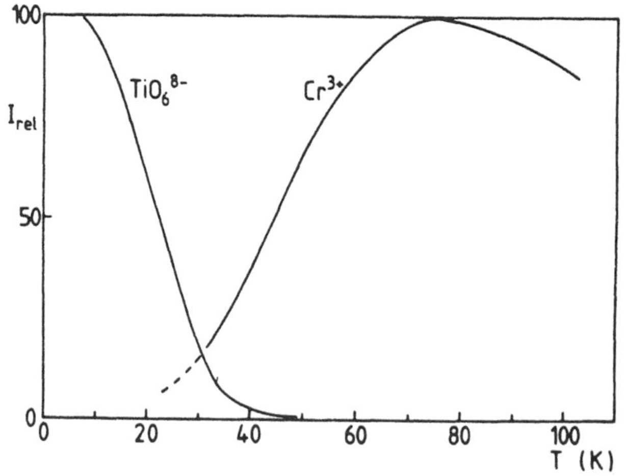

Fig. 2. Temperature dependence of the intensity of the titanate and the Cr^{3+} emission of $SrTiO_3:Cr^{3+}$ under band-gap excitation. Source: see ref. 8.

a crystal structure which is an ordered variant of corundum, those of $NaNbO_3$ and $KNbO_3$ are distorted perovskite modifications.

The flexibility of the perovskite structure permits us to go from $M(II)TiO_3$ via $M(I)NbO_3$ to $\square WO_3$ ($\square$: vacancy) without drastic structure change: the units "$M(d^0)O_3$" show still perovskite-like build up and are isoelectronic, since Ti^{4+}, Nb^{5+} and W^{6+} have a noble-gas configuration (here indicated as d^0). Covalency will increase in this series from titanate to tungstate in view of the increase of the formal charge. This implies that the delocalization becomes more important relative to the localization (determined by electron-lattice coupling). In fact WO_3 is yellow, i.e. the optical absorption has shifted to the visible due to band broadening. It shows a very weak luminescence with relatively small Stokes shift (15). All this is compatible with more pronounced delocalization (6,16).

Another advantage of the perovskite structure is the possibility to have the optically-active centers discussed in this section separated from each other in a 1:1 ordered perovskite. Relevant examples are La_2MgTiO_6, Ba_2GdNbO_6 and Ba_2MgWO_6. In this superstructure of perovskite the $M(d^0)O_6$ octahedra contain metal ions (Mg^{2+}, Gd^{3+}) of which the optical transitions occur at much higher energy. In this way wave-function overlap between the $M(d^0)O_6$ octahedra becomes of negligible importance.

The luminescence of La_2MgTiO_6 (4,17) and Ba_2GdNbO_6 (18) is only weak. Of much more interest are the properties of the ordered perovskites A_2BWO_6 (4,19) where A^{2+} and B^{2+} are alkaline-earth metal ions. The lowest optical absorption of these compounds is situated in the ultraviolet between 300 and 320 nm, i.e. at much higher energy than for WO_3. The luminescence depends strongly on the choice of A and B. In this way this perovskite system serves as a model to investigate nonradiative transitions. This will be illustrated below for the uranium-doped tungstates.

In the case of B = Mg (A_2MgWO_6) there is a blue as well as a yellow emission present. The blue emission is due to the intrinsic tungstate group. The yellow emission occurs as a consequence of Mg-W disorder on the octahedral sublattice. It is most probably due to a center consisting of two edge-sharing tungstate octahedra. This disorder does not occur for the larger B ions.

It is possible to dilute the tungstate ions by using tellurate: systems $A_2BTe_{1-x}W_xO_6$ show complete miscibility. Interestingly enough, the quantum efficiency of the tungstate luminiscence at 300 K and the quenching temperature goes up for lower values of x. This shows that in the x = 1 composition the tungstate luminescence shows concentration quenching, or in other words, the tungstate octahedra have interaction with each other although they are separated from each other by optically innocent ions. This is an example of nonradiative energy transfer between the tungstate octahedra (2,5,6).

Finally we draw attention to the fact that these ordered perovskites are model compounds for the vibrational spectroscopy of the octahedron with the higher charged metal ion (WO_6^{6+} in A_2BWO_6, for example). This is due to their O_h site symmetry. As a more or less exotic example we give in Fig. 3 the vibrational spectra of Ba_2NaIO_6 (20). The infrared spectrum shows clearly the allowed ν_3 and ν_4 modes, the Raman spectrum the allowed ν_1, ν_2 and ν_5 modes. Many other examples are available in literature.

Let us now turn to the optical properties of transition metal ions with partly filled d-shell in the perovskite structure.

III. Transition Metal Ions with Partly-Filled d-Shell

Since a large amount of research has been performed in this field, we restrict ourselves to a few examples. Starting with some older cases, we subsequently pay attention to results induced by developments in photoelectrochemistry and in instrumentation.

Ions with d^3 configuration have always been peculiarly popular. The Cr^{3+} spectroscopy has been studied at lengths in

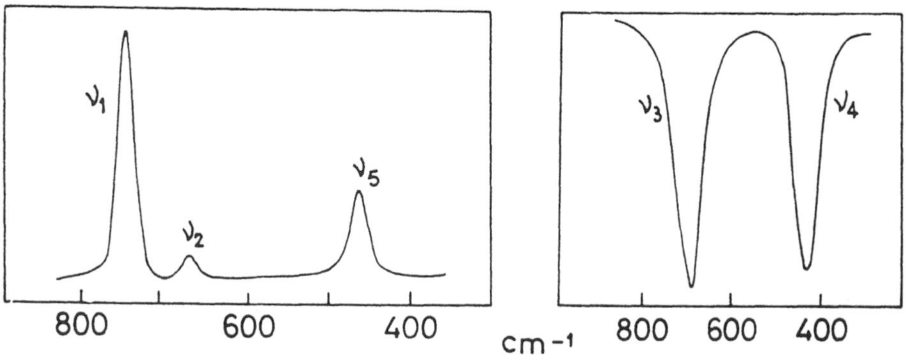

Fig. 3. Infrared (ν_3 and ν_4) and Raman (ν_1, ν_2, and ν_5) spectra of Sr_2NaIO_6 at 300 K. Modified from ref. 20.

perovskites $MTiO_3$ (M = Ca, Sr, Ba) and ABO_3 (A = La, Gd, Y; B = Al, Ga). It is imperative to realize that the Cr^{3+} ion may show two completely different types of emissión, viz. narrow-line and spin-forbidden $^2E \rightarrow {^4A_2}$ emission and broad-band and spin-allowed $^4T_2 \rightarrow {^4A_2}$ emission. The latter occurs only for relatively weak crystal fields.

In $LaAlO_3:Cr^{3+}$ (21) and $GdAlO_3:Cr^{3+}$ (22) the Cr^{3+} ion shows its characteristic crystal-field absorption bands. The luminescence is from the ^2E level, with a dominating zero-phonon line. In $CaTiO_3:Cr^{3+}$ the emission is also from the ^2E level, but the amount of phonon assisted emission has strongly increased (23) (see Fig. 4). The isoelectronic Mn^{4+} ion shows in $LaAlO_3$ (21) as well as in $CaZrO_3$ (24) ^2E emission with mainly phonon-assisted lines. This shows that in the perovskite structure the electron-lattice coupling is strong apart from the case of the aluminates.

The Cr^{3+} luminescence has come into focus again when it appeared possible to design tunable lasers using the broad-band $^4T_2 \rightarrow {^4A_2}$ emission (2). In the perovskite field $KZnF_3:Cr^{3+}$ has become a successful material in this respect (25). From a chemical point of view Reinen's work is of interest (26). Here the elpasolite-type of order in the perovskite structure was used to investigate the influence of the surroundings of a given transition metal ion on its spectra. Examples are Ba_2NiWO_6 in comparison with Ba_2NiTeO_6. The stronger π bonding in the tungstate reduces the π bonding in the Ni-O bond, so that the Ni^{2+} crystal field in the tungstate is 25 % larger than in the telurate, though the Ni-O distances are the same.

The spectroscopy of $SrTiO_3$:Fe,Mo is interesting from a different point of view (27). In this composition the photo-excited charge carriers are trapped by the impurities and charge

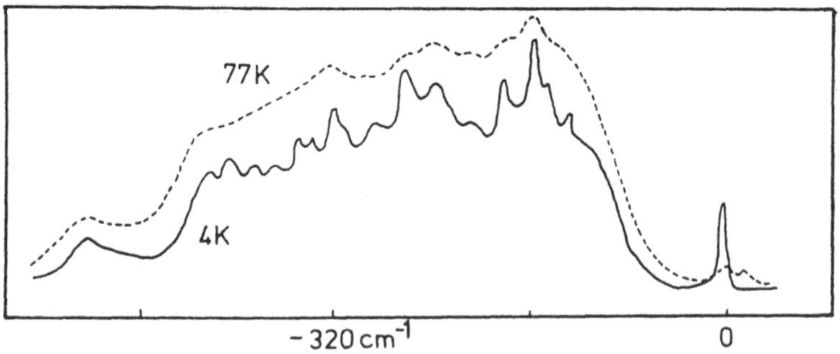

Fig. 4. Emission spectra of $CaTiO_3:Cr^{3+}$ at 4.2 and 77 K with respect to the no-phonon lines. After ref. 23.

separation can be maintained during some time. This results in a color change. In the dark state Fe and Mo are present as Fe^{3+} and Mo^{6+} and the material is white. After 400 nm irradiation, Fe^{4+} and Mo^{5+} are created which absorb in the visible. The Mo^{6+} ion plays the role of an electron trap. At 300 K the thermal decay lasts for several minutes.

The transition-metal doped perovskite-titanates became again of interest after the invention of photoelectrochemical water-splitting by titanates (28). It appeared possible to sensitize titanate electrodes (like $SrTiO_3$) for visible light by doping with transition metal ions (29). The optical transitions involved were shown to be of the metal-to-metal charge-transfer type, for example $Cr^{3+} + Ti^{4+} + h\nu \rightarrow Cr^{4+} + Ti^{3+}$. They are responsible for the brownish color of such compositions (30).

Exciting spectroscopy has been reported in recent years using new types of instrumentation. First we mention results obtained by McClure et al. (31) on $Cs_2GeF_6:Mn^{4+}(3d^3)$ using two-photon spectroscopy. The MnF_6^{2-} octehedron in Cs_2GeF_6 has perfect octahedral symmetry. In one-photon spectroscopy the transitions within the $3d^3$ shell are forbidden. However, in two-photon spectroscopy they are allowed. Indeed strong zero-phonon lines were observed making a complete analysis of the vibrational structure possible. The $^4A_2 \rightarrow {}^2E$ transition shows that the expansion in the ^2E level is only 0.003 nm (weak coupling scheme). For the $^4A_2 \rightarrow {}^4T_2$ transition this value is much larger, viz. 0.0053 nm. The coupling is not only with the symmetrical ν_1 mode, but also with the Jahn-Teller active ν_2 and ν_5 modes indicating a Jahn-Teller distortion in the excited state.

Güdel and coworkers have reported during recent years many cases of (near) infrared emission from several transition-metal ions in the perovskite structure (mainly chlorides and bromides). This was only possible by using suitable detectors of infrared radiation. Examples are the following: (i) $CsMg_{1-x}Ni_xCl_3$ (32) with Ni^{2+} emission at about 5000 cm^{-1}; (ii) $Cs_2NaM^{3+}X_6:Cr^{3+}$ (M = In, Y and X = Cl, Br), a model system for the study of nonradiation transitions (33); (iii) several V^{3+}-doped halide elpasolites which nicelly illustrate complete understanding of the optical processes in these materials (34).

IV. Ions with s^2 Configuration

Spectroscopy of s^2 ions in oxidic perovskites is relatively rare. On the larger A-sites these ions may show a weak luminescence. An extensively discussed example is $CaZrO_3$:Pb (36). However, the chloro-elpasolites are extremely useful as model systems to investigate the spectroscopy of these ions. A first example is Cs_2NaYCl_6:Bi^{3+} (37). The $^1S_0$ -> $^3P_1$ and $^1P_1$ absorption transitions occur at 30,600 and 46,000 cm^{-1}. These bands are very narrow and show extended vibrational structure at low temperatures. This structure can be interpreted in terms of the vibrational modes of the $BiCl_6^{3-}$ octahedron. There is a clear progression in the ν_1 mode. The emitting $^3P_1$ level is not the lowest excited state. This is the $^3P_0$ level. It can be only populated via the $^1P_1$ level. The nonradiative decay from $^3P_1$ to $^3P_0$ is extremely slow since their energy difference (1150 cm^{-1}) is much larger than the highest vibrational frequency of the $BiCl_6^{3-}$ octahedron (290 cm^{-1}). The Stokes shift of this emission is very small. This has consequences for samples with bismuth concentrations above 0.1 mole %. Energy transfer between Bi^{3+} ions becomes now possible. Actually $Cs_2NaBiCl_6$ shows only emission from defects (Bi^{3+}-O^{2-} centers) which are fed by excitation energy migration (5) over the bismuth lattice in spite of the large Bi-Bi distance (~ 0.8 nm).

Another example is the spectroscopy of $Sb^{3+}(5s^2)$. Figure 5 gives as an example the excitation and emission spectra of Sb^{3+} in $Cs_2NaScCl_6$ at 4.2 K. These spectra are of the broad band type with considerable Stokes shift. The splitting of the absorption bands is due to a dynamic Jahn-Teller effect in the excited state (38).

In the case of A_2SnCl_6:Te^{4+} (A = Cs, Rb, K) the emission of the Te^{4+} ion shows even an impressive progression in the Jahn-

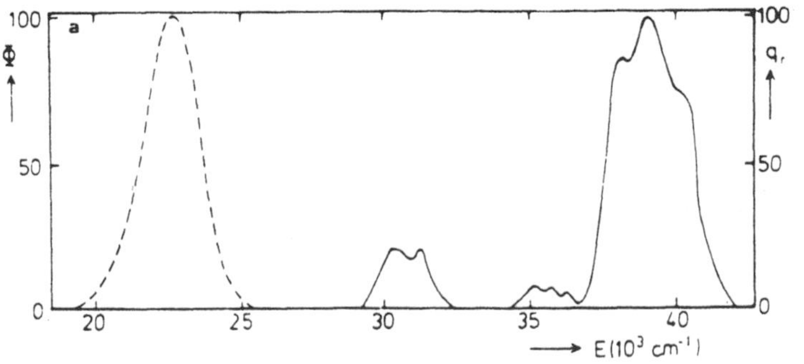

Fig. 5. Emission (broken line) and excitation (solid line) spectra of the luminescence of $Cs_2NaScCl_6$:Sb^{3+} at 4.2 K. After E.W.J.L. Oomen, Thesis, Utrecht, 1987.

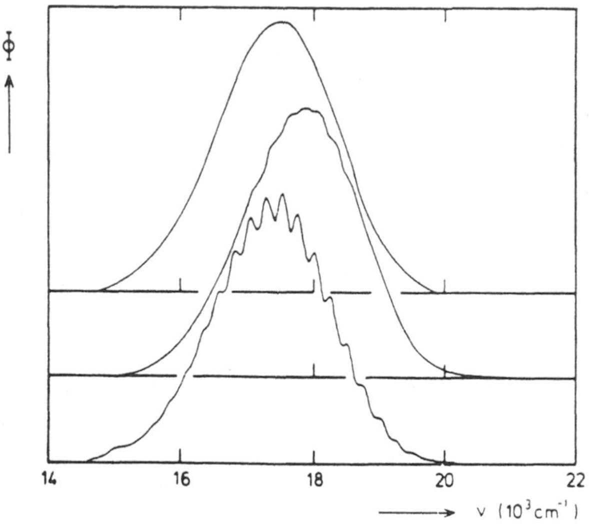

Fig. 6. The emission spectra of the Te^{4+} ion in A_2SnCl_6 (A = Cs, Rb, K) at 10 K. Top curve A = K, lowest curve A = Cs. After H. Donker, Thesis, Utrecht, 1989.

Teller active ν_2 mode, pointing to a pronounced distortion in the excited state of this ion (39). This phenomenon depends on the choice of the A ion (see Fig. 6). The success of the elpasolite lattice in the spectroscopy of s^2 ions is closely related to the perfect octahedral symmetry shown by this lattice for the octahedrally coordinated ions. As a contrast with this spectroscopy of localized S^2 ion centers, we would like to mention the perovskite $CsPbCl_3$ (40). It is clearly a semiconductor. The spectroscopy does not reveal the properties of isolated Pb^{2+} ions. In contrast to the elpasolites considerable delocalization took place (5).

V. The Uranate (UO_6^{6-}) Center

With the uranyl (UO_2^{2+}) group having a long history in spectroscopy it has sometimes been assumed that this is the only oxidic U^{6+} species which exists. The ordered perovskites have shown that also the UO_6^{6-} complex is worthwhile studying (41). Compounds A_2BUO_6 (A, B alkaline earth metal) have ordered perovskite structure and form solid solution series with the analogous tungstates and tellurates A_2BWO_6 and A_2BTeO_6. In fact, compositions of the type $A_2B(W,Te)_{0.99}U_{0.01}O_6$ often show efficient luminescence and present a wealth of spectroscopical information

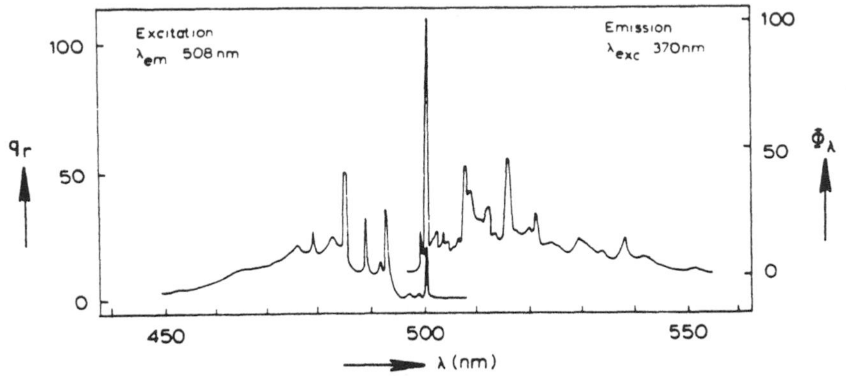

Fig. 7. Emission (right-hand-side) and excitation (left-hand-side) spectra of the luminescence of Ba_2MgWO_6:U at 4.2 K.

at low temperatutes due to the presence of vibrational structure in the spectral bands (41-44) (Fig. 7).

These vibrational structures can be fully analyzed. The magnetic-dipole zero-phonon line is followed by many vibronic lines the greater part of which are due to coupling with internal UO_6^{6-} vibrational modes; there is a clear progression in the totally symmetric ν_1 mode. This spectroscopy can also be used to analyze the constitution of luminescent uranate centres. Runciman at al. (45), for example, have shown that in the perovskite $KZnF_3:U^{6+}$ the optically active center is a tetragonally distorted UO_4F_2 octahedron.

The diluted compositions $A_2BW_{0.99}U_{0.01}O_6$ are very suitable model compounds to study the dependence of structure and composition on the nonradiative transitions in a luminescent center (6,46). Theories based on the simple configuration coordinate model show that the nonradiative rate is strongly restricted if the expansion of the excited state relative to that of the ground state is small. This can be shown nicely on the tungstate and uranate luminescence of ordered perovskites (see Table 1). The table shows the thermal quenching temperature of the uranate luminescence of A_2BWO_6:U as a function of A and B. These temperatures are used as a measure of the radiationless transition rate. It is clear that the rate does not depend on the nature of A, whereas that of B determines the value of this rate strongly. The smaller the B^{2+} ion, the smaller the nonradiative rate. The expansion of the luminescent center is actually hardly counteracted by the A^{2+} ions which are in the [111] directions of the luminescent octahedron. However, the B^{2+} ions are immediately

Table 1. Thermal Quenching Temperature T_q of the Uranate Luminescence of Ordered Perovskites $A_2BWO_6:U^{6+}$

$A_2BWO_6:U^{6+}$		T_q (K)	Δr (au)*
A = Ba	B = Ba	180	10.9
Ba	Sr	240	10.6
Ba	Ca	310	10.2
Ba	Mg	350	10.0
Sr	Mg	350	10.0
Ca	Mg	350	10.0

* Δr is the expansion of the excited state relative to the ground state in arbitrary units, calculated from a single configurational coordinate diagram.

behind the oxygen ions of the octahedron, the angle U(W)-O-B being 180°. The stronger the B-O bond, the smaller the expansion, and the lower the nonradiative rate. In Table 1 the total change in the expansion Δr is less than 10 % (<0.001 nm) which results in drastic changes in the nonradiative rates. It is hard to imagine a better structure for these types of studies than the perovskite structure.

The concentrated systems A_2BUO_6 show another phenomenon of interest, viz. energy migration of the excited state to traps and quenching centers (47). This is comparable to what has been described above for $Cs_2NaBiCl_6$ and the tungstates. Excitation into the intrinsic uranate group is followed by a sequence of energy transfer steps (energy migration) between the intrinsic uranate groups. Finally the migrating energy is trapped by uranate traps. These are UO_6^{6-} groups with an excited state which is slightly below that of the intrinsic groups. Consequently the excitation energy cannot leave the trap at low temperatures and trap emission is observed. At temperatures which are high enough to escape from the traps, energy migration brings the excitation energy to quenching centers, so that the luminescence is quenched. This is shown schematically in Fig. 8. In Ba_2CaUO_6 the traps are uranate groups near Ca^{2+} and Ba^{2+} ions which have been interchanged, and the quenching centers are U^{5+} ions. The scheme

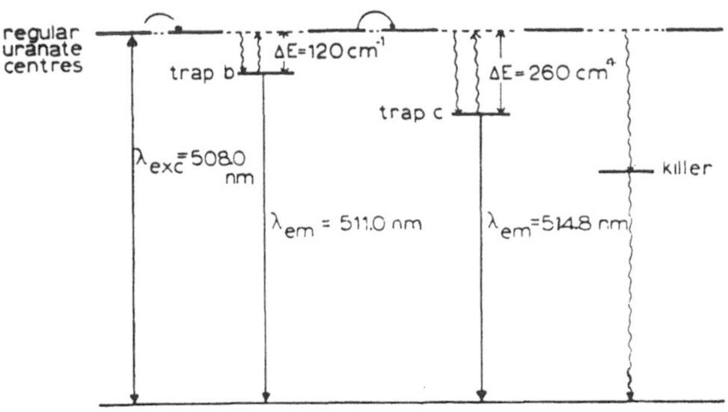

Fig. 8. Schematic energy level scheme of Ba_2CaUO_6 (see text).

of Fig. 8 has a very general meaning and can be applied to many other compounds with high concentrations of optically active centers (5,6).

VI. Rare Earth Ions

A tremendous amount of work has been performed on the optical properties of rare earth ions in perovskites. Elaborate studies have been performed on the elpasolites of the type Cs_2NaRCl_6 (R = (mixture of) rare earth(s)) by Richardson's group (48). These authors have not only measured the spectra in great detail, but have also presented full theoretical analysis. The elpasolites have perfect cubic symmetry, so that the rare earth ions are in sites with O_h symmetry. Consequently, the electronic electric-dipole transitions are parity forbidden and only the magnetic-dipole and vibronic transitions are observed. The dominant contributions to the vibronic structure can be assigned to coupling with the three ungerade vibrational modes of the RCl_6^{3-} octahedron.

The gadolinium member ($Cs_2NaGdCl_6$) was investigated elsewhere (49). Fig. 9 shows as an example of the type of spectra in these elpasolites the absorption spectrum of the $^8S_{7/2} \rightarrow {}^6P_{7/2}$ transition of Gd^{3+} in $Cs_2NaGdCl_6$ at 4.2 K. Lines 1, 2 and 3 are the three electronic magnetic-dipole transitions. There are three lines, since the cubic crystal field splits the $^6P_{7/2}$ level into three components. The distance between lines 1 and 3 correspond therefore to the cubic crystal-field splitting of the $6P_{7/2}$ level. It amounts to 29 cm^{-1} only. The subsequent lines are vibronic in nature and due to coupling with (mainly) the ν_3, ν_4 and ν_6 modes

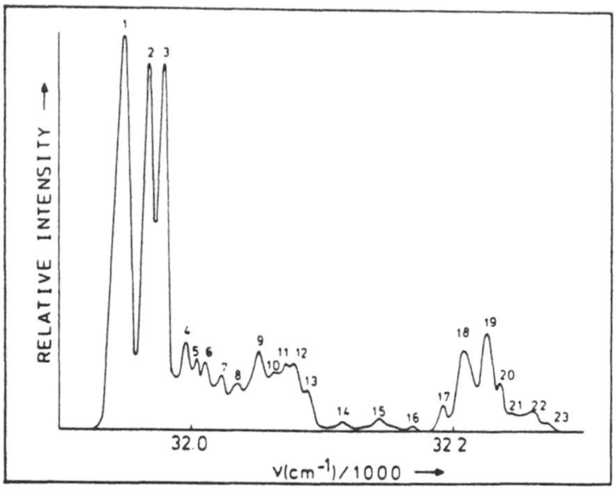

Fig. 9. The $^8S_{7/2}$ -> $^6P_{7/2}$ excitation (absorption) spectrum of the Gd^{3+} ion in $Cs_2NaGdCl_6$ at 4. 2 K. From A.J. de Vries, Thesis, Utrecht, 1988.

of the $GdCl_6^{3-}$ octahedron (49).

Let us now turn to rare earth ions in oxidic perovskite compounds. In early times there have been studies of the spectra of these ions in compounds of the type $RAlO_3$ (50). Another interesting lattice is Ba_2RNbO_6. In the former the R^{3+} ions are in the larger metal ion sites and lack perfect cubic site symmetry; in the latter the R^{3+} ions are in the smaller metal ion sites and have perfect octahedral site symmetry. Fig. 10 shows how the site symmetries influence the spectra on the example of the Eu^{3+} ion. The $^5D_0$ -> $^7F_2$ emission of this ion is allowed as magnetic-dipole transition and forbidden as electric-dipole transition; the $^5D_0$-> $^7D_2$ emission is forbidden as magnetic-dipole transition, but allowed as electric-dipole transition in the absence of inversion symmetry. Fig. 10 shows that, apart from weak vibronic lines, the latter emission is absent in Ba_2GdNbO_6:Eu^{3+} (inversion symmetry at the Eu^{3+} site), that it starts to grow in $Ba_2Gd(Nb,Sb)O_6$:Eu^{3+} (the $Nb^{5+/Sb}5+$ mixture implies a very small deviation from inversion symmetry), that it is clearly present in $LaAlO_3$:Eu^{3+} (deviation from inversion symmetry), and that it is very strong in La_2MgTiO_6:Eu^{3+} (pronounced deviation from inversion symmetry due to the Mg/Ti superstructure) (51).

Extremely strong vibronic lines have been reported for the emission spectrun of the Eu^{3+} ion in $SrTiO_3$ (52,53). This is similar to the case of $CaTiO_3$:Cr^{3+} mentioned above. This phenomenon

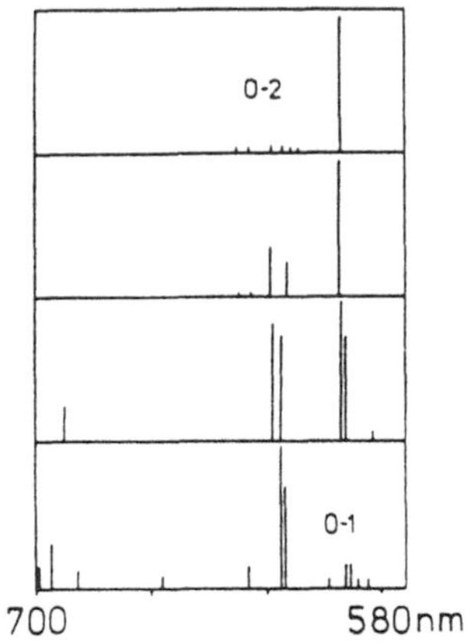

Fig. 10. Schematic emission spectra of the Eu^{3+} ion in (from top to bottom) Ba_2GdNbO_6, $Ba_2Gd(Nb,Sb)O_6$, $LaAlO_3$, La_2MgTiO_6. The $^5D_0$ -> $^7F_1$ transition is at about 595 nm, the $^5D_0$ -> $^7F_2$ transition at about 615 nm. See also text. Modified from ref. 51.

suggests that the emitting state has a larger amount of opposite-parity admixture than the ground state. This might be due to the low-energy position of the absorption edge of the host lattice. In the $LaAlO_3$ perovskite this phenomenon does not occur and the vibronic lines have their usual weak intensity.

Another type of vibronic lines have been reported for the Gd^{3+} ion in perovskites Ba_2RTaO_6 (R = La, Gd, Y, Lu) (54). The emission spectrum of the Gd^{3+} ion in these lattices consists of the electronic $^6P_{7/2}$ -> $^8S_{7/2}$ transition (allowed as magnetic-dipole transition) and a couple of vibronic lines which are due to coupling with the internal vibrational modes of the TaO_6 octahedron. Since the electronic transition occurs on the Gd^{3+} ion and, simultaneously, the vibrational transition in the tantalate octahedron, these vibronic lines are called cooperative vibronic transitions (55). It has been observed that the vibronic intensity in these host lattice increases from R = La to Lu by about 50 %. This has been ascribed to an increase in covalency as is to be expected from theory (54,56). It is an example of fine tuning of distances by changing the radii of a given type of ions in the perovskite structure.

Another example of using the high site symmetry in the perovskite structure is the absorption spectrum of the Eu^{3+} ion in $BaZrO_3$ (57). The absorption spectrum of the Eu^{3+} ion is due to the $4f^7$ -> $4f^65d$ transitions. In a cubic field the 5d level splits into two components. This can be nicelly observed for $BaZrO_3:Eu^{2+}$ where the cubic crystal-field splitting of the Eu^{2+} ions is 13,000 cm^{-1} (57).

Nowadays interest has changed from isolated-ion spectroscopy to the properties of concentrated systems where energy migration occurs. In the case of the trivalent rare-earth ions with highly-forbidden transitions within the $4f^n$ configuration and near-to-cubic site symmetry in the perovskite lattice, the interaction responsible for energy transfer between two ions of one kind is expected to be the exchange interaction (2,5-7). The other possible interaction mechanism is the electric-multipole-multipole interaction.

Let us return to the system mentioned above, viz. $Ba_2(Gd,Eu)NbO_6$. It turns out that the concentration quenching occurs at the composition with 20 % Eu and 80 % Gd (58,59). This implies that for this Eu concentration the excitation energy on the Eu^{3+} ion can be transferred to others (migration), so that quenching centers are reached. The shortest Eu-Eu distance in this lattice is 0.6 nm, since the Eu^{3+} ions are separated from each other by a niobate octahedron. For the isostructural $Cs_2NaEuCl_6$ there is no report of considerable concentration quenching (60), and the same holds true for $Cs_2NaGaCl_6$ (49). In these ordered perovskites all electronic electric-dipole transitions are forbidden, so that exchange is expected to be the interaction responsible for transfer. Since exchange interaction is expected to be negligible for distances of about 0.5 nm, it is to be expected that in these ordered perovskites no energy transfer occurs between the rare earth ions. In the chlorosystems this is the case. In the niobate, however, the occurrence of concentration quenching and energy transfer points to the importance of a certain exchange interaction. This is of the superexchange type using the niobate group as an intermediary (Eu-O-Nb-O-Eu).

Much faster migration occurs in compounds of the type $RAlO_3$ (R = Eu, Gd, Tb) where the R-R distance is shorter than in the ordered perovskites (61-63). Also in these compounds the transfer mechanism is of the exchange type. The transfer rates are of the order of 10^7 s^{-1}, to be compared with the radiative rates of ~ 10^3 s^{-1}. This means that in a pure crystal many steps are made over the rare-earth sublattice before radiative emission occurs.

Of special interest is the influence of magnetic order on the energy migration in these systems (62,63). The compounds $GdAlO_3$ and $TbAlO_3$ show an antiferromagnetic order at 3.9 and 3.8 K, respectively. Below the Néel temperature the energy migration slows down, because the nearest neighbor Gd^{3+} (or Tb^{3+}) ions are oriented antiparallel. This makes energy transfer by exchange interaction impossible, so that the migration can occur only by energy transfer between next-nearest Gd^{3+} (or Tb^{3+}) neighbors. This effect becomes clear by considering the diffusion constant of the excited state: at 4.4 K it amounts to 1.6×10^{-9} cm^2s^{-1} for $GdAlO_3$ and $TbAlO_3$ both, but only 8×10^{-12} cm^2s^{-1} and 8×10^{-14} cm^2s^{-1} at 1.6 K for $GdAlO_3$ and $TbAlO_3$, respectively. In $EuAlO_3$ such an effect does not occur. The ground state of the Eu^{3+} ion ($^7F_0$) does not have a magnetic moment. Fast energy migration occurs down to very low temperatures due to exchange interaction (61).

In $GdAlO_3$ this energy migration does not only take place for the lowest excited level ($^6P_{7/2}$), but also for the higher $^6I_{7/2}$ level (61). This is possible because the nonradiative $^6I_{7/2} \to {}^6P_{7/2}$ decay is slow compared to the transfer rate to neighboring ions: 8×10^3 s^{-1} and 10^7 s^{-1}, respectively. Excitation into the $^6I_{7/2}$ level of Gd^{3+} in $GdAlO_3$ at 4.2 K results a.o. in the $^6I_{7/2} \to {}^8S_{7/2}$ emission shown in Fig. 11. The intrinsic emission (intr.) is very weak. The emissions indicated by T_i are due to Gd^{3+} traps which are fed by energy migration over the Gd^{3+} sublattice. Their

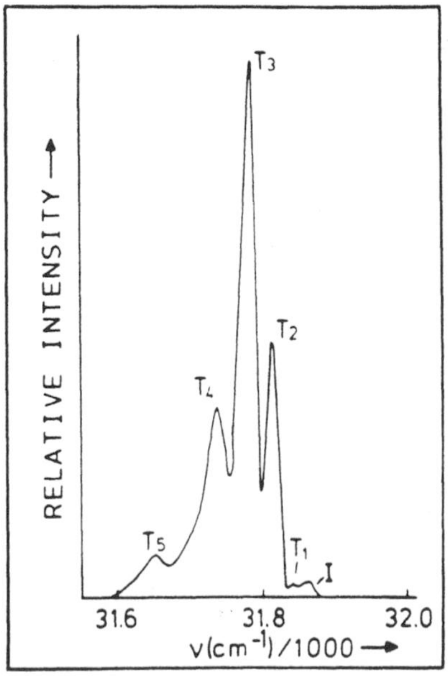

Fig. 11. The Gd^{3+} emission spectrum of $GdAlO_3$ at 4.2 K upon excitation into the $^6I_{7/2}$ level. Source: see Fig. 9.

dominant presence in the emission spectrum proves the occurrence of energy migration. The trap emissions disappear at higher temperatures: detrapping brings the excitation energy back into the Gd^{3+} sublattice and migration to quenching centers starts to prevail. Fig. 12 shows the temperature dependence of the decay time of the Gd^{3+} trap T_2.

The low-temperature decay time corresponds to the radiative decay rate of this trap (375 s^{-1}). The decrease above 5 K points to detrapping. From this curve a trap depth of 55 cm^{-1} is derived. From the spectrum in Fig. 10 this trap depth is found to be 48 cm^{-1}. The agreement is very good in view of the experimental error (about 10 %).

The host $GdAlO_3$ is also very suitable to investigate the trapping of the migrating excitation energy by impurities. These can be introduced on the Gd^{3+} sites (rare earth ions) or on the Al^{3+} sites (transition metal ions) (22). After excitation into the Gd^{3+} ions, the ratio of the amount of impurity emission to that of the Gd^{3+} emission is measured. It turns out that this ratio decreases in the sequence $Cr^{3+} > Tb^{3+} \approx Eu^{3+} > Sm^{3+} > Dy^{3+} > Tm^{3+} > Er^{3+}$. The Cr^{3+} ion is an exceptionally efficient trap for the migrating excitation energy, the Er^{3+} is very inefficient. These results have been analyzed using the theories on energy transfer. It is possible to obtain very reasonable agreement.

We hope to have illustrated clearly that the amount and diversity of spectroscopical studies on rare earth ions in the perovskite structure are impressive.

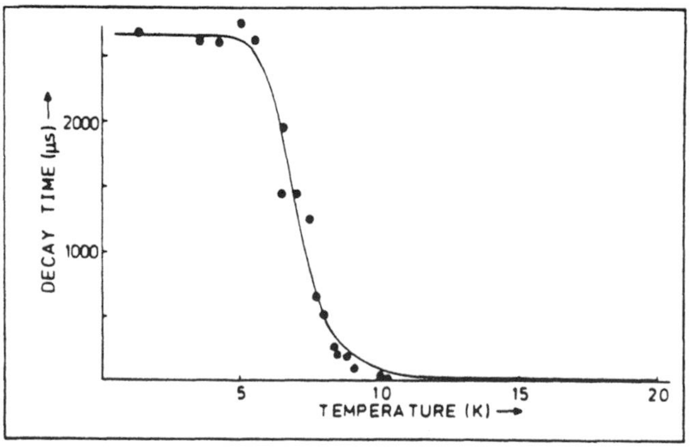

Fig. 12. Temperature dependence of the decay time of the Gd^{3+} trap T_2 (See Fig. 11). Source: see Fig. 9.

VII. Conclusion

Practically all kinds of optical centers have been studied in the perovskite structure, not only in low, but also in high concentrations. Due to the high symmetry of this crystal lattice and its chemical flexibility, many of the compositions studied served as model systems. The results obtained on these model systems have deepened our insight into the relevant fundamental issues. These systems are also a successful starting point in the interpretation of the optical properties of more complicated systems. In this way the optical studies on perovskite systems have contributed considerably to the development of many optical materials of importance (laser materials, phosphors for several applications, materials for solar energy conversion, excintillators).

References

1. Goodenough, J. B. and J. M. Longo, Crystallographic and magnetic properties of perovskites and perovskite-related compounds, Landolt-Börnstein, *Numerical data and functional relationships in science and technology*, New Series, Group III, Vol. 4a, Springer Verlag, Berlin (1970).
2. Henderson, B. and G. F. Imbusch, *Optical spectroscopy of inorganic solids*, Clarendon Press, Oxford (1989).
3. Lever, A. B. P., *Inorganic electronic spectroscopy*, 2nd , Elsevier, Amsterdam (1984).
4. Blasse, G., The luminescence of closed-shell transition-metal-complexes. New developments, *Structure and Bonding* **42:** 1 (1980).
5. Powell, R. C. and G. Blasse, Energy transfer in concentrated systems, *Structure and Bonding* **42:** 43 (1980).
6. Blasse, G., Luminescence of inorganic solids: from isolated centers to concentrated systems, *Prog. Solid State Chem.* **18:** 79 (1988).
7. Blasse, G., Interaction between optical centers and their surroundings: an inorganic chemist's approach, *Adv. Inorg. Chem.* **35:** 319 (1990).
8. De Haart, L. G. J., A. J. de Vries and G. Blasse, On the photoluminescence of semiconducting titanates applied in photoelectrochemical cells, *J. Solid State Chem.* **59:** 291

(1985).
9. Macke, A. J. H., Investigations on the luminescence of titanium-activated stannates and zirconates, *J. Solid State Chem.* **18:** 337 (1976).
10. Aguilar, M. and F. Agulló-López, X-ray induced processes in $SrTiO_3$, *J. Appl. Phys.* **53:** 9009 (1982).
11. Sihvonen, Y. T., Luminescence of strontium titanate, *J. Appl. Phys.* **38:** 4431 (1967).
12. Grabner, N., Photoluminescence in $SrTiO_3$, *Phys. Rev.* **117:** 1315 (1969).
13. Fugol, J. Y., Excitons in rare-gas crystals, *Adv. Phys.* **27:** 1 (1978); Schwentner, N., E. E. Koch and J. Jortner, Energy transfer in solid rare gases, *Energy transfer processes in condensed matter* (B. Di Bartolo, ed.), Plenum Press, New York (1983).
14. Blasse, G. and L. G. J. de Haart, The nature of the luminescence of niobates $MNbO_3$ (M = Li, Na, K), *Mat. Chem. Phys.* **14:** 481 (1986).
15. Paracchine, C. and G. Schianchi, Luminescence of WO_3, *Phys. Stat. Sol.* (a) **72:** K 129 (1983).
16. Blasse, G., On the luminescence of $SrTiO_3$ and related titanates, *Mat. Res. Bull.* **18:** 525 (1983).
17. Macke, A. J. H., Luminescence and energy transfer in the ordered perovskite system $La_2MgSn_{1-x}Ti_xO_6$, *Phys. Stat. Sol.* (a) **39:** 117 (1977).
18. Blasse, G. and G. J. Dirksen, Luminescence of the isolated niobate octahedron in $MgNb_2(P_2O_7)_3$, *Inorg. Chim. Acta* **157:** 141 (1989).
19. Van Oosterhout, A. B., Tungstate luminescence in ordered perovskites, *Phys. Stat. Sol.* (a) **41:** 607 (1977); An ab initio calculation on the WO_6^{6-} octahedron with an application to its luminescence, *J. Chem. Phys.* **67:** 2412 (1977).
20. Hair, J. Th. W., A. F. Corsmit and G. Blasse, Vibrational spectra and force constants of periodates with ordered perovskite structure, *J. Inorg. Nucl. Chem.* **36:** 313 (1974).
21. Tkachuk, A. M. and Z. M. Zonn, Optical spectroscopy of chromium in lanthanum aluminate, *Opt. Spectrosc.* **26:** 322 (1969); Zonn, Z. N., V. A, Ioffe and P. P. Feofilov, Luminescence of Cr^{3+} in $LaAlO_3$, *Opt. Spectrosc.* **19:** 541 (1965).
22. Vries, A. J. de, W. J. J. Smeets and G. Blasse, The trapping of Gd^{3+} excitation energy by Cr^{3+} and rare earth ions in $GdAlO_3$, *Mat. Chem. Phys.* **18:** 81 (1987).

23. Grabner, L. and S. E. Stokowski, Photoluminescence of Cr-doped $CaTiO_3$, *Phys. Rev.* **28:** 4351 (1970).
24. Blasse, G. and P. H. M. de Korte, The luminescence of tetravalent manganese in $CaZrO_3$:Mn, *J. Inorg. Nucl. Chem.* **43:** 4305 (1981).
25. Brauch, U. and U. Dürr, $KZnF_3:Cr^{3+}$ - A tunable solid state near infrared laser, *Opt. Comm.* **49:** 61 (1984); Room temperature operation of the vibronic $KZnF_3:Cr^{3+}$ laser, *Opt. Lett.* **9:** 441 (1984).
26. Reinen, D., Anwendung der Ligandenfeld-Spektroscopie auf Probleme der chemischen Bindung in Festkörpern, *Ang. Chem.* **83:** 991 (1971).
27. Faughnan, B. W. and Z. J. Kiss, Photoinduced reversible charge-transfer processes in transition-metal-doped single-crystal $SrTiO_3$ and TiO_2, *Phys. Rev. Lett.* **21:** 1331 (1968).
28. See e.g. Maruska, H. P. and A. K. Ghosh, Photocatalytic decomposition of water at semiconductor electrodes, *Solar Energy* **20:** 443 (1978); Butler, M. A. and D. S. Ginley, Principles of photoelectrochemical solar energy conversion, *J. Mat. Sci.* **51:** 1 (1980).
29. Blasse, G., P. H. M. de Korte and A. Mackor, The coloration of titanates by transition-metal ions in view of solar energy applications, *J. Inorg. Chem.* **43:** 1499 (1981).
30. Blasse, G., Optical electron transfer between metal ions and its consequences, in press.
31. Chien R. L., J. M. Berg, D. S. McClure, P. Rabinowitz and B. N. Perry, Two-photon electronic spectroscopy of $Cs_2GeF_6:Mn^{4+}$, *J. Chem. Phys.* **84:** 4168 (1986).
32. Reber, C. and H. U. Güdel, Luminescence properties of $CsNiX_3$ and $CsMg_{1-x}Ni_xX_3$ (X = Cl, Br), *Inorg. Chem.* **25:** 1196 (1986).
33. Knochenmuss, R., C. Reber, M. V. Rajasekharan and H. U. Güdel, Broad-band near infrared luminescence of Cr^{3+} in the elpasolite lattices Cs_2NaYCl_6 and Cs_2NaYBr_6, *J. Chem. Phys.* **85:** 4280 (1986).
34. Reber, C. and H. U. Güdel, Near infrared luminescence spectroscopy and relaxation behavior of V^{3+}-doped in $Cs_2NaYCl_{6-m}Br_m$ (m = 0, 0.3, 3, 6), *J. Luminescence* **42:** 1 (1988).
35. Reber, C., H. U. Güdel, G. Meyer, T. Schleid and C. A. Daul, Optical spectroscopic and structural properties of V^{3+}-doped fluoride, chloride and bromide elpasolite lattices, *Inorg. Chem.* **28:** 3249 (1989).
36. Braam, A. W. M. and G. Blasse, Luminescence of Pb^{2+} ions in

$CaZrO_3$ ($CaZrO_3$), *Solid State Chem.* **20:** 717 (1976).
37. Steen, A. C. van der, Luminescence of Cs_2NaYCl_6:Bi($6s^2$), *Phys. State Solid* (b) **100:** 603 (1980).
38. Oomen, E. W. J. L., W. M. A. Smit and G. Blasse, On the luminescence of Sb^{3+} in Cs_2NaMCl_6 (with M = Sc, Y, La): a model system for the study of trivalent s2 ions, *J. Phys. C: Solid State Phys.* **18:** 3263 (1986).
39. Wernicke, R., H. Kupka, W. Ensslin and H. H. Schmidtke, Low temperature luminescence spectra of the $d^{10}s^2$ complexes Cs_2MX_6 (M = Se, Te and X = Cl, Br). The Jahn-Teller effect in the γ_4^- ($^3T_{1u}$) excited state, *Chem. Phys.* **47:** 235 (1980); Ackerman, J. F., Preparation and luminescence of some K_2PtCl_6 materials, *Mat. Res. Bull.* **19:** 783 (1984); Donker, H., W. M. A. Smit and G. Blasse, On the luminescence of Te^{4+} in A_2ZrCl_6 (A = Cs, Rb) and A_2SnCl_6 (A = Cs, Rb, K), *J. Chem. Phys.* **50:** 603 (1989).
40. Belikovich, B. A., I. P. Pashchuk and N. S. Pidzyrailo, Luminescence of $CsPbCl_3$ single crystals, *Opt. Spectrosc.* **42:** 62 (1977); Heidrich, K., H. Künsel and J. Treusch, Optical properties and electronic structure of $CsPbCl_3$ and $CsPbBr_3$, *Solid State Comm.* **25:** 887 (1978).
41. Bleijenberg, K. C., Luminescence properties of uranate centers, *Structure and Bonding* **42:** 97 (1980).
42. Bleijenberg, K. C. and H. G. M. de Wit, On the vibrational structure in the luminescence spectra of uranium-doped tungstates with ordered perovskite structure, *J. Chem. Soc., Faraday Trans. II* **76:** 872 (1980).
43. Hair, J. Th. W. de and G. Blasse, Luminescence of the octahedral uranate group, *J. Luminiscence* **14:** 307 (1976).
44. Hair, J. Th. W. de and G. Blasse, The luminescence properties of the octahedral uranate group in oxides with perovskite structure, *J. Solid State Chem.* **19:** 263 (1976).
45. Manson, N. B., A. P. Radlinski and W. A. Runciman, Excitation and emission of hexavalent uranium in $KMgF_3$, *Solid State Comm.* **52:** 1011 (1984).
46. Blasse, G., The influence of the ligands on the luminescence of metal ions, *Photochem. and Photobiol.* **52:** 417 (1990).
47. Krol, D. M. and G. Blasse, Luminescence and energy migration in Ba_2CaUO_6, *J. Chem. Phys.* **69:** 3124 (1978).
48. Richardson, F. S., M. F. Reid, J. J. Dallara and R. D. Smith, Energy levels of lanthanide ions in the cubic

$Cs_2NaLnCl_6$ and Cs_2NaYCl_6:Ln^{3+} (doped) systems, *J. Chem. Phys.* **83:** 3813 (1985), and references therein.
49. Vries, A. J. de and G. Blasse, On the luminescence of $Cs_2NaGdCl_6$, *J. Chem. Phys.* **88:** 7312 (1988).
50. See e.g. Weber, M. J., T. E. Varitimos and B. H. Matsinger, Optical intensities of rare-earth ions in yttriun ortho-aluminate, *Phys. Rev.B* **47:** 726 (1973); Delsart, Ch., Déexcitation radiative et nonradiative des ions Eu^{3+} dans LaAlO3, *J. Physique* **34:** 711 (1973).
51. Blasse, G, A. Bril and W. C. Nieuwpoort, On the Eu^{3+} fluorescence in mixed metal oxides. I. The crystal structure sensitivity of the intensity ratio of electric and magnetic dipole emission, *J. Phys. Chem.* **27:** 1587 (1966); Blasse, G., The Eu^{3+} luminescence as a measure for chemical bond differences in solids, *Chem. Phys. Lett.* **20:** 573 (1973).
52. Weber, M. J. and R. F. Schaufele, Vibronic spectrum of Eu^{3+} in strontium titanate, *Phys. Rev. A* **138:** 1544 (1965).
53. Blasse, G., The intensity of vibronic transitions in the spectra of the trivalent europium ion, *Inorg. Chim. Acta* **169:** 33 (1990).
54. Blase, G. and L. H. Brixner, The intensity of vibronic transitions in the emission spectrum of the trivalent gadolinium ion, *Inorg. Chim. Acta* **169:** 25 (1990).
55. Stavola, M., L. Isganitis and M. J. Sceats, Cooperative vibronic spectra involving rare earth ions and water molecules in hydrated salts and diluted aqueous solutions, *J. Chem. Phys.* **74:** 4228 (1981).
56. Judd, B. R., Vibronic contributions to ligand-induced pseudo-quadrupole absorption of rare-earth ions, *Phys. Scripta* **21:** 543 (1980).
57. Blasse, G., W. L. Wanmaker, J. W. ter Vrugt and A. Bril, Fluorescence of Eu^{3+}-activated silicates, *Philips Res. Repts.* **23:** 189 (1968).
58. Blasse, G., Concentration quenching of Eu^{3+} fluorescence, *J. Chem. Phys.* **46:** 2583 (1967).
59. Vliet, J. P. M. van, D. van der Voort and G. Blasse, Luminescence and energy migration in Eu^{3+}-containing scheelites with different anions, *J. Luminescence* **42:** 305 (1989).
60. Serra, O. A. and L. C. Thompson, Emission spectra of $Cs_2NaEuCl_6$ and $Cs_2Na(Eu,Y)Cl_6$, *Inorg. Chem.* **15:** 504 (1976).
61. Vries, A. J. de, J. P. M. van Vliet and G. Blasse,

Mechanism of concentration quenching of Gd^{3+} and Eu^{3+} emission in perovskite aluminates, *Phys. Stat. Sol. (b)* **149:** 391 (1988).
62. Salem, Y., M. F. Joubert, C. Linares and B. Jacquier, Effect of magnetic order on the fluorescence dynamics of $LnAlO_3$ (Ln = Gd, Tb), *J. Luminescence* **40/41:** 694 (1988).
63. Joubert, M. F., B. Jacquier, C. Linares, J. P. Chaminade and B. M. Wanklyn, Electronic exchange interaction effect on the fluorescence properties of terbium magnetic insulators, *J. Luminescence* **37:** 239 (1987).

Chapter 7

Poisoning of Perovskite Oxides by Sulfur Dioxide

Li Wan

Department of Chemistry and Environmental Engineering, Beijing Polytechnic University, Beijing 100871, People's Republic of China

I. Introduction

Perovskite oxides have been widely studied on their crystal structure, electric, magnetic, high-temperature superconductive, and catalytic properties (1-3). One of the early studies of perovskite as catalyst was conducted by Parravano (4,5), who found that the CO oxidation activity of $NaNbO_3$, $KNbO_3$, $LaFeO_3$, and $La_{1-x}Sr_xMnO_3$ was related to their electronic properties. In catalysis, since the early seventies, the main interest in perovskite has been focused on their potential use as oxidation and reduction catalysts for air pollution control. Suggestions by Meadowcroft (6) and Libby (7) on the use of cobaltate perovskites as substitutes of platinum for fuel cell electrodes and for treatment of auto exhaust has greatly stimulated the study of these materials (8-17). It was observed that under most of the conditions relevant to automobile operations, these base metal perovskite were significantly less active but far more sensitive to SO_2 poisoning than platinum catalyst. However, excellent results with regard to SO_2 resistance and activity for both CO oxidation and NO_x reduction were given by catalysts in which small amount of noble metals were added to or incorporated in the perovskite structure (18-20). Noble metal promoted perovskites have been receiving considerable attention and may well be developed into a practical device for auto exhaust catalysis. Moreover, some perovskites can be applied as catalyst to control air pollutants from stationary sources and to be used in alcohol sensors and O_2 sensors (21-23). Some perovskites have been found to be active catalysts for reactions like hydrogenation and hydrogenolysis of hydrocarbons (24), reduction of SO_2 by CO (25), selective oxidation of C_3H_8 (26), CH_4 combustion (27), photodecomposition of water (28), and thermal decomposition of NO (29).

Perovskite is also an ideal model structure for the study of structure-activity relation (30,31) and for catalyst design, since this structure can tolerate multiple ion substitution. Partial substitution of the A cation with non-equivalent ions may cause the stabilization of an unusual valence state and mixed valence state of transition metal ions in the B site as well as the variation of lattice defect density. These changes in electronic properties will in turn modify the catalytic performance of the resultant perovskite. On the other hand, proper choice of B site cations may bring about novel properties. The SO_2 resistance has been shown to be improved by substitution of both A and B cations.

In spite of the highly technological importance of SO_2 effect on oxide catalysts, literature on this subject remains quite limited. Many problems are left open for exploration.

II. Adsorption of SO_2 on Metal Oxides

The readiness of SO_2 poisoning of metal oxide catalysts has been generally attributed to low temperature stable sulfate formation (32), although there are cases in which SO_2 poisoning has been reported to be reversible due to blocking of surface sites by chemisorbed SO_2.

Adsorption and oxidation of SO_2 on metal oxides can be regarded as the prerequisite to surface (or even bulk) sulfate formation. To know how SO_2 is adsorbed and the route of their conversion may be of help for developing perovskite catalysts with improved SO_2 resistance. But it will be seen from the text that studies on these problems have only just been started.

A. Nature and Reactivity of the Adsorbed Species

SO_2^- Anion

Kazansky et al. found by EPR technique that SO_2, which has a higher electron affinity than O_2, adsorbs at room temperature in the form of SO_2^- and SO^- anions on partially reduced oxides such as TiO_2 (33-35), ZnO (34), and V_2O_5/SiO_2 (36, 37). For instance, when SO_2 is introduced to a TiO_2 (anatase) sample pretreated at 773 K under a vacuum of 10^{-5} Torr (1 Torr = 133.3 Nm^{-2}), the EPR signals that appear (g = 2.005, g = 2.0017) resemble that of the SO_2^- anion formed in the decomposition of $K_2S_2O_5$. Works of Lunsford et al. showed that SO_2 adsorbs as SO_2^- on thermally activated and UV-irradiated MgO (38) and on reduced V_2O_5/SiO_2 (39) at room temperature. SO_2^- formation during SO_2 adsorption at temperature above 373 K has been observed on dehydrated Al_2O_3 by Ono et al. (40) and on Al_2O_3 supported Mo and Mo-Co by Khulbe and Mann (41).

The adsorption sites for SO_2^- are believed to be oxygen ion vacancies. High temperature outgassing of metal oxide is essential for creating high energy defect centers. The maximum amount of SO_2^- was obtained on a 973 K pretreated Al_2O_3 at an adsorption temperature of 613 K, reported by Ono. Kazansky observed that no EPR signals of SO_2^- appeared on insufficiently reduced V_2O_5/SiO_2. The SO_2^- anions obtained on various oxides are stable at room temperature under a dynamic vacuum of 10^{-5} Torr, but will vanish at elevated temperatures.

SO_2^- may be an active intermediate in the oxidation and reduction of SO_2 on metal oxides. Evidence was provided by Kazansky (35) that reaction of SO_2^- with molecular O_2 (or O_2^- with molecular SO_2) on reduced TiO_2 leads to the formation of O_2^- (or SO_2^-) and a diamagnetic species, very likely to be adsorbed sulfate ion. This surface sulfate may then decompose to SO_3 and O^{2-}. Lunsford reported that SO_2^- can react with O_2 to form SO_3^- on MgO (42) and SO_4^- anion on V_2O_5/SiO_2. Photooxidation of SO_2 on MgO in the presence of water and O_2 may occur via SO_3^-, which reacts with surface sulfite species to form SO_2^- and SO_4^{2-} (43).

Chemisorbed SO_2 Species

It is known that SO_2 has a lone pair of electrons and can act as a Lewis base. It may also act as a Lewis acid. Therefore, SO_2 can be linked in several ways with metal ions to form a SO_2 complex, which can be identified by infrared technique (44). Chemisorbed SO_2 on metal oxides may take the form of a SO_2 complex.

SO_2 adsorption is affected by the number of OH^- and CO_3^{2-} groups on surfaces of non-transition metal oxides. On well outgassed CaO, MgO, and Al_2O_3, SO_2 chemisorbs immediately at room temperature. The IR bands at 1250-1100 cm^{-1} on CaO, 1323 and 1140 cm^{-1} on MgO were assigned as surface SO_2 complex by Low et al. (45,46). On Al_2O_3, the band at 1326 cm^{-1} observed by Chang (47), and bands at 1327-1378 and 1148 cm^{-1} observed by Yao (48) were assigned to chemisorbed SO_2. These chemisorbed SO_2 species react very rapidly with lattice oxygen to form sulfite species.

Sulfite and Sulfate Species

Conversion of chemisorbed SO_2 to sulfite species on freshly outgassed CaO and MgO occured so fast that IR bands due to S-bonded sulfite (e.g. 992, 929, and 632 cm^{-1} on CaO, 1088, 1041, and 956 cm^{-1} on MgO) were immediately observed when the sample was exposed to even a quite small amount of SO_2. Bands due to surface SO_2 complex appeared only at moderate surface coverage, when sterically and energetically suitable sites usable for sulfite formation were diminished. Upon introducing the first dose of SO_2 to well dehydrated Al_2O_3, an IR band at 1060 cm^{-1} appeared immediately, which could only be removed at 873 K, and was assigned to sulfite species by Chang. Lunsford (49) observed two forms of sulfite species after heating the sample covered with chemisorbed SO_2 at 573 K under vacuum; the one assigned as O-bond gave bands at 925 and 875 cm^{-1}; the other yielding bands at 1070, 1040, and 875 cm^{-1} was ascribed to S-bond sulfite.

These sulfite species can be oxidized to sulfate species by reacting with O_2 at temperatures above 673 K. On heating in a vacuum above 773 K, some of the sulfite species may decompose, some may convert to sulfate species via disproportionation. IR bands assigned to sulfate were reported to be 1150 and 981 cm^{-1} on CaO, 1300, 1050, 1250, and 1092 cm^{-1} on MgO, 1365 and 1060 cm^{-1} on Al_2O_3. Deo et al. (50) observed sulfate-like species on Al_2O_3 giving bands at 1375 and 1110 cm^{-1}.

Transition metal oxides like MnO_2 (51), Co_3O_4 (52), V_2O_5 (53), and CuO without pretreatment can adsorb SO_2 to form sulfate at room temperature. IR spectra displaying sulfate formation on these oxides are given in Fig. 1. The one for CuO was recorded after exposure to a gas mixture of 400 ppm SO_2 in air at 473 K (54). The bands obtained are comparable with those of bulk $CuSO_4$ observed by Pechkovsky (53). $La2O_3$, basic in nature shows its surface fully covered by hydroxyl and carbonate groups, therefore it does not adsorb SO_2 at room temperature without vacuun pretreatment, as shown by the IR spectra in Fig. 1 d. When unpretreated TiO_2 (52) and ZrO_2 (55) were exposed to SO_2 at 983 K, there were some nearly unperceptable weak bands between 1199 and 1089 cm^{-1} on TiO_2 and weak bands on ZrO_2, probably due to SO_2

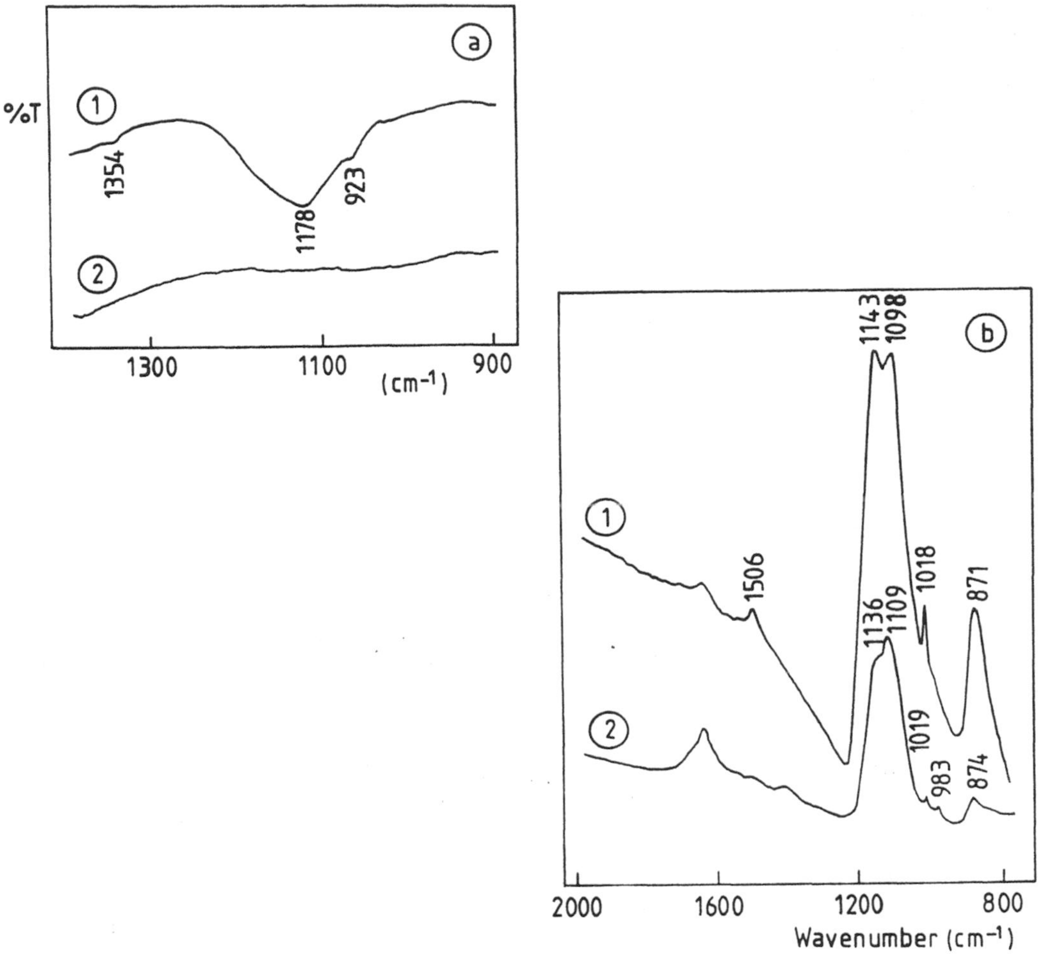

Fig.1. (a) Infrared spectra of SO_2-treated MnO_2: 1, SO_2 adsorption at room temperature; 2, background. (b) Infrared spectra of SO_2-treated Co_3O_4: 1, exposed to SO_2 at room temperature; 2, heated in air at 513 K.

complexes, as shown in Figure 2. The spectrun for TiO_2, pretreated at 298 K under a vaccum of ca. 10^{-5} Torr for two hours, then exposed to 80 Torr SO_2 and allowed to stand for three hours, showed a band at 1110 cm^{-1}, which could be assigned to SO_4^{2-}.

These infrared observations are in accord with kinetics studies on SO_2-metal oxide interaction by DeBerry and Sladek (56). Reacting with a gas mixture containing 0.35% SO_2 and small amounts of O_2, CO_2, H_2O, balance N_2, some oxides like CuO and Co_3O_4

displayed measurable rates at temperatures between 473 and 773 K, and were converted to sulfates, while rates were immeasurably small in a temperature range from 298 K to 1073 K for Al_2O_3, ZnO, TiO_2, and ZrO_2. The IR data of TiO_2 is also consistent with studies on single crystals by Henrich et al. (57,58). Henrich reported that oxides of transition metals in their highest valence sates, e.g. TiO_2, ZnO, NiO, and V_2O_5, when cleaved under UHV, provide surfaces that are quite inert to adsorption of O_2, CO, and SO_2 at 298 K. However, UHV cleaved nearly a perfect surface of suboxides such as Ti_2O_3 and V_2O_3 interacts strongly with these molecules, as revealed by UPS, XPS, and LEED studies. SO_2 was found to be adsorbed dissociatively on the Ti_2O_3(1012) plane, and resulted in completely oxidizing the surface to TiO_2 and TiS_2 (59).

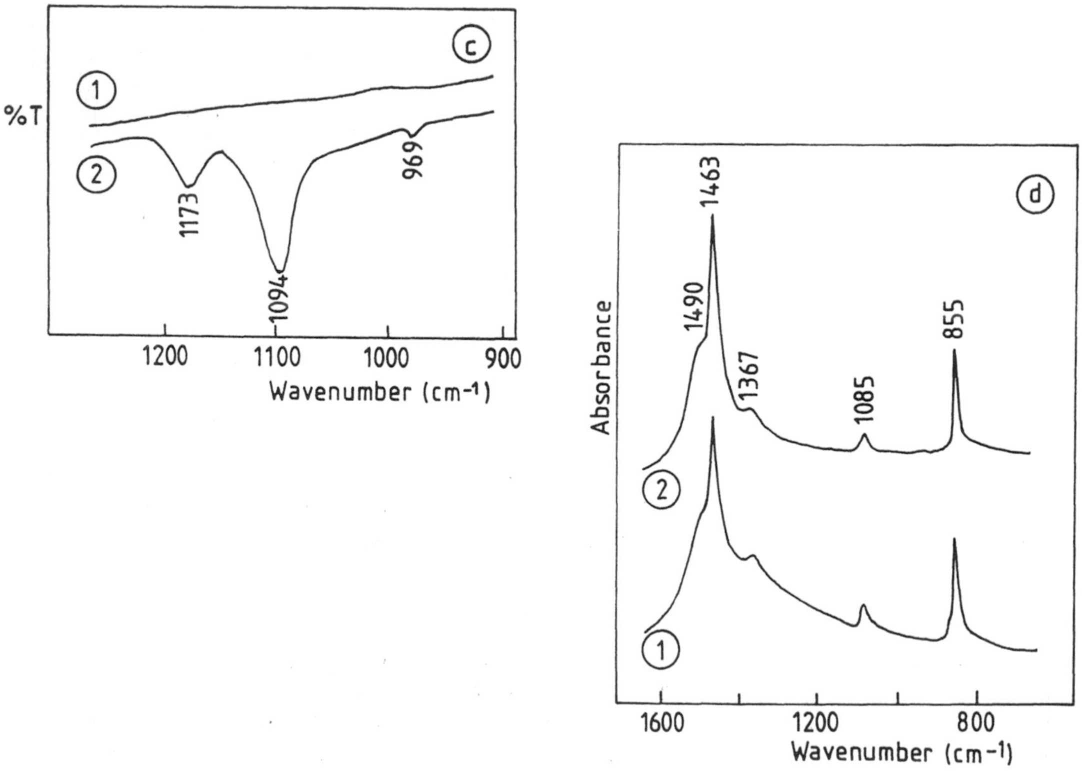

Fig. 1. (c) Infrared spectra of SO_2-treated CuO: 1, background; 2, exposed to a mixture of 400 ppm SO_2 in air at 473 K. (d) Infrared spectra of SO_2-treated La_2O_3: 1, background; 2, exposed to SO_2 at room temperature.

B. Thermal Stability of Metal Sulfate

Thermodynamic functions (heat of formation, heat of decomposition) and sulfate decomposition temperatures both provide a basis for evaluating the tendency of formation and the stability of various metal sulfates. Useful information is available in Lowell's review (60), Kellog's review (61), and Kubaschewski's book (62). In evaluating metal oxide for use as a sorbent for removing SO_2 from flue gas, Lowell had discussed and collected decomposition behavior of sulfate, and sulfite of 46 elements. Recently, Tagawa et al (63,64) measured anew the initial decomposition temperatures of 14 metal sulfates. The results were found to be well related with thermodynamic functions. From a survey of the heats of sulfate decomposition data, Tagawa remarked that trivalent metal sulfates will

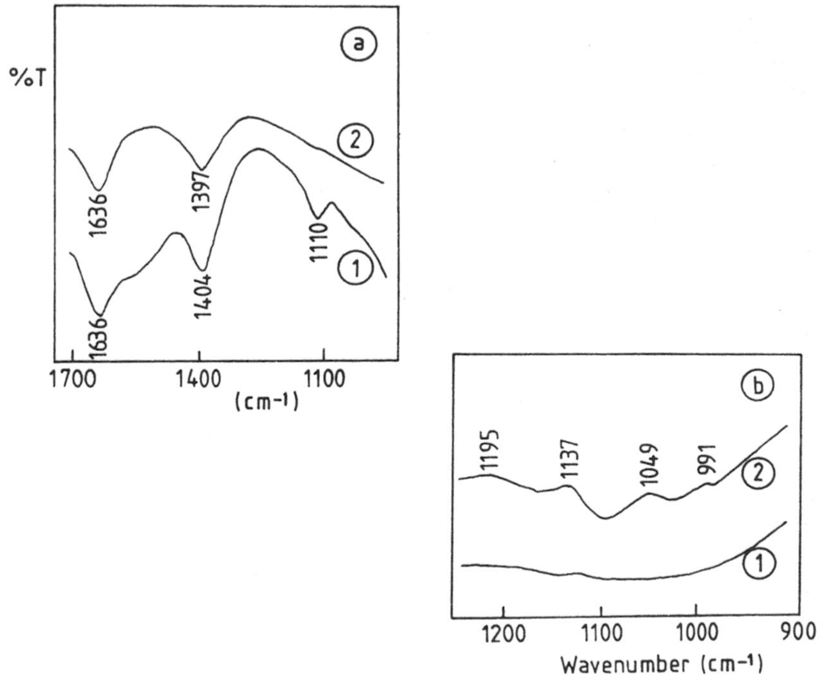

Fig. 2. (a) Infrared spectra of SO_2-treated TiO_2: 1, outgassed at 298 K under 10^{-5} Torr for 2 h and then exposed to 80 Torr SO_2; 2, prereduced in H_2 at 673 K and then exposed to 80 Torr SO_2. (b) Infrared spectra of SO_2-treated ZrO_2: 1, background; 2, exposed to SO_2 at room temperature.

decompose at lower temperatures than divalent metal sulfates, if the determination begins at the same SO_3 pressure. Factors possibly influencing thermal stability of sulfates were discussed by Ostroff and Sanderson (65) with special consideration on the effect of the sulfate structure. They concluded that in the literature data there existed a tendency toward greater stability, the lower the electronegativity of the metal and the larger its crystal radius.

III. SO_2 Effect

Base metal perovskites, as reported, are subject to SO_2 poisoning as are nearly all oxide catalysts. In the case of CO and C_2H_4 oxidation, a study by YaO (15) showed that at 673-773 K a few ppm of SO_2 in the gas stream containing 1% CO (or 0.1 % C_2H_4), 1% O_2, balance N_2, caused practically complete and irreversible poisoning of seven perovskites, namely, $LaCoO_3$, $BaCoO_3$, $LaMnO_3$, $La_{.5}Sr_{.5}MnO_3$, $La_{.9}Sr_{.1}MnO_3$, and $La_{.7}Pb_{.3}MnO_3$. The amount of SO_2 passed to cause complete poisoning in CO oxidation was approximately 0.2 ml/m^2.

Differential thermal analysis is a useful technique for quick evaluation of catalyst activity and poisoning effect (66-68). Johnson et al (69,70) observed that the CO oxidation activity of $La_{.5}Sr_{.5}MnO_3$ declined to nil when it was heated in a DTA cell up to 773 K in a gas blend of 2% CO, 2% O_2, 96% N_2 plus 150 ppm SO_2. In other experiments, $La_{.5}Sr_{.5}MnO_3$ and $La_{.5}Pb_{.5}MnO_3$ were initially exposed to the above gas blend at 298 K for four hours and then heated up to 773 K in SO_2 free feedstreams. It was found that this mild treatment was sufficient to nearly totally poison these two perovskites. The activity of these two catalysts could only be partially restored by heating to 873 K in air. A similar test on $LaCu_{.5}Mn_{.5}O_3$ revealed that this perovskite also lost its CO reactivity when heated in the presence of 150 ppm SO_2 at 673 K (71).

Voorhoeve et al. (11) studying perovskite-like catalysts Ba_2CoWO_6 and Ba_2FeNbO_6, found that 50 ppm SO_2 at 728 K caused a dramatic reduction in activity for both CO oxidation and NO_x reduction. Partial regeneration could be achieved by heating in a SO_2-free gas stream at temperatures ranging from 673 to 728 K.

Studies by Huang et al. (72,73) on SO_2 poisoning of $La_{1-x}A_xMnO_3$ system where A = Sr, Ba, Ca, and Pb, x = 0.3 or 0.5, showed that these catalysts were severely and irreversibly

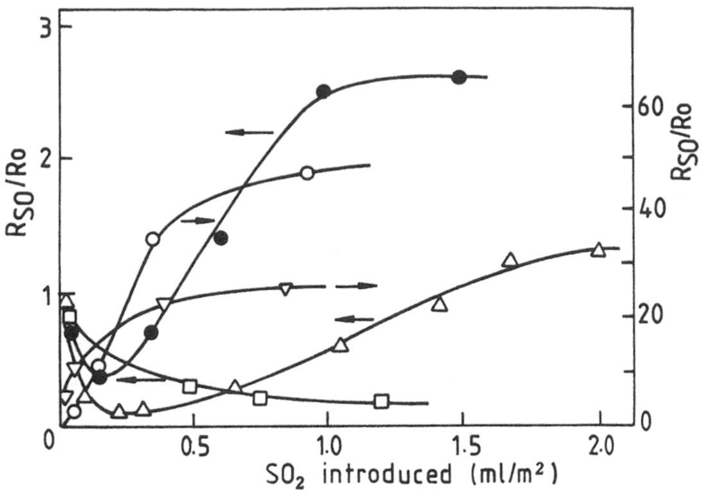

Fig. 3. Rate of oxidation as a function of the amount of SO_2 introduced : ●, oxidation of CO on $La_{0.7}Pb_{0.3}MnO_3$ (100 ppm Pt); ○, oxidation of C_2H_4 on the same catalysts; △, oxidation of CO on $La_{0.7}Pb_{0.3}MnO_3$ (30 ppm Pt); ▽, oxidation of C_2H_4 on the same catalyst; □, oxidation of CO on $La_{0.5}Sr_{.5}MnO_3$ (Pt). (Reproduced by permission from ref.15).

poisoned by 50 ppm SO_2 at 513-573 K. Infrared spectra revealed the presence of sulfate in $La_{.7}Pb_{.3}MnO_3$ and $La_{.5}Sr_{.5}MnO_3$ after poisoned at 673 and 873 K, respectively (Fig. 6).

The degree of poisoning of a base metal oxide catalyst depends on several factors, namely: SO_2 concentration in the gas phase, gas flow rate, temperature, formulation, and surface area of the catalyst. In general, the degree of poisoning increases with decreasing temperature, increasing SO_2 concentration, and increasing space velocity. E.g. with a space velocity of 20000 h^{-1} and SO_2 concentration of 200 ppm in the gas stream of 1% CO in air, the time and amount of SO_2 passed to reduce the CO conversion over $La_{.6}Sr_{.4}CoO_3$ to below 5% at 473 K and 673 K are 40 min-0.23 ml/m^2 and 60 min-0.35 ml/m^2, respectively. At a higher concentration of SO_2, e.g. 400 ppm, this catalyst is deactivated in 30 min with 0.35 ml/m^2 SO_2 passed (55).

YaO (74-77) reported that CuO and $CuCr_2O_4$ were more SO_2 tolerant than oxides like NiO, Cr_2O_3, Co_3O_4, MnO_2, Fe_2O_3, SnO_2, and ZrO_2. In the presence of 20 ppm SO_2, at 773 K, $CuCr_2O_4$ lost about 80% and 60% of its activity for oxidation of CO and C_2H_2, respectively. The total amount of SO_2 passed were about 2 ml/m_2 in both cases. While Co_3O_4 and MnO_2 were the oxides most sensitive to SO_2, they were totally deactivated by a few ppm of SO_2 at 773

K for both CO and C_2H_2 oxidation. The amount of SO_2 passed was about 0.1 ml/m^2 which was less than that required to cover one monolayer of the surface, taking 30 $Å^2$ as the area occupied by one SO_2 molecule for calculation.

Studies on SO_2 poisoning of CuO, Cr_2O_3, and $CuCr_2O_4$ by Sultanov (78) indicated that SO_2 adsorbed on CuO forming sulfate, which decomposed at 873-973 K to form oxysulfate and CuO, thereby its activity for CO restored. Farrauto and Wedding (78) provided evidence that SO_2 poisoning of $CuCr_2O_4$ depended strongly on temperature and SO_2 concentration. A sample poisoned by 100 ppm SO_2 at 773 K could partially recover its activity by shutting off the SO_2 supply, indicating some degree of reversibility. In other sample exposed to 800 ppm SO_2 at 1073 K, no change in CO and hexane reactivity was observed, while those exposed at 973 K and below were all severely poisoned. These samples poisoned by high SO_2 concentration could be regenerated by heating at 1073 K for one hour. The possibility of sulfate formation at high SO_2 concentration (e.g. 10000 ppm) but not at low concentration (100 ppm) was demonstrated by water washing and analyzing the ions dissolved in the water leachate.

A marked SO_2 induced activation effect over $La_{.7}Pb_{.3}MnO_3$ crystals containing 30 ppm of Pt was reported by Yao (15). Upon introducing 20-40 ppm SO_2, the activity of the sample first decreased then increased with time to a value higher than the original activity. At 773 K this effect was very significant and

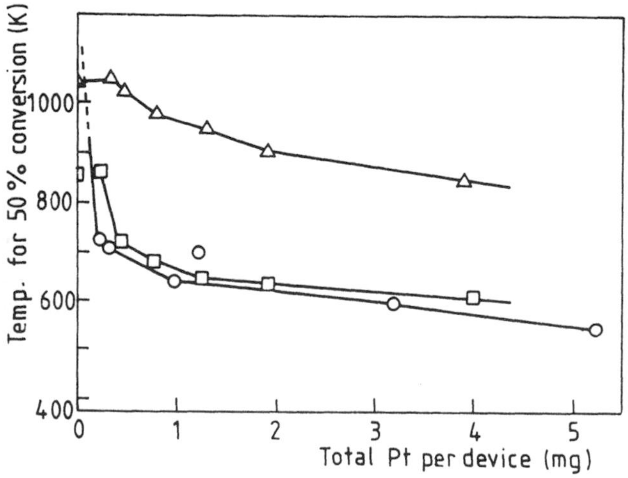

Fig. 4. Activity of supported catalysts for CO oxidation with SO_2 (1.1-1.7 g perovskite) over: △, $LaMn_{0.5}Cu_{0.5}O_3$; □, $La_{0.5}Sr_{0.5}MnO_3$; ○, Pt/Al_2O_3 catalysts. (Reproduced by permission from ref. 82).

the final activity reached eight times the initial value for CO and 50 times for C_2H_2. The kinetics over the SO_2 treated sample followed that over Pt. This effect was also observed with polycrystalline preparation doped with 100 ppm Pt. The role of SO_2, according to Yao, was to free Pt from the surface Pt-Pb alloy through the formation of $PbSO_4$. Single crystals of this composition were known to undergo thermal deactivation due to migration of Pb on the surface at 673 K (80). Gallagher et al (81,82) reported similar SO_2 induced activation with $La_{.7}Pb_{.3}MnO_3$ and $La_{.5}Sr_{.5}MnO_3$. Doping with Pt tended to degrade the activity of $LaCu_{.5}Mn_{.5}O_3$ in the absence of SO_2, perhaps due to a surface Pt-Cu alloy formation. Introducing 50 ppm SO_2, the activity of this perovskite increased but in a lower level than that for the other two perovskites, as illustrated in Figure 3 and Figure 4.

It is very interesting to quote here the result obtained recently by Sachtler et al (83) in studying the effect of SO_2 on the oxidation of C_3H_8 by Pb-poisoned Pt/Al_2O_3 catalyst. An increased activity due to SO_2 exposure was attributed to the formation of large $PbSO_4$ crystals which effectively removed Pb from Pt crystallites. The C_3H_8 oxidation activity of Pb free Pt/Al_2O_3 was also enhanced because of $Al_2(SO_4)_3$ formation. SO_2 poisoning of CO oxidation offsets the enhancement due to removal of Pb from Pt crystals, thus the net effects of SO_2 on CO conversion by Pb poisoned Pt catalyst was small.

Angelov et al (84,85) reported that SO_2 tolerance of a spinel system was related to the fine structure of these catalysts. They found that the resistance of $Cu_xCo_{3-x}O_4$ to SO_2 increased substantially when x > 0.3. This fact was explained by the increase in surface acidity due to formation of electrophilic O^- species, which resisted the adsorption of acidic SO_2. The O^- anions were generated to compensate the positive charge on Cu (II) cation in octahedral site of the spinel and such Cu ion was formed at x > 0.3. It was found that $Cu_{.8}Co_{2.2}O_4$ retained all its activity in the first four hours of reaction at 573 K in a gas stream from a pulse flame reactor burning a blend containing 0.03 wt% S, added as ethanethiol.

IV. Regeneration

Adsorption poisoning is reversible. By reversible we take the strict definition here that when SO_2 is removed from the feedstream and the reaction is allowed to continue at the same temperature and inlet conditions, the activity is restored with time. A reversible poisoning of CuO/SiO_2 by 350 ppm SO_2 at 573 K and 673 K in CO oxidation was observed by Mehanrev et al. (86). The activity restored to its original value when the SO_2 supply was shut off. At 773 K this catalyst remained unpoisoned in 300 minutes on stream. In this case, the adverse effect of SO_2 was attributed to the blocking of surface sites by chemisorbed SO_2 complex.

Poisoning is irreversible when the adsorption of SO_2 readily leads to sulfate formation. When sulfate formation is involved, as in the case of Cu-Cr-Al-O catalyst studied by Fishel et al. (87) and Jenson et al. (88), the poisoned catalyst could be regenerated by heating at higher temperature in air or in N_2 to decompose the sulfate. In fact, a concomittant improvement in the CO oxidation activity and a decline in the sulfate content accumulated on this catalyst from vehicle exhaust was observed.

The removal of sulfate formed on a $CuCr_2O_4$ catalyst by water washing was demonstrated by Farrauto and Wedding (79). This catalyst lost virtually all of its ability to catalyze CO oxidation below 473 K after poisoned at 773 K by 1000 or 10000 ppm SO_2. It recovered 50%-60% of its original activity by water washing. Meanwhile, cations and soluble sulfur were detected in the water leachate. Sultanov et al. (88) has analyzed the water leachate of eight transition metal oxides poisoned by 1000 ppm of SO_2 at 773 K in CO oxidation reaction. From the amount of sulfate ions detected, it was concluded that SO_2 was oxidized to sulfate on all of the oxides tested. The extent of reaction of SO_2 with Co_3O_4, CuO, and Mn_3O_4 was higher than that with oxides of Sn, Ni, Fe, Cr, and Ce. It was further observed that the regeneration temperatures for SnO_2, Mn_3O_4, and CeO_2 were higher than the decomposition temperatures of the corresponding sulfates, probably due to the strong adsorption of the decomposition product or due to the formation of free acid during the heat treatment.

Yao (77) indicated that the rate of regeneration of some SO_2 poisoned oxides could be increased by the introduction of 0.2-0.5 % of water without altering the final activity attained. Besides, catalysts poisoned by injection of a few microliters of SO_2 to the inlet gas at a time (high concentration, short time) were much more difficult to regenerate than those poisoned at a constant concentration of below 100 ppm. Furthermore, over 70 % of the SO_2 introduced at 80 ppm could be recovered by heating in a N_2 stream at 773 K while only 20 % of that introduced at 1000 ppm was removed under the same condition.

V. Improving SO_2 Resistance of Perovskite Catalyst

As stated above, an ideal catalyst formulation should contain active constituent which either does not form a sulfate or forms a sulfate that decomposes at low temperature. A survey on the decomposition data for metal sulfates indicated that very few base metal ions meet this requirement.

Under oxidizing conditions, noble metals are not affected by

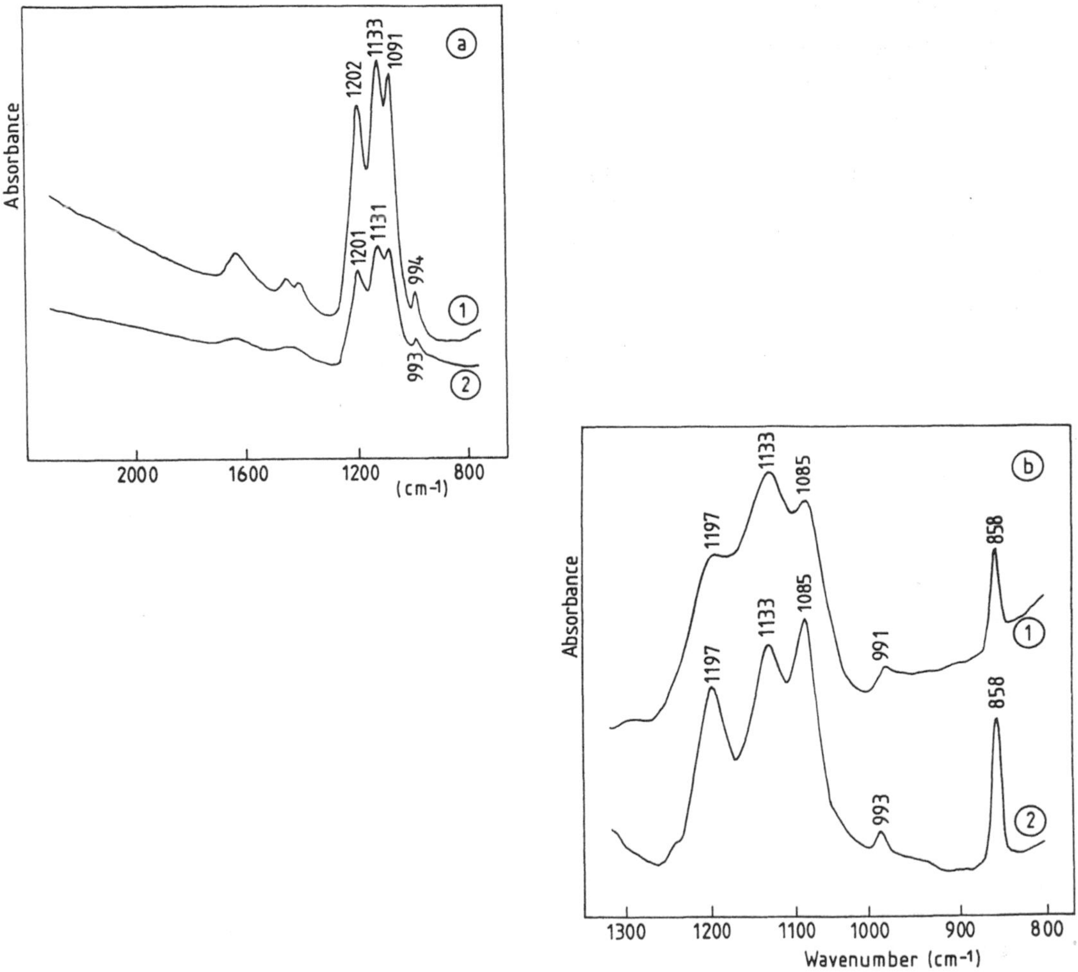

Fig. 5. (a) Infrared spectra of SO_2-poisoned $La_{0.6}Sr_{0.4}Co_{0.8}Ti_{0.2}O_3$: 1, poisoned at 473 K; 2, then calcined in air at 673 K. (b) Infrared spectra of So_2-poisoned $La_{0.6}Sr_{0.4}Co_{0.85}Zr_{0.15}O_3$: 1, poisoned at 473 K; 2, then calcined in air at 673 K.

SO_2 in a level of 20 ppm, normal to that in the exhaust from gasoline containing 0.03 wt % S (90,91), although SO_2 in high concentration (e.g. 10 % in air) may transform Pt, Pd, and Rh to their sulfates (92). The three-way catalysts are inhibited by SO_2 (93-95). Hunter in 1972 (96) reported a loss in CO activity of noble metal catalyst in a car test burning 0.03 wt % S-containing gasoline. Recently, Sodowski (97), Yao et al. (48) showed the loss in CO activity was a result of surface sulfate formation on γ-Al_2O_3, which suppresses both CO and alkene oxidation by decreasing the associative CO and alkene adsorption and enhances alkane oxidation by increasing the dissociative alkane adsorption on Pt (98). Studies on the effect of SO_2 on noble metals have been receiving more attention recently (99-101).

A. Noble Metal Stabilized in Perovskite Structure

Snyder et al. (102) claimed that noble metal promoted base metal catalysts did not lose any activity after 5000 equivalent miles of exposure to the exhaust from fuel containing 0.1 wt % S. Similar observations were reported by Roth and Gambell (103).

The effect of Pt on $La_{.7}Pb_{.3}MnO_3$ in the form of crushed single crystals grown from lead borate flux in Pt crucibles had been studied extensively. This perovskite was reported to be unusal among base metal oxides in being resistant to SO_2. However, latter studies showed that single crystals prepared in this way contain Pt as a contaminant (104), and it is this Pt impurity that imparts SO_2 resistance to the perovskite (18, 80,81,105). Pt free single crystal of the same composition prepared in silica (80) or MgO (105) crucibles were readily poisoned by SO_2 while polycrystalline preparation doped with Pt showed high resistance to SO_2.

Johnson et al. showed that as little as 200 ppm of Pt can dramatically enhance both the SO_2 resistance and the CO reactivity of $La_{.7}Pb_{.3}MnO_3$ (70). Katz et al. (12) observed $La_{.7}Pb_{.3}MnO_3$ with 300 ppm Pt was resistant to SO_2 and its specific activity was roughly 100 times higher than the Pt-free polycrystalline preparation. The dramatic role played by such a small amount of Pt was ascribed to a 100-fold surface enrichment of Pt observed by Croat et al. (105) using XPS technique. Johnson et al. (82) put forward an explanation also based on XPS observation that in the perovskite structure Pt exists as Pt (IV) which exhibits higher activity than Pt (0) in Pt/Al_2O_3.

A preparation of $La_{.8}Ba_{.2}Mn_{.5}Cu_{.4}Pd_{.1}O_3$ was found to withstand

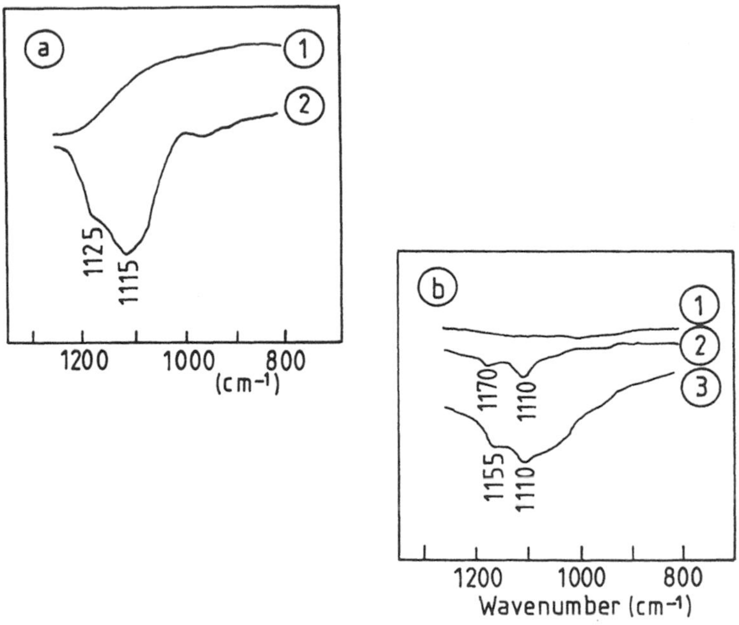

Fig. 6. (a) Infrared spectra of SO_2-poisoned $La_{0.5}Sr_{0.5}MnO_3$: 1, background; 2, SO_2-poisoning at 873 K. (b) Infrared spectra of SO_2-poisoned $La_{0.7}Pb_{0.3}MnO_3$: 1, background; 2, SO_2-poisoning at 539 K; 3, SO_2-poisoning at 673 K.

300 ppm of mercaptan or 350 ppm SO_2 at 673 K. No decline in hexane oxidation activity was observed within 100 hours reaction on stream (106).

Works of Trimble (20) on NO reduction indicated that the total NO reduction was poisoned less and recovered more quickly on a $La_{.8}K_{.2}Mn_{.94}Ru_{.06}O_3$ and $SrRuO_3$ than Ru/Al_2O_3. The SO_2 poisoning caused a decrease in NH_3 formation below 673 K over all three catalysts. $La_{.8}Sr_{.2}Co_{.9}Ru_{.1}O_3$ on Torvex Al_2O_3 honeycomb was claimed by Lauder to withstand exposure to the exhaust from a single cylinder engine, with leaded gasoline, for 1000 hours with only minimal change in NO reduction activity (107).

B. Base Metal Perovskite with Improved SO_2 Resistance

Several transition metal elements such as Ti, Zr, V, Sn, Cu were chosen to substitute partially the Co ions in $La_{.6}Sr_{.4}CoO_3$, in view of the relativity low heat of formation, low rate of sulfate formation, and low temperature of sulfate decomposition of these elements. Introduction of Sr was aimed at pushing the

Table 1. SO_2 Effect on the CO Oxidation Activity of Some Perovskites

Catalyst	T (K)	SO_2 (ppm)	X_s/X_o (min)	ml/m^2	X'/X_o (min)
Pt/γ-Al_2O_3	383	20	0.05(45)	--	0.05(60)
	573	100	1.00(60)	--	1.00
$LaCoO_3$	533	20	0.05(20)	0.13	0.05(60)
		100	" (8)	0.27	" "
	673	20	" (57)	0.38	" "
		100	" (14)	0.45	" "
$La_{.6}Sr_{.4}CoO_3$	473	20	" (54)	0.17	" "*
		100	" (12)	0.20	" "*
	673	20	" (101)	0.33	0.96(25)
		100	" (30)	0.48	1.00(25)
$La_{.6}Sr_{.4}Co_{.8}Cu_{.2}O_3$	473	20	" (44)	0.20	0.90(71)
		100	" (18)	0.41	0.85(12)
	673	20	0.85(227)	1.03	0.97(22)
		100	0.98 (76)	1.73	1.00(.5)
60000 h^{-1}	673	380	0.05 (7)	1.12	0.05(60)
$La_{.6}Sr_{.4}Co_{.8}V_{.2}O_3$	473	20	" (55)	0.32	" "*
		100	" (14)	0.40	" "*
	673	20	0.60(165)	0.97	0.98(33)
		100	0.05 (80)	2.40	1.00(12)
60000 h^{-1}	673	380	" (5)	1.12	0.05(60)
$La_{.6}Sr_{.4}Co_{.8}Sn_{.2}O_3$	473	20	" (70)	0.29	" "*
		100	" (15)	0.31	" "*
	673	20	0.20(134)	0.56	0.90(30)
		100	0.05 (14)	0.28	0.80(19)
$La_{.98}Ce_{.02}CoO_3$	473	20	" (72)	0.21	0.05(60)*
		100	" (23)	0.33	" "
	673	20	0.93(110)	0.31	--
		100	0.05 (58)	0.83	0.05(60)
60000 h^{-1}	673	380	" (13)	1.41	" "
$La_{.6}Sr_{.4}Co_{.8}Ti_{.2}O_3$	473	50	" (10)	0.09	" (30)
56000 h^{-1}	673	170	" (300)	4.30	1.00(30)
2% CO in air		400	" (120)	4.00	0.90(60)
$La_{.6}Sr_{.4}Co_{.85}Zr_{.15}O_3$	473	200	" (40)	0.18	0.05(30)*
20000 h^{-1}	673	200	1.00 (60)	0.28	--
1% CO in air		400	0.05 (80)	0.73	1.00(30)
$LaMnO_3$	573	50	" (10)	0.03	0.05(60)
$La_{.8}Sr_{.2}MnO_3$	573	50	" (10)	0.04	" "
$La_{.8}Sr_{.2}Mn_{.8}Ti_{.2}O_3$	673	50	" (19)	0.10	" "

*Restores their original activity at 573 K with SO_2.
Except specified, all run at 30000 h^{-1}, with 2% CO in air.

B site cations to their highest valence state and to increase the number of oxygen ion vacancies, thereby increasing the mobility of lattice oxygen. It was expected by these measures the SO_2 bonding with the surface might be weakened, thereby increasing the SO_2 resistance. SO_2, in virtue of its high electron affinity, may scramble the seat of adsorption for both oxygen and CO. Studies of Tascon and Tejuca (108-110) showed that centers for oxygen adsorption on $LaCoO_3$ are metallic ions; CO can adsorb either as carbonyl on metallic ions or as carbonate species on surface oxide ions. The results of some of the perovskites designed are given in Table 1. A Pt/Al_2O_3 catalyst with 0.1 % Pt loading is also listed. In this table x_s represents % conversion of CO in the presence of SO_2, x_o- that in the absence of SO_2, x'- the conversion observed after shutting off the SO_2 supply at otherwise the same conditions. The time for achieving a certain value or x_s/x_o and x'/x_o is given in parenthesis. Some samples lost their activity at 473 K, could not resume their activity within 60 minutes after cutting off the SO_2 supply, and were found to display their original activity at 573 K in the presence of SO_2. This may be due to the long adsorption lives of SO_2 species at low temperature while higher temperature accelerates their desorption.

These perovskites are all single phase by XRD examinations. The IR spectra of the Ti and Zr substituted cobaltates showed bands due to adsorbed SO_2 complex. Some of these species desorbed when heated in air at 773 K as shown in Figure 5. IR spectra of some manganate perovskites (73) after poisoned by SO_2 are given in Figure 6. Thermogravimetry observation indicated that Ti or Zr substituted perovskites do have higher lattice oxygen mobility. The mechanism of CO oxidation over these perovskites was proposed to occur between adsorbed CO and lattice oxygen (52-55,115).

Results on manganate perovskites were quite different (51,111). Substitution of Ti to a $La_{1-x}Sr_xMnO_3$ ($x = 0.1 - 0.9$) system did not result in improved SO_2 resistance. These manganates were severly poisoned by 50 ppm SO_2 at temperatures from 473 to 673 K at a space velocity of 56000 h^{-1}. The amount of SO_2 passed was around 0.1 - 0.2 ml/m^2. Temperature programmed reduction results showed that the mobility of lattice oxygen decreased with increasing Sr substitutuion (51,111). These results can be explained by the fact, contrary to the $LaCoO_{3-x}$ system, introduction of Sr to the $LaMnO_{3+x}$ system decreases the number of cation vacancies and excess oxygen. Besides, the Mn

(IV) formed is rather stable (2,3,112,113).

To conclude this section, the results obtained on cobaltate base metal perovskites and manganate perovskites with noble metal stabilized in the structure are encouraging, indicating that by proper design, the susceptibility to SO_2 poisoning of these compounds is not an unsurmountable barrier. But many works remain to be conducted such as the effect of sulfur compound on the behavior of these perovskites under reducing atmosphere, the surface structure of these perovskites, the valence states of the transition ions, and the mechanism of the interaction between SO_2 and the surface, etc.

References

1. Galasso, F. S., *Structure, Properties, and Preparation of Perovskite-type Compounds*, Pergamon Press, Oxford (1969).
2. Voorhoeve, R.J.H., Perovskite related oxides as oxidation-reduction catalyst, *Advanced Materials in Catalysis*, J.J. Burton and R.L. Garten (Eds.), Academic Press, 129-180 (1977).
3. Happel, J., M. Hnatow, and L. Bajars, *Base Metal Oxides Catalysts*, Marcel Dekker Inc., 117-156 (1977).
4. Parravano, G., Ferroelectric transition and heterogeneous catalysis, *J. Chem. Phys.*, **20**: 342 (1952).
5. Parravano, G., *J. Amer. Chem. Soc.*, **75**: 1497 (1953).
6. Meadowcroft, D.B., Low cost oxygen electrode material, *Nature (London)*, **226**: 847-848 (1970).
7. Libby, W.F., Promising catalyst for auto exhaust, *Science*, **171**: 499-500 (1971).
8. Voorhoeve, R.J.H., J.P. Remeika, P.E. Freeland, and B.T. Matthias, Rare-earth oxides of manganese and cobalt rival Pt for the treatment of CO in auto exhaust, *Science*, **177**: 353-354 (1972).
9. Voorhoeve, R.J.H., J.P. Remeika, and D.W. Johnson, Jr., *Science*, **180**: 62-64 (1973).
10. Schlatter, J.C., R.L. Klimisch, and K.C. Taylor, Exhaust catalysts: Appropriate condition for comparing Pt and base metal, *Science*, **179**: 797-800 (1973).
11. Voorhoeve, R.J.H., L.E. Trimble, and C.P. Khattak, *Mat. Res. Bull.* **9**: 655-666 (1974).
12. *Amer. Ceram. Soc. Bull.*, **52**: 440 (1973).

13. Chien, M.N., I.M. Pearson, and K. Nobe, Reduction and adsorption kinetics of nitric oxide on cobalt perovskite catalysts, *I.E.C. Prod. Res. Dev.*, **14**: 131-134 (1975).
14. Jhaverl, N.C., L.S. Caretto, and K. Nobe, Propylene oxidation on cobalt perovskite catalysts, *I.E.C. Prod. Res. Dev.*, **14**: 142 (1975).
15. Yao, Y.F.Y., The oxidation of hydrocarbons and CO over metal oxides. IV. Perovskite-type oxides. *J. Catal.*, **36**: 266-275 (1975).
16. Katz, S., J.J. Croat, and J.V. Laukonis, Lanthanum Lead manganite catalyst for carbon monoxide and propylene oxidation, *I.E.C. Prod. Res. Dev.*, **14**: 274 (1975).
17. Katzman, H., L. Pandolfi, L.A. Pederson, and W.F. Libby, Pb-tolerant auto exhaust catalyst, *CHEMTECH June*: 369-372 (1976).
18. Shelef, M., and H.S. Gandhi, The reduction of NO_x in automobile emissions, stabilization of catalyst containing ruthenium, *Plat. Met. Rev.*, **18**: 2 (1974).
19. (a) Trimble, L.E., Effect of SO_2 on nitric oxide reduction over Ru-containing perovskite catalysts, *Mat. Res. Bull.*, **9**: 1405-1412 (1974).
19. (b) *Chem. Eng. News*, *Aug.*, **4**: 8 (1975).
20. Croat, J.S., and G.G. Tibbetts, RE-manganites: Surface segregated Pt increases catalytic activity, *Science*, **194**: 318-319 (1976).
21. Zhao, F.H., C.Y. Lu, and W. Li, Rare earth containing perovskite type catalysts for catalytic deodorization of pyridine, *Environ. Chem.(Chinese)*, **6**: 16-20 (1978).
22. Arakawa, T., et al., The catalytic activity of rare earth manganites. *Mater. Res. Bull.*, **15**: 269 (1980).
23. Seiyama, T. et al., Perovskite oxides having semiconductivity as oxygen sensors. *Chem. Lett.*, 377-380 (1985).
24. Ichimura, K., Y. Inoue, and I. Kojima, Comparative studies of mixed perovskite catalysts $LaCoO_3$, $LaFeO_3$, and $LaAlO_3$ for alkene hydrogenation and alkane hydrogenolysis, *New Horizons in Catalysis*, K. Tanabe and T. Seiyama (Eds.), Kodansha-Elsevier, 1281 (1981).
25. Happel, J., M.A. Hnatow, L. Bajars, and M. Kandrath, Lanthanum titanate catalyst-SO_2 reduction, *I.E.C. Prod. Res. Dev.*, **14**: 154 (1975).
26. Conner, W.C., and S. Soled, Propane oxidation over metal

oxides, perovskite trirutiles, and columbites, *New Horizons in Catalysis*, K. Tanabe and T. Seiyama (Eds.), Kodansha-Elsevier, 840 (1981).

27. McCarty, J.G., M.A. Quinlan, and H. Wise, Catalytic combustion of methane by complex oxides, *Catalysis, Theory to Practice*, Elsevier, **4**, 1818 (1988).
28. Nasby, R.D., and R.K. Quinn, Photoassisted electrolysis of water using a $BaTiO_3$ electrode, *Mat. Res. Bull.*, **11**: 985-992 (1976).
29. Raj, S.L., and V. Srinivasan, Decomposition of NO on RE-manganites, *J. Catal.*, **65**: 121-126 (1981).
30. Balasubramanian, M.R., R. Natesan, and P. Rajendran, Correlation between catalytic activities and physico-chemical properties of perovskite oxides, *J. Sci. Ind. Res.*, **43**: 500-506 (1984).
31. Voorhoeve, R.J.H., D.W. Johnson, Jr., J.P. Remeika, and P.K. Gallagher, Perovskite oxides: Materials science in catalysis, *Science*, **195**: 827-833 (1977).
32. Shelef, M., K.Otto, and N.C. Otto, Poisoning of automotive catalysts, *Advances in Catalysis*, D.D. Eley, H. Pines and P.B. Weisz (Eds.), Academic Press, V. **27**: 311-365 (1978).
33. Kasansky, V.B., and G.B. Pariisky, The ESR study of free radicals adsorbed on catalysts, *Proc. 3rd Int. Cong. Catal.*, 367-376 (1964).
34. Mashchenko, A.E., G.B. Pariisky, and V.B. Kazansky, ESR study of radicals formation during SO_2 adsorption on partially reduced TiO_2 and ZnO. *Kinet. Catal. (Russian)*, **8**: 704-705 (1967).
35. Mashchenko, A.E., G.B. Parrisky, and V.B. Kazansky, Studies on radicals formed during chemisorption of electron acceptor molecules on the surfaces of n- semiconductors. III. Formation of radicals during SO_2 adsorption on reduced TiO_2 surface and interaction with oxygen, *Kinet. Catal. (Russian)*, **9**: 151-157 (1968).
36. Kolosov, A.K., V.A. Shvets, and V.B. Kazansky, ESR of SO_2^- and SO^- anion radicals adsorbed on supported vanadium silica gel catalyst, *J. Catal.*, **37**: 387 (1975).
37. Worotinzev, V.M., V.A. Shvets, and V.B. Kazansky, EPR studies on changes of coordinative states of V^{4+} ions on the surface of vanadium catalyst during adsorption of NH_3, SO_2, and N_2, *Kinet. Catal. (Russian)*, **12**: 678-684 (1971).
38. Schoonheydt, R.A., and J.H. Lunsford, An electron

paramagnetic resonance study of SO_2 on MgO, *J. Phys. Chem.*, **76**: 323-328 (1972).

39. Rao, K.V.S., and J.H. Lunsford, An EPR study of SO_2^- and SO_4^- ions on vanadium oxide supported on silica gel, *J. Phys. Chem.*, **78**: 649-651 (1974).
40. Ono, Y., H. Takagiwa, and S. Fukuzami, Formation of SO_2^- anions on Al_2O_3, *J. Catal.*, **50**: 181 (1977).
41. Khulbe, K.C., and R.S. Mann, ESR studies of SO_2 and H_2S adsorption on Al_2O_3 and Al_2O_3 supported Mo and Mo-Co, *J. Catal.*, **51**: 364-371 (1978).
42. Taarit, Y.B., and J.H. Lunsford, EPR evidence for the formation of SO_3^- by the oxidation of SO_2^- on MgO, *J. Phys. Chem.*, **77**: 1365-1367 (1973).
43. Lin, M.J., and J.H. Lunsford, Photooxidation of SO_2 on the surface of MgO, *J. Phys. Chem.*, **79**: 892 (1975).
44. Nakamoto, K., *Infrared and Raman Spectra of Inorganic and Coordination Compounds*, 3rd ed., Wiley, New York, 337-338, 247-249, 239-241 (1978).
45. Low, M.J.D., A.J. Goodsel, and N. Takezawa, Reactions of gaseous pollutants with solids. I. Infrared study of the sorption of SO_2 on CaO, *Environ. Sci. Tech.*, **5**: 1191-1195 (1971).
46. Goodsel, A.J., M.J.D. Low, and N. Takezawa, Reactions of gaseous pollutants with solids. II. IR study of sorption of SO_2 on MgO, *Env. Sci. Tech.*, **6**: 268 (1972).
47. Chang, C.G., Infrared studies of SO_2 on alumina, *J. Catal.*, **53**: 374 (1978).
48. Yao, H.C., H.K. Stepien, and H.S. Gandhi, The effect of SO_2 on the oxidation of hydrocarbons and carbon monoxide over Pt/Al_2O_3 catalysts, *J. Catal.*, **67**: 231 (1981).
49. Deo, A.V., I.G. Dalla Lana, and H.W. Habgood, IR studies of the adsorption and surface reactions of H_2S and SO_2 on some aluminas and zeolites, *J. Catal.*, **21**: 270-281 (1971).
50. Schoonheydt, R.A., and J.H. Lunsford, Infrared spectroscopic investigation of the adsorption and reactions of SO_2 on MgO., *J. Catal.*, **26**: 261-271 (1972).
51. Wu, *N.*, *Thesis*, Beijing Polytechnic University, Studies on SO_2 effect on Ti substituted manganite perovskites in CO and Hydrocarbons oxidation (1988).
52. Li, W., Q. Huang et al., Improving the SO_2 resistance of perovskite type oxidation catalysts, *Catalysis and Automotive Pollution Control*, A. Crucq, A. Frennet, (Eds.), Elsevier,

405-415 (1987).
53. Pechkowsky, V.V., and M.S. Gaisinnobich, IR spectra of thermal decomposition products of sulfates of Cu, Zn, Cd and Al, *J. Inorg. Chem. (Russian)*, **9:** 2299-2302 (1964).
54. Hongxing, Dai, *M.E. Thesis*, Beijing Polytechnic University, Improving SO_2 resistance of cobaltate perovskite substituted with Cu, V, Sn, Ce, and Pd (1990).
55. Weidong, Yin, *M.E. Thesis*, Beijing Polytechnic University, Activity for CO oxidation and SO_2 resistance of perovskite $La_{1-x}Sr_xCo_{1-y}Zr_yO_3$ catalysts (1988).
56. DeBerry, D.W., and K.J. Sladek, Rates of reaction of SO_2 with metal oxides, *Canadian J. Chem. Eng.*, **49:** 781 (1971).
57. Henrich, V.E., Electronic and geometric structure of defects on oxides and their role in chemisorption, *Surface and Near Surface Chemistry of Oxide Materials*, J. Nowotny, and L.C. Dufour (Eds.), Elsevier, 189-217 (1988).
58. Henrich, V.E., The surface of metal oxides, *Rep. Prog. Phys.*, **48:** 1481 (1985).
59. Smith, K.E., and V.E. Henrich, Interaction of SO_2 and CO with the Ti_2O_3 (1012) surface, *Phys. Rev.*, **B32:** 5384 (1985).
60. Lowell, P.S., K. Schwitzgebel, T.B. Parsono, and K.S. Sladek, Selection of metal oxides for removing SO_2 from flue gas, *I.E.C. Proc. Des. Dev.*, **10:** 384-390 (1971).
61. Kellogg, H.H., A critical review of sulfation equilibria, *Trans. Metal. Soc. of AIME*. **230;** 1622-1634 (1964).
62. Kubachewski, O., E.L.L. Evans, and C.B. Alcock, *Metallurgical Thermochemistry*, Pergamon, New York, 268-323, 378-384 (1967).
63. Tagawa, H., and H. Saijo, Kinetics of the Thermal decomposition of some transition metal sulfates, *Thermochim. Acta*, **91:** 67-77 (1985).
64. Tagawa, H., Thermal decomposition temperatures of metal sulfates, *Thermochim. Acta*, **80:** 23-33 (1984).
65. Ostroff, A.G., and R.T. Sanderson, Thermal stability of some metal sulfates, *J. Inorg. Nucl. Chem.*, **9:** 45-50 (1959).
66. Locke, C.E., and H.F. Rase, D.T.A.-rapid definition of catalysts characteristics, *I.E.C.*, **52:** 515-516 (1960).
67. Wedding, B., and R.J. Farrauto, Rapid evaluation of automotive exhaust oxidation catalysts with a differential scanning calorimeter, *I.E.C. Proc. Des. Dev.*, **13:** 45 (1974).
68. Johnson, Jr., D.W., and P.K. Gallagher, Studies of some perovskite oxidation catalysts using DTA technique, *Thermochim. Acta*, **7:** 303-309 (1973).

69. Johnson, Jr., D.W., P.K. Gallagher, E.M. Vogel, and F. Schrey, Effect of SO_2 upon the catalytic activity of Pt doped oxidation catalysts by DTA, *Thermal Analysis*, I. Buzas (Ed.), **3**: 181-192 (1975).
70. Gallagher, P.K., D.W. Johnson, Jr., and E.M. Vogel, Examples of the use of DTA for the study of catalytic activity and related phenomena, *Catalysis in Organic Synthesis*, Academic Press, 113-136 (1976).
71. Gallagher, P.K., D.W. Johnson, Jr., and E.M. Vogel, Preparation, structure, and selected catalytic properties of the system $LaMn_{1-x}Cu_xO_{3-y}$, *J.Am., Ceram. Soc.* **60**: 28-31 (1977).
72. Huang, M.M., H.X. Yang, Q.W. Wang, P.Y. Lin, and G.F. Rung, SO_2 poisoning of $La_{1-x}A_xMnO_3$ catalysts, *Cuehua Huebao (Chinese)*, **3**: 277-282 (1982).
73. Huang, M.M., H.X. Yang, Q.W. Wang, P.Y. Lin, and G.F. Rung, Studies on the active centres of $La_{.5}Sr_{.5}MnO_3$, *Cuehua Huebao (Chinese)*, **4**: 311-314 (1983).
74. Yao, Y.F.Y., and J.T. Kummer, The oxidation of hydrocarbons and CO over metal oxides, I. NiO crystals, *J. Catal.*, **28**: 124-138 (1973).
75. Yao, Y.F.Y., The oxidation of HC and CO over metal oxides II. α-Cr_2O_3, *J. Catal.*, **28**: 139-149 (1974).
76. Yao, Y.F.Y., The oxidation of HC and CO, III. Co_3O_4, *J. Catal.*, **33**: 108-122 (1974).
77. Yao, Y.F.Y., The oxidation of HC and CO. V. SO_2 effect, *J. Catal.*, **39**: 104-114 (1975).
78. Sultanov, M.U., E.S. Altachel. Z.Z. Mahmudova, and T.F. Ganievat, SO_2 poisoning of CuO, Cr_2O_3, and $CuCr_2O_4$ in CO oxidation, *Kinet. Catal. (Russian)*, **23**: 754 (1982).
79. Farrauto, R.J., and B. Wedding, Poisoning by SO_x of some base metal oxide auto exhaust catalysts, *J. Catal.*, **33**: 249-255 (1974).
80. Gallagher, P.K., D.B. Johnson, Jr., J.P. Remeika, F. Schrey, L.E. Trimble, E.M. Vogel, and R.J.H. Voorhoeve, The activity of $La_{.7}Sr_{.3}MnO_3$ without Pt and $La_{.7}Pb_{.3}MnO_3$ with varying Pt contents for the catalytic oxidation of CO, *Mater. Res. Bull.*, **10**: 529-538 (1975).
81. Gallagher, P.K., D.W. Johnson, Jr., E.M. Vogel, and F. Schrey, Effect of the Pt content of $La_{.7}Pb_{.3}MnO_3$ on its catalytic activity for the oxidation of CO in the presence of SO_2, *Mat. Res. Bull.*, **10**: 623-628 (1975).

82. Johnson, Jr., D.W., P.K. Gallagher, G.H. Wertheim, and E.M. Vogel, The nature and effect of platinum in perovskite catalysts, *J. Catal.*, **48**: 87-97 (1977).
83. Sachtler, J.W.A., I. Onal, and R.E. Marinangeli, Reactivation of lead poisoned Pt/Al_2O_3 catalysts by SO_2, *Catalysis and Automotive Pollution Control*, Elsevier, Amsterdam, 267-274 (1987).
84. Angelov, S., D. Mehandziev, B. Piperov, V. Zarkov, A. Terlecki-Baicevic, D. Jovanovic, and Z. Jovanovic, CO oxidation on mixed spinels $Cu_xCo_{3-x}O_4$ ($0<x<1$) in the presence of S compounds, *Appl. Catal.*, **16**: 431-437 (1985).
85. Terlecki-Baricevic, A., B. Grbic, D. Jovanovic, S. Angelov, D. Mehandziev, C. Marinova, and P. Kirilov-Stefanov, Activity and sulfur tolerance of monophase spinels in CO and C_xH_y oxidation, *Appl. Catal.*, **47**: 145-153 (1989).
86. Mehanrev, D., and B. Piperov, Deactivation of metal oxide in CO oxidation catalysts, *Kinet. Catal. (Russian)*, **21**: 1051-1055 (1980).
87. Fishel, N.A., R.K. Lee, and F.C. Wilhelm, Poisoning of vehicle emission control catalysts by sulfur compounds, *Env. Sci. Techn.*, **3**: 260-267 (1974).
88. Jenson, J.V., and E.J. Newson, A deactivation reactor for catalyst screening and evaluation, *Proc. 6th Intern. Cong. Catal.*, **2**: 796-805 (1976).
89. Sultanov, M.U., E.S. Altschel, and Z.Z. Mahmudova, Deactivation of metal oxides in CO oxidation in relation to their interaction with SO_2, *Kinet. Catal. (Russian)*, **23**: 754 (1987).
90. Giacomazzi, R.A., and M.F. Homfeld, The effect of Pb, S and P on the deterioration of two oxidizing bead type catalysts, *SAE* 730595 (1973).
91. Neal, A.H., E.E. Wigg, and E.L. Holt, Fuel effect on oxidation catalyst and catalyst equipped vehicles, *SAE* 730593 (1973).
92. Wang, T., A. Vazquez, A. Kato, and L.D. Schmidt, Sulfur on noble metal catalyst particles, *J. Catal.*, **78**: 306-318 (1982).
93. Williamsen, W.B., H.S. Gandhi, M.E. Heyde, and G.A. Zawacki, Deactivation of three way catalysts by fuel contaminants: Pb, P. ans S, *SAE* 790942 (1979).
94. Williamsen, W.B., H.K. Stepien, and H.S. Gandhi, Poisoning of Pt-Rh automotive TWC's: Behavior of single component

catalyst and effect of S and P, *Env. Sci. Techn.*, **14:** 319-324 (1980).
95. Schlatter, J.C., and P.J. Mitchell, Three way catalysts response to transients, *I.E.C. Prod.Res. Dev.*, **19:** 288-293 (1980).
96. Hunter, J.E., Studies of catalyst degradation in automotive emission control system, *SAE* 720122 (1972).
97. Sadowski, G., and D. Reibmann, Zum Einfluss von SO_2 auf die katalytische Totaloxydation verschiedener Verbindungen an Pt/γ-Al_2O_3, *Z. Chem.*, **19:** 189 (1979).
98. Yao, Y.F.Y., Oxidation of alkanes over noble metal catalysts, *I.E.C. Prod. Res. Dev.*, **18:** 293-298 (1980).
99. Summers, J.C., and K. Baron, The effect of SO_2 on the performance of noble metal catalysts in automobile exhaust, *J. Catal.*, **57:** 380-389 (1979).
100. Su, E.C., W.L.H. Watkins, and H.S. Gandhi, Sulfur poisoning of the oxidation of H_2 and CO over a (Pd + CeO_2)/γ-Al_2O_3 catalyst, *Appl. Catal.*, **12:** 59-68 (1984).
101. Blinova, N.A., G.S. Yablonski, G.L. Koval, G.P. Korneichuk, and V.I. Filinov, Kinetic models of CO oxidation on Pd-containing catalyst (with and without SO_2), *React. Kinet. Catal. Lett.*, **19:** 207-211 (1982).
102. Snyder, P.W., W.A. Stover, and H.G. Lassen, Status report on HC/CO oxidation catalysts for exhaust emission control, *SAE* 720479 (1972).
103. Roth, J.F., and J.W. Gambell, Control of automotive emission by particulate catalysts, *SAE* 730277 (1973).
104. Croat, J.J., G.C. Tibbetts, and S. Katz, Surface and catalytic properties of lanthanum lead manganite ($La_{.7}Pb_{.3}MnO_3$) containing traces of platinum, Preprints, Div. Petrochem., *A.C.S.*, **21:** 830-840 (1976).
105. Robertson, J.M., J.P.M. Damen, H.D. Jonker, M.J.G. van Hout, W. Kamminga, and A.B. Voermans, The growth of oxide single crystal-use and care of Pt apparatus, *Plat. Met. Rev.*, **18:** 15 (1974).
106. Xi, Z.S., and W. Li, unpublished data.
107. Lauder, A., *U.S. Pat.*, 3,897,367 (1975).
108. Tascon, J.M.D., and L.G. Tejuca, Adsorption of oxygen on the perovskite type oxide $LaCoO_3$, *Zeit. Phys. Chem. Neue Folge*, **121:** 79-93 (1980).
109. Tascon, J.M.D., and L.G. Tejuca, Adsorption of CO on the perovskite type oxide $LaCoO_3$, *Zeit. Phys. Chem. Neue Folge*, **121:**

63–78 (1980).

110. Tascon, J.M.D., J.L.G. Fierro, and L.G. Tejuca, Kinetics and mechanism of CO oxidation on $LaCoO_3$, *Zeit. Phys. Chem. Neue Folge*, **124**: 249–257 (1981).

111. Chen, Q., *Thesis*, Beijing Polytechnic University, The SO_2 effect on Co and Mn perovskites (1990).

112. (a) Nakamura, T., M. Misono, and Y. Yoneda, Catalytic properties of perovskite type mixed oxide $La_{1-x}Sr_xCoO_3$, *Bull. Chem. Soc. Jpn.*, **55**: 394–399 (1982).

112. (b) Misono, M., and T. Nitadori, Redox and catalytic properties of perovskite type mixed oxides. A comparative Study of $Ln_{1-x}Sr_xBO_3$ (Ln = RE, B = Co, Fe, Mn), *Adsorption and Catalysis on Oxide Surfaces*, M. Che and G.C. Bond (Eds.), Elsevier, Amsterdam, 409–420 (1985).

113. Seiyama, T., *Surface and Near Surface Chemistry of Oxide Materials*, J. Nowotny, and L.C. Dufour (Eds.), Elsevier, Amsterdam, 189–217 (1988).

114. Teraoka, Y., S. Furukawa, N. Yamazoe, and T. Seiyama, Oxygen-sorptive and catalytic properties of defect perovskite type $La_{1-x}Sr_xCoO_{3-\delta}$, *J. Chem. Soc. Jpn. (Japanese)*, **8**: 1529–1534 (1985).

115. Geng, Q.B., X.L. Huang, Q. Huang. and W.Li, Studies on the SO_2 resistance of Cobaltate perovskite oxidation catalysts, *Cuehua Huebao (Chinese)*, **10**: 79–82 (1989).

Chapter 8

Infrared Spectroscopy and Temperature-Programmed Desorption of Perovskite Surfaces

J.M.D. Tascón[1], J.L.G. Fierro[2] and L.G. Tejuca[2]

[1] Instituto Nacional del Carbón y sus Derivados, CSIC, Apdo. 73, 33080 Oviedo, Spain

[2] Instituto de Catálisis y Petroleoquímica, CSIC, Serrano 119, 28006 Madrid, Spain

I. Introduction

Temperature-programmed desorption (TPD) and infrared (IR) techniques have often been used in recent years to study the surface properties of ABO_3 perovskite-type oxides, especially in order to gain a deeper understanding of their catalytic activity for various reactions. Consequently, this review will be limited to the application of these two techniques to perovskites exhibiting remarkable catalytic properties, namely those oxides containing transition metal ions in position B and rare-earth or alkaline-earth ions in position A of the perovskite structure.

Large differences may be observed in the relative number of results obtained by using each of these techniques for the various gases studied. These differences merely reflect the distinct performance of TPD and IR techniques as a function of the nature of the adsorptive molecule under study.

Infrared spectra of bulk perovskites, whose knowledge is necessary prior to carrying out IR studies of adsorbed molecules, were already described in detail in the early seventies by Subba in the 4000-80 (1) or 800-40 cm^{-1} (2,3) ranges for perovskites by

Rao et al. (1) and Couzi and Huong (2,3). Spectra were recorded containing rare-earth ions in position A, and Al, Cr, Mn, Fe and Co in position B of the structure. Subba Rao et al. (1) discussed their results in terms of crystallography, magnetic properties and covalency of bonds. Couzi and Huong (2,3)interpreted their results in connection with the crystal structure of these compounds by means of factor group analysis. Also, correlations were established between the spectral evolution and the crystal deformations of these ionic compounds.

As the surface of perovskites may be related to their catalytic properties, the nature and concentration of the surface sites can be evaluated by the analysis of the interactions of suitable probe molecules and the surface of these materials. Along the current revision, TPD and IR spectroscopic results reported up to date for perovskites as a function of the nature of the adsorbed gas molecule will be sequentially discussed. The case of homomolecular simple molecules (H_2, O_2) will be first examined, then the adsorption of some oxygen-containing molecules (NO, CO, CO_2) will be considered, and finally the interactions of less-commonly studied molecules (SO_2, H_2O, pyridine, hydrocarbons) with the surfaces of perovskites will be reviewed.

II. Adsorption of Simple Molecules

A. Hydrogen and Deuterium

The hydrogen (or deuterium) probe molecule has largely been used to study the nature of the adsorption sites in a great variety of perovskites. Quantitative data have been obtained almost exclusively by TPD because the H_2 (or D_2) molecule is highly symmetric, viz. difficult to be polarized, and therefore inadequate to be studied by IR spectroscopy.

Ichimura et al. (4) were the first to report TPD spectra upon hydrogen adsorption on perovskites. Adsorption of hydrogen was carried out at 298 K on $LaFeO_3$, $LaCoO_3$ and $LaAlO_3$. The TPD spectra were found to be complex, but some common features can be drawn: according to the desorption temperature they may be grouped into two categories. The H-species desorbing above 340 K was considered (4) to be responsible for hydrogenolysis of C_2H_6 and hydrogenation of C_2H_4 and C_2H_2.

Thermal desorption of deuterium from $LaRhO_3$ was later studied by Watson and Somorjai (5,6). D_2 was desorbed from the fresh oxide at 450 K, this temperature dropping to 377 K in the case of a $LaRhO_3$ sample used in the syngas (CO + H2) reaction. In order to obtain a clearer picture of TPD results, the heats of desorption were calculated and compared with those determined for metallic Rh, Rh_2O_3 and La_2O_3. High values for the desorption heat of deuterium (ca. 117 kJ/mol) were obtained for La_2O_3 and fresh $LaRhO_3$, presumably due to hydroxyl formation upon adsorption. Lower desorption heats were measured on used $LaRhO_3$ and Rh_2O_3 oxides, and an even lower value (ca. 84 kJ/mol) was calculated for metallic rhodium. In addition, it was concluded (5,6) that much of Rh in the used $LaRhO_3$ catalyst was in a partially reduced (probably Rh^+) state.

In recent years, Tejuca et al. (7-12) have carried out a comprehensive study by TPD of adsorbed hydrogen on $LaMO_3$ (M = Cr, Mn, Fe, Co, Ni) perovskite-type oxides as a function of the adsorption temperature and of the reduction temperature of the oxide. For $LaCrO_3$ and $LaNiO_3$ systems, activated adsorption was observed when the adsorption was effected above room temperature. A desorption peak between 340 and 380 K was observed after hydrogen adsorption at 298 K on prereduced $LaMO_3$ oxides. Indeed, reduced $LaNiO_3$ exhibited a second desorption peak at 610-615 K. As all these peaks increased in intensity with the reduction temperature of sample, it was concluded (7) that TPD peaks observed upon H_2 adsorption on prereduced $LaCoO_3$ and $LaNiO_3$ are ascribable to hydrogen adsorbed on metallic Co and Ni. On the contrary, for reduced $LaMO_3$ (M = Cr, Mn, Fe), more oxidized metal centers (M^{n+}, $0<n<3$) should be involved in hydrogen adsorption (7).

Tejuca et al. have also studied by TPD the interactions between hydrogen and carbon monoxide adsorbed on $LaCrO_3$ (8), $LaMnO_3$ (9), $LaFeO_3$ (10), $LaCoO_3$ (11) and $LaNiO_3$ (12). The TPD spectra were recorded after sequential H_2-CO or CO-H_2 adsorption treatments. As an illustration, Figure 1 shows hydrogen TPD spectra for successive adsorptions on $LaCrO_3$ reduced at 923 K. The TPD spectrum of hydrogen upon hydrogen adsorption on the same reduced sample (Fig. 1, dashed line) is also included as a reference. It can be seen that the desorption peak at 395 K for the latter system almost disappears from TPD spectrum upon sequential CO-H_2 adsorption (spectrum a), and is also severely depressed upon the sequential H_2-CO adsorption (spectrum b). In addition, new desorption peaks appear at 560 K for the CO-H_2 sequence (spectrum a), and at 575, 635 and above 800 K for the H_2-CO sequence (spectrum b).

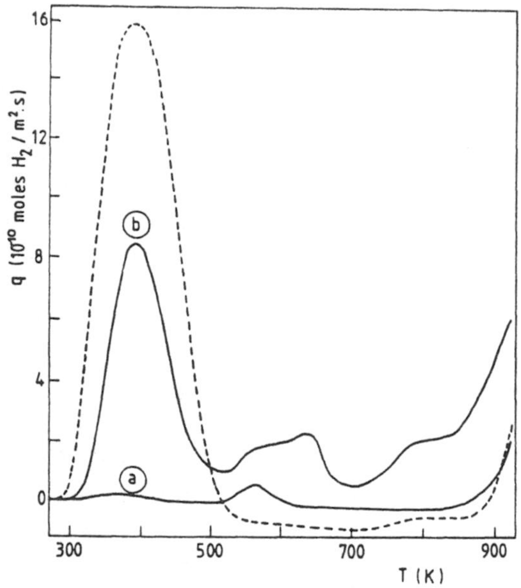

Fig. 1. TPD spectra of H_2 after CO-H_2 (a) and H_2-CO (b) successive adsorption runs at 298 K on $LaCrO_3$ reduced at 923 K. Dashed line, TPD spectrum of H_2 after H_2 adsorption at 298 K on the same reduced sample. Reproduced with kind permission of Elsevier Science Publishers from Appl. Surf. Sci. (Ref. 8).

TPD results of H_2 adsorption from reduced $LaCrO_3$ indicate, as stated above, that hydrogen adsorbs on partially reduced metallic centers (Cr^{n+}, $0 < n < 3$) (7). On the other hand, it has been shown (8) that CO adsorption on reduced $LaCrO_3$ takes place as linear or bridged species on reduced metallic ions. Thus, both gases adsorb on the same type of surface sites. Results from Fig. 1 (8), as well as similar studies with other perovskites (9,10), indicate that the interactions of these reduced oxides with CO are stronger than with H_2. CO preadsorption on the reduced oxides inhibits subsequent hydrogen adsorption more severely than preadsorbed hydrogen inhibits subsequent CO adsorption. Also, CO adsorbed after H_2 causes a displacement of the TPD peaks of H_2 towards higher temperatures. These results could suggest that adsorbed CO displaces adsorbed hydrogen from the sites involved in its low-temperature desorption peak (from fresh perovskites) to another type of site giving rise to a H_2-higher temperature desorption peak. However, a more plausible explanation for these results is the formation of an O-containing mixed surface species after any of the sequential CO-H_2 or H_2-CO adsorption runs even at room temperature.

Spectral changes involved in the formation of this oxygenated species are much more evident on $LaCrO_3$ (8), $LaMnO_3$ (9) or $LaFeO_3$ (10) than on $LaCoO_3$ (11) or $LaNiO_3$ (12). In the case of $LaNiO_3$ no interaction at all between CO and H_2 in the adsorbed state was observed. Since $LaCoO_3$ and $LaNiO_3$ are very easily reducible perovskites, while bulk reduction of $LaCrO_3$ is negligible even at temperatures as high as 1273 K, it was concluded (8) that partially reduced metal sites act as effective coadsorption centers and thus may play an important role in the synthesis of oxygenates from syngas. On the contrary, $LaCoO_3$ and $LaNiO_3$ easily reduced to metallic Co and Ni, and these sites should be less active for this reaction.

The difference observed among $LaMO_3$ (M = Cr, Mn, Fe, Co, Ni) perovskites are especially conspicuous between the extremes of the series. Results of hydrogen adsorption on reduced $LaNiO_3$ (7) showed that a substantial fraction of adsorbed hydrogen (desorption peaks above 610 K) interacts with the surface of $LaNiO_3$ reduced at 773 K much more strongly than with the surface of $LaCrO_3$ reduced at 923 K. This stronger adsorption bond of hydrogen may hinder its interaction with CO adsorbed on $LaNiO_3$ reduced at 773 K and impede the subsequent formation of oxygenated species at the surface of the reduced perovskite. Thus, the different behaviors of reduced perovskites should be related to the relative strengths of the interactions of CO and H_2 with the surfaces of these materials (12).

B. Oxygen

An interesting potential application of perovskites as catalysts is their use for redox reactions in connection with purification of automobile exhaust gases. Active catalysts for total oxidation operate through adsorbed oxygen and thus the study by TPD of oxygen adsorption on these oxides has attracted considerable interest, particularly in relation to total oxidation of hydrocarbons.

Yamazoe et al. (13) and Ichimura et al. (14) reported, almost simultaneously, the first oxygen TPD results from perovskites. In both cases, the objective was to study the influence of partial substitution of La^{3+} by Sr^{2+} in $La_{1-x}Sr_xMO_3$ (M = Mn, Fe, Co) oxides on their surface and catalytic properties. Both research teams concluded that the amount of desorbed oxygen from

$La_{1-x}Sr_xCoO3$ increased with increasing x-substitution. Since A-site substitution with a divalent ion is known to induce the formation of oxygen vacancies, these authors associated the low-temperature O_2 TPD peak with oxygen vacancies. The different behaviors of $La_{1-x}Sr_xMnO3\pm\lambda$, exhibiting in some cases oxidative nonstoichiometry, and $La_{1-x}Sr_xFeO3-\lambda$ and $La_{1-x}Sr_xCoO_{3-\lambda}$, showing only reductive nonstoichiometry, have been explained by Seiyama et al. (15) on the basis of the nonstoichiometric character of these oxides. However, a high temperature oxygen desorption peak has been associated with the B cation (15). Oxygen adsorption and desorption were indeed higher for the perovskites exhibiting a higher catalytic activity for total oxidation, namely perovskites of Mn, Co and Ni. A detailed account of Seiyama et al.'s work in this field is given elsewhere in this book (16), and thus no additional details will provided in this chapter.

On the other hand, the effect of A-site substitution wivth a tetravalent (Ce^{4+}) cation on TPD of oxygen from Fe and Co perovskites has been investigated by Nitadori and Misono (17). The effect of Ce^{4+} substitution was quite different in these two systems, the corresponding TPD spectra being shown in Fig. 2a. In the case of $La_{1-x}Ce_xCoO_3$, the amount of desorbed oxygen remarkably increased with increasing x, and the desorption peaks of oxygen appearing at about 573 K were similar to those obtained for $La_{1-x}Sr_xCoO_3$. On the contrary, the oxygen adsorption decreased distinctly upon Ce substitution in $La_{1-x}Ce_xFeO_3$. This was interpreted as due to the reduction of Fe^{3+} to Fe^{2+} as Ce^{4+} is introduced, thus making oxygen vacancies difficult to be formed. Such a reduction justifies the relative ease with which oxygen can be desorbed with increasing Ce substitution (Fig. 2a). However, in the case of Co perovskites the formation of an A site vacancy becomes significant and a part of the Co^{3+} ion in the B site changes to Co^{4+}. Therefore, the effect of Ce^{4+} substitution in the $LaCoO_3$-derived system is expected to be similar to that of Sr^{2+} substitution.

To confirm the effect of A-site vacancies, Nitadori and Misono (17) compared TPD spectra of oxygen from $LaCoO_3$, $La_{0.9}Ce_{0.1}CoO_3$ and $La_{0.9}\phi_{0.1}CoO_3$ (ϕ = A-site cation vacancy). The corresponding results are shown in Fig. 2b. Almost identical TPD spectra were obtained for the two last preparations, in parallel with their catalytic activities for propene oxidation that were, as expected, higher than that of $LaCoO_3$. It was concluded (17) that it is very likely that only a small fraction of Ce^{4+} is actually introduced to an A site of $LaCoO_3$, and most of the A

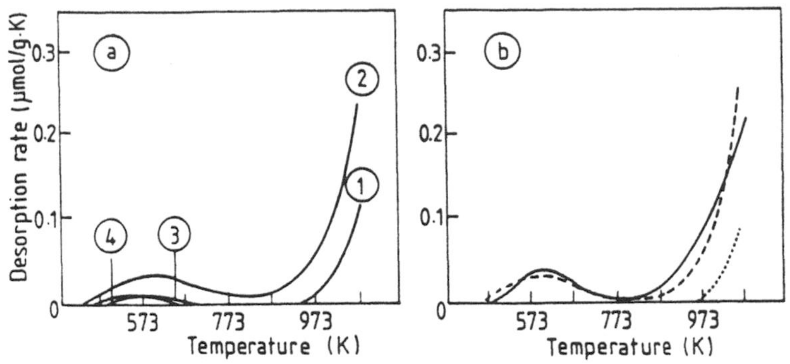

Fig. 2. (a) TPD spectra of oxygen after oxygen adsorption at room temperature on $LaCoO_3$ (1); $La_{0.9}Ce_{0.1}CoO_3$ (2); $LaFeO_3$ (3) and $La_{0.9}Ce_{0.1}FeO_3$ (4). (b) The same on $La_{0.9}CoO_3$ (solid line), $La_{0.9}Ce_{0.1}CoO3$ (broken line) and $LaCoO_3$ (dotted line). Reproduced with kind permission of Academic Press, Inc. from J. Catal. (Ref. 17).

sites which are expected to be occupied be Ce^{4+} form cation vacancies.

Min and Pei-yan (18) have also studied thermal desorption of oxygen from perovskites in connection with their catalytic activity for oxidation of hydrocarbons. They reported the existence of four TPD peaks after preadsorption of this gas on $Nd_{0.7}Sr_{0.3}MnO_3$; all these surface species were able to react with ethane at different temperature regions.

More recently, TPD of oxygen from perovskite-type oxide catalysts has attracted interest in relation to the oxidative coupling of methane. Amorphous $LaAlO_3$ oxides are active and selective for this reaction, but C_2 hydrocarbon selectivity is strongly inhibited by the growth of crystalline phases (19). Thus, these catalysts do not crystallize in a perovskite structure, but their oxygen TPD behavior will be discussed here since their composition is analogous to that of many perovskite catalysts. Two desorption peaks were observed for various mixed La-Al oxides by Imai et al. (19). When the samples were air-pretreated after recording TPD spectra up to 983 K these desorption peaks appeared again, indicating that the surface sites adsorb oxygen reversibly (20). However, the oxygen desorption peaks were severely decreased when $LaAlO_3$ was heated at 1373 K for 16 h. A comparison of the amounts of oxygen desorbed from the monolayer suggested that oxygen adsorption sites in amorphous $LaAlO_3$ catalysts are responsible for the oxidative coupling of methane. Later on, Tagawa and Imai (21) have indicated that the oxygen species responsible for the high temperature TPD peak shows a much greater activity and a high selectivity to C_2 hydrocarbons than does that arising from the continuous-flow reaction. Consequently both, oxygen from the low-temperature peak (weakly-bound oxygen) and gaseous oxygen are active for combustion, an undesired reaction.

Results of subsequent studies by Shamsi and Zahir (22) point to an opposite trend in the case of $A_{1-x}B_xMnO_3$ (A = Gd, Sm, La, Ho; B = Na, K) perovskites, also tested for the oxidative coupling of methane. All oxygen TPD spectra exhibited two O_2 peaks: the maximum temperatures for the oxygen peak desorbing at the lower temperature correlated well with the corresponding C_2 selectivities. Shamsi and Zahir (22) concluded that in the selective oxidation of methane to higher hydrocarbons, the most important factor may be the binding energy of oxygen to surface sites of the catalysts. A strong binding of oxygen atoms thus seems essential to selectively produce olefins from methane as opposed to complete oxidation leading to undesirable carbon dioxide.

III. Adsorption of Oxygen-Containing Molecules

The chemisorption of oxygen-containing molecules such as NO, CO and CO_2, has been studied by means of IR spectroscopy to reveal the nature of surface and catalytic sites in perovskite-type oxides. This has been made feasible because of the high polarizability of these molecules. An overview of the IR data concerning these molecules to investigate the nature of the adsorption sites in perovskites reveals that two major topics, namely the symmetry of surface ions (or atoms) and the degree of coordination, have been examined. A brief account of the interactions of these molecules on the surface of perovskites is given in the next sections.

A. Nitric Oxide

Infrared spectroscopic studies of NO adsorption have been carried out on $LaMO_3$ (M = Cr, Mn, Fe, Co, Ni) perovskites. Fig. 3 shows the IR spectra of NO adsorbed on $LaFeO_3$ as a function of temperature (23). Bands in the spectral region 2000-1600 cm^{-1}

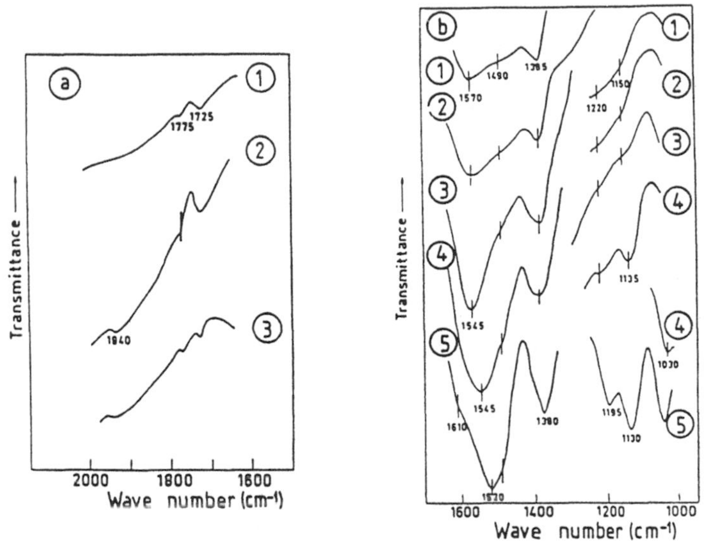

Fig. 3. (a) IR spectra (2000-1650 cm^{-1}) of adsorbed NO on $LaFeO_3$ at various temperatures: (1) 298 K; (2) 373 K; (3) 473 K. (b) IR spectra (1650-1000 cm^{-1}) of adsorbed NO on $LaFeO_3$ at various temperatures: (1) 298 K; (2) 373 K; (3) 473 K; (4) 573 K; (5) 673 K. Reproduced with kind permission of Gauthier Villars from Nouv. J. Chim. (Ref. 23).

(Fig. 3a) did not increase noticeably in intensity with increasing temperature. The band at 1940 cm^{-1} was assigned to a dinitrosyl complex chemisorbed on Fe^{2+} or Fe^{3+} ions. Bands at 1775 and 1725 cm^{-1} were ascribed to mononitrosyls adsorbed on Fe^{2+} and Fe^{o}, respectively. On the other hand, bands in the region 1650-1000 cm^{-1} (Fig. 3b) shifted, in general, to lower wavenumbers and increased remarkably in intensity for increasing adsorption temperatures, showing a progressive chemisorption character. Other bands at 1610, 1570, 1220 and 1030 cm^{-1} were assigned to bridged and bidentate nitrates, and those at 1490 and 1150 cm^{-1} to monodentate nitrates, all these species interacting with surface iron ions. Finally, the band at 1385 cm^{-1} was ascribed to nitrite structures associated to the La^{3+} ion (23). On the other hand, IR spectra of the gaseous phase in contact with the sample between 298 and 573 K exhibited weak bands at 2235 and 2210 cm^{-1} of N_2O gas, that became intense at 673-773 K. The formation of N_2O suggests that NO adsorbs in dissociative as well as molecular forms, indicating that this gas interacts with both cations and anions on the surface of $LaFeO_3$.

The IR spectrum recorded upon adsorption of NO on $LaMnO_3$ (24) exhibited bands typical of the same species found on $LaFeO_3$. However, formation of N_2O was not detected. The simultaneous presence of nitosyl and nitrate bands again evidenced the interaction of NO with both metal cations and oxygen anions. The relative constancy of the extent of NO adsorption with temperature on various perovskites (23-25) and the strength of its bond with these surfaces suggested the possible use of this molecule for determining surface metallic centers, as in the case of simple metal oxides. However, the IR evidence of nitrates formation shows that NO does not have any particular specificity for adsorption on metal or oxide ions. Consequently, the assumption of a 1:1 correspondence between adsorbed NO molecule and surface sites, as that reported for simple oxides, will not give a proper estimate of the surface concentration of transition metal ions because of the complexity of the interactions of NO with perovskites (25).

Infrared spectra obtained after simultaneous adsorption of NO+CO on $LaMnO_3$, $LaFeO_3$ and $LaCoO_3$ at 573 and 773 K provided evidence for the presence of N_2O, isocyanate species, and NO adsorbed with a donor-type or coordinative bond (25). An additional band of nitrosyl groups was detected on $LaCoO_3$. These results may provide some clues for the mechanism of the NO+CO reaction on these oxides, since N_2O and isocyanate species have been previously suggested as intermediates for this reaction on simple oxides. The chemisorption of NO thus seems to play an important role in this reaction catalyzed by perovskites (25).

On the other hand, IR spectra were obtained after sequential NO-CO and CO-NO adsorption runs. Infrared bands of carbonates formed upon CO adsorption at 298 K on $LaMnO_3$ containing preadsorbed NO were weak and could be removed by outgassing at temperatures as low as 573 K (24). This contrasts with CO adsorption on fresh $LaMnO_3$ perovskite, for which a much higher stability was observed, strong carbonate bands being present even at 673 K (26). Contrary to this, NO bands were not modified upon subsequent CO adsorption on $LaFeO_3$ (25). The extent of NO and CO adsorptions upon sequential NO-CO and CO-NO runs on $LaCrO_3$ and $LaNiO_3$ at 298 K showed an inhibiting effect of NO on subsequent CO adsorption much larger than the one of CO on NO adsorption (26). Thus, NO appears to be more strongly adsorbed than CO on the perovskite surface: both molecules compete for O^{2-} anions, NO interacting with the most reactive ones. This has been already confirmed by comparison of the heats and entropies of adsorption of NO and CO on $LaCrO_3$ (27). Briefly, all these results indicate, in agreement with earlier findings, that CO should be a less useful probe-molecule than NO for characterization of perovskite surfaces.

Nitric oxide TPD has been used much more scarcely than IR. To our knowledge, the only NO TPD results were reported by Voorhoeve et al. as early as in 1975 (28,29). To establish the relative importance of molecular and dissociative adsorption, NO was adsorbed at 423 K on $La_{0.8}K_{0.2}MnO_3$ (28) and $La_{0.8}K_{0.2}Mn_{0.94}Ru_{0.06}O_3$ (29), previously used as catalysts for the reduction of NO with a $CO-H_2$ mixture. After this, the catalysts were studied by TPD. Desorption of molecular NO resulted in a peak at 423-473 K indicating the presence of moderatly strongly chemisorbed NO in molecular form, as a nitrosyl group, presumably on low-valency metal ions. A binding energy of 117 kJ/mol was estimated in the case of $La_{0.8}K_{0.2}MnO_3$ (28). Evolution of NH_3, N_2 and CO at higher temperatures indicated the additional presence of tightly-bound nitrogen containing radicals, likely as an isocyanate compound since CO and N_2 desorbed simultaneously at ca. 903 K in a N:CO ratio close to 1. It was also concluded that the high concentration of molecularly-adsorbed NO correlates with an enhanced NO conversion at 473-573 K observed on many manganite catalysts (28).

B. Carbon Monoxide and Carbon Dioxide

Results concerning adsorption of these two molecules on perovskites will be reviewed together here since CO and CO_2 usually form similar adsorbed species on metal oxides. In addition, each of these molecules frequently gives rise to TPD peaks of the other one as a consequence of oxidation or reduction reactions between the adsorbed species and the perovskite surface at the temperatures required for desorption of the former one.

IR spectroscopy has shown (26) that CO interacts with both surface M^{3+} and O^{2-} ions of perovskites to yield linear and/or bridged CO species as well as various types of carbonates. As an illustration of the former type of adsorbed species, Fig. 4 shows IR bands in the spectral range 2200-1800 cm^{-1} after adsorption of CO on various perovskite-type oxides. Bands above 2000 cm^{-1} are generally assigned to linear CO species, while bands below 2000 cm^{-1} can be ascribed to bridged structures. From Fig. 4, one may conclude that multiple types of adsorbed species coexist at the perovskite surfaces. Various types of IR bands ascribable to monodentate and/or bridged or bidentate carbonates have also been found for various perovskites (8,24-26,30), it being usual to find both types of species simultaneously on the same oxide. Consequently, the complexity of surface-adsorbate interactions precludes the use of CO as a probe molecule for the characterization of surface centers in these solids (26,31). This parallels the findings discussed already in the preceding section for NO.

A similar picture holds for CO_2 adsorption on perovskites. Both monodentate and bidentate carbonates were detected by IR on $BaTiO_3$ (32) as well as $LaMO_3$ (M = Cr, Mn, Fe, Co, Ni) (8, 10, 26, 30, 33-35) oxides. Formation of monodentate carbonates was favored at low temperatures, and transformation of this type of species into bidentate carbonates was observed in some cases with increasing adsorption temperature (34). The large width of the IR bands, together with the quantitative description of the equilibrium results by the Freundlich model of adsorption (36) was interpreted as indicating heterogeneity of the perovskite surface.

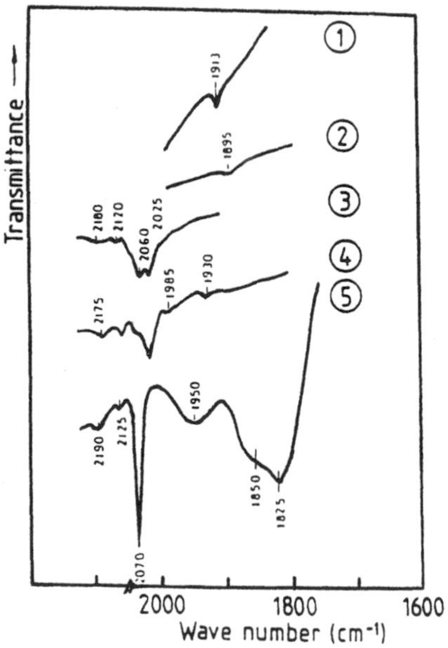

Fig. 4. IR spectra between 2200 and 1800 cm^{-1} of adsorbed CO on $LaCrO_3$ at 673 K (1); $LaMnO_3$ at 673 K (2); $LaFeO_3$ at 298 K (3) and 473 K (4); $LaCrO_3$ (reduced in H_2 at 773 K) at 298 K (5). Reproduced with kind permission of The Royal Society of Chemistry from J. Chem. Soc., Faraday Trans. I (Ref. 26).

To illustrate the types of carbonate bands appearing upon adsorption of CO_2 as well as the influence of reductive pretreatment on the nature of species formed, Fig. 5 shows IR spectra of CO_2 adsorbed on $LaFeO_3$ reduced to various extents (10). Bands of carbonate species at 1600, 1325, 1218, 1050 and 840 cm^{-1} are observed upon CO_2 adsorption on the fresh sample (Spectrum 1). While the wide bands at 1600 and 1325 cm^{-1} decrease in intensity with increasing reduction temperature of the oxide (spectra 2-4), the opposite trend can be observed for the band at 1050 cm^{-1}. The decrease in intensity of the two former bands suggested that they are due to monodentate carbonate (10). Likewise, the band at 1050 cm^{-1} was ascribed to the symmetric CO stretching of a carbonate species of a higher thermal stability, such as a bidentate carbonate, this being in agreement with TPD results. Finally, the band at 1218 cm^{-1} is indicative of the presence of a very labile carbonate species since it was removed by pumping at 298 K (10).

Watson and Somorjai (5,6) first studied by TPD the adsorption of CO on $LaRhO_3$. CO desorbed from the fresh oxide in a single, broad peak centered at ca. 498 K. Adsorption of CO on $LaRhO_3$ catalyst used for the $CO+H_2$ reaction resulted in two desorption

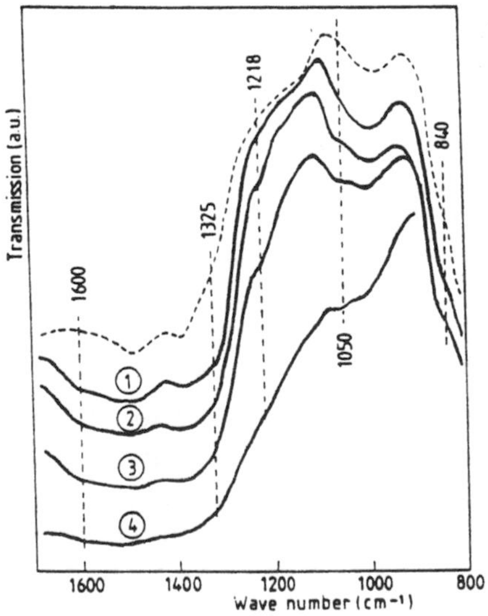

Fig. 5. IR spectra of adsorbed CO_2 at 298 K on oxidic $LaFeO_3$ (1), and on $LaFeO_3$ reduced in H_2 at 523 K (2); 723 K (3); 773 K (4). Broken curve, background spectrum. Reproduced with kind permission of Chapman and Hall from J. Mat. Sci. (Ref. 10).

peaks at 423 and 525 K. Heats of desorption were calculated from peak temperatures and compared with previous results for other compounds. Fresh $LaRhO_3$ behaved similarly to La_2O_3, while adsorbed CO was stabilized only slightly by including Rh^{3+} in the crystal lattice instead of the presence of Rh metal. One of the binding states of CO on a spent $LaRhO_3$ sample appeared to correlate with the CO desorbed from the fresh oxide (ca. 134 kJ/mol), except that it shifted to an even higher heat of desorption (138 kJ/mol). The change in CO desorption behavior for the spent catalyst seemed to reflect a change in the oxidation state of rhodium at the catalyst surface, since the multiplicity of bonding states of CO is indicative of at least two oxidation states for Rh at the active $LaRhO_3$ surface.

A comprehensive TPD study of the adsorption of CO and CO_2 on $LaMO_3$ (M = Cr, Mn, Fe, Co, Ni) perovskites has been carried out by Tejuca et al. (8-12). Figure 6a shows TPD spectra of CO after CO adsorption at 298 K on $LaMnO_3$ (8). CO desorption from $LaMnO_3$ oxidized at 873 K shows a wide peak centered at 473 K, another poorly-resolved shoulder at 540 K, and a tail at ca. 773 K (spectrum 1). For $LaMnO_3$ reduced in H_2 at 573 K (spectrum 2), two well-defined peaks appear at 360 and 550 K. CO desorption above 773 K was also observed. On $LaMnO_3$ reduced at 873 K, the three desorption peaks appearing now at 395, 545 and 800 K (spectrum 3) increase substantially in intensity; in addition, a shoulder at 860 K can be seen. Based on literature results from TPD and IR spectroscopic studies, it was concluded (8) that the peak at 473 K from the oxidized catalysts is associated with a carbonate species.

The above assignment is in agreement with some features in the TPD spectra of CO_2 obtained form CO adsorbed on the same $LaMnO_3$ sample (Fig. 6b). The presence of CO_2 should be due to CO oxidation via formation and decomposition of carbonates. On the oxidized surface a wide peak at 473 K and a tail centered at 773 K were observed (spectrum 1). The former was assigned to monodentate carbonates. Its high intensity and total diappearance from reduced surface at 573 K (spectrum 2) and 873 K (spectrum 3) point to interaction of these carbonates with Mn^{3+}. The tail at 773 K is associated with carbonates of higher thermal stability, i.e., bidentate carbonates. On the other hand, the shoulder at 540 K and the tail centered at 773 K of CO desorption (Fig. 6a, spectrum 1), which become well-resolved peaks on reduced $LaMnO_3$ surfaces (Fig. 6a, spectra 2 and 3) and increase in intensity with the degree of reduction, should be associated

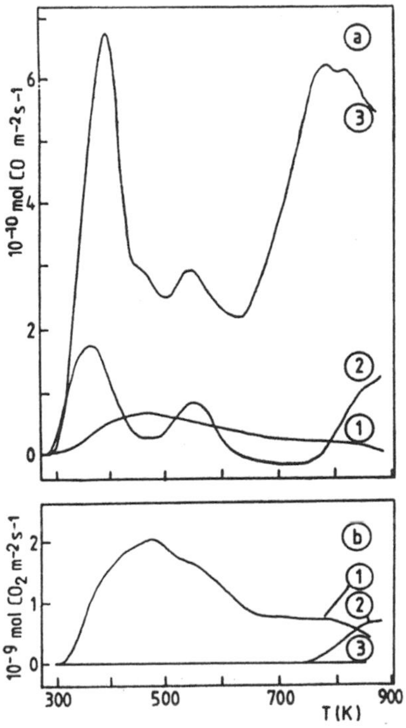

Fig. 6. (a) TPD spectra of CO after CO adsorption at room temperature on $LaMnO_3$ oxidized at 873 K (1); reduced at 573 K (2) and 873 K (3). (b) TPD spectra of CO_2 after CO adsorption at room temperature on $LaMnO_3$ oxidized at 873 K (1); reduced at 573 K (2) and 873 K (3). Reproduced with kind permission of The Royal Society of Chemistry from J. Chem. Soc., Farad. Trans. I (Ref.9).

to linear or bridged CO adsorbed on reduced Mn^{n+} ($0<n<3$) ions. IR spectra showing bands at 2130 and 2080 cm^{-1}, ascribable to linear CO on Mn^{2+}, confirms this assignment (8). This agrees, in turn, with the absence of CO_2 peaks in spectra 2 and 3 of Fig. 6b (as indicated above, CO_2 peaks are due to decomposition of adsorbed carbonate species).

Similarly, a progressive increase of the intensity of TPD peaks of CO associated to linear and bridged CO species adsorbed on reduced Cr^{n+} ($0<n<3$) centers was observed for $LaCrO_3$ (8) with increasing severity of the reducing conditions. This was paralleled with a decrease of the intensity of CO_2 peaks around 500 K appearing after CO adsorption, although unlike the case of $LaMnO_3$, CO_2 peaks were not totally suppressed. A similar behavior was observed for $LaFeO_3$ (10). However, more reduced iron centers were assumed to participate in the adsorption process in sight of the higher reducibility of $LaFeO_3$ as compared with $LaCrO_3$ and $LaMnO_3$ counterparts (27).

A markedly different behavior was observed for $LaCoO_3$ (11) and $LaNiO_3$ (12), for which the intensity of peaks due to linear or bridged CO adsorbed on reduced metal ions was at a maximum for intermediate reduction temperatures (473 K for $LaCoO_3$ and 573 K for $LaNiO_3$). This is not surprising in view of TPR results showing that both Co^{2+} and Ni^{2+} ions are stable intermediates during the H_2-reduction of these perovskites (31). In these two cases, the intensity of peaks due to CO_2 desorbing after CO adsorption decreased with increasing temperature, while the peak maximum shifted towards higher temperatures. This was associated (12) to the fact that O^{2-} ions remaining on the surface after reduction at increasing temperatures should be strongly bound to the oxide lattice, and therefore the carbonates formed are removed at higher temperatures than those needed for elimination of carbonates formed on less reduced samples.

As for CO, TPD has been used to study CO_2 adsorption on $LaCrO_3$ (8), $LaMnO_3$ (9), $LaFeO_3$ (10), $LaCoO_3$ (11) and $LaNiO_3$ (12) perovskites. Results corresponding to $LaMnO_3$ and $LaNiO_3$ are shown in Figs. 7a and 7b, respectively. In addition to CO_2 peaks, peaks of CO desorption were found for $LaMnO_3$ reduced at 723 and 873 K (Fig. 7a, spectra 3 and 4) (8). A similar behavior was shown by $LaCrO_3$ reduced at 623-923 K (8) and $LaFeO_3$ reduced at 823 K (10). These CO peaks should be due to CO_2 reduction on the surfaces. No desorption peaks of CO were, however, observed after CO_2 adsorption on reduced $LaNiO_3$ (Fig. 7b) (12) or reduced $LaCoO_3$ (11). This could be related to the higher catalytic activity for CO oxidation of the latter two perovskites with respect to the other members of $LaMO_3$ series (31).

On the other hand, CO_2 adsorption on $LaMO_3$ oxides yielded two main CO_2 TPD peaks at 340-425 K and 540-940 K (8-12). The low-temperature desorption peak was ascribed to decomposition of monodentate carbonates (the less stable adsorbed species). The intensity of this peak decreased with the extent of oxide reduction in $LaMnO_3$ (9) and $LaFeO_3$ (10), the opposite trend being observed for the other perovskites studied. The CO_2 desorption peak appearing at the higher temperature increased in intensity with increasing reduction temperature for $LaCrO_3$ (8), $LaMnO_3$ (9) and $LaFeO_3$ (10), this being ascribed to the formation and decomposition of bidentate and/or bridged carbonates, whose presence was indeed confirmed by IR (see above). Formation of this type of species takes place on surface sites composed of a lattice oxygen and an anion vacancy; consistent with this interpretation, concentration of these carbonates increased with the extent of oxide reduction. This effect was more notorious as the oxide reducibility increased (8-10).

The evolution of the high-temperature TPD peak found upon CO_2 adsorption on reduced $LaCoO_3$ (11) and $LaNiO_3$ (12) was very different from the other perovskites studied. The intensities of

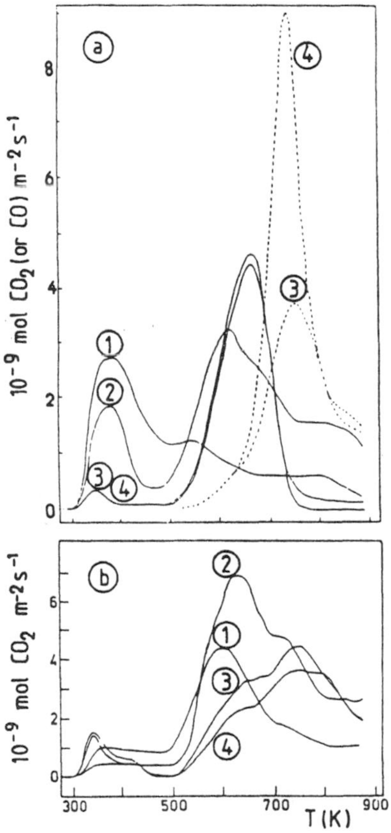

Fig. 7. (a) TPD spectra of CO_2 (solid line) and CO (broken line) after CO_2 adsorption at 298 K on $LaMnO_3$ oxidized at 873 K (1); reduced at 573 K (2); 723 K (3); and 873 K (4). Reproduced with kind permission of The Royal Sciety of Chemistry from J. Chem. Soc., Faraday Trans. I (Ref.9). (b) TPD spectra of CO_2 after CO_2 adsorption at 298 K on H_2-reduced $LaNiO_3$ at 473 K (1); 573 K (2); 673 K (3); and 773 K (4). Reproduced with kind permission of Elsevier Science Publishers from Thermochim. Acta (Ref.12).

TPD peaks showed maxima for $LaCoO_3$ and $LaNiO_3$ samples reduced at 573 K and decreased for higher reduction temperatures. This indicates interaction of bidentate or bridged carbonates with intermediate reduction states of cobalt and nickel, likely Co^{2+} and Ni^{2+}. The number of pair sites required for adsorption appears to increase in these latter perovskites up to a reduction temperature of 573 K. At higher reduction temperatures the concentration of anion vacancies is expected to increase drastically thus decreasing the concentration of pair sites and, therefore, hindering the formation of bidentate or bridged carbonates.

IV. Adsorption of Other Less Common Molecules

Wan et al. (37) studied by IR spectroscopy the adsorption of SO_2 on partially-substituted La-Co perovskites. The $La_{0.6}Sr_{0.4}Co_{1-x}M_xO_3$ sample exposed to SO_2 at 473 K gave rise to bands at 1133 and 994 cm^{-1}, ascribable to a bridged structure of adsorbed SO_2. In the same conditions, adsorption of SO_2 on $La_{0.6}Sr_{0.4}CoO_3$ showed bands at 1294 and 1128 cm^{-1}, which were assigned to adsorbed SO_2 on a single metallic site. This linear species could be converted into the bridged one upon heating the sample in air to 673 K. Wan et al. (37) compared the behavior of these perovskites with that of their single oxide constituents. While no bands of adsorbed SO_2 at room temperature were observed on La_2O_3, the Co_3O_4 oxide exposed to SO_2 at room temperature gave a spectrum with bands positioned closely to those of bridged and linear species. The above findings agree with the catalytic behavior for CO oxidation of La-Co perovskites: the poisoning effect by SO_2 takes place through adsorption of this molecule on B sites of the structure and/or oxygen anions, that are the centers responsible for catalytic activity of these compounds. SO_2 may also interact with cations in position A, but this process does not result in catalyst deactivation (31).

The adsorption of water at either oxidic or reduced perovskites has been used to reveal changes in the hydroxyl groups concentration. Fierro and Tejuca (38) studied the IR hydroxyl bands obtained after water adsorption at 423 K on oxidic and H_2-reduced $LaCrO_3$. Two bands at 3680 cm^{-1} of isolated OH groups (centers b), and at 3550 cm^{-1} (centers a) were found in both cases. The O-H bond in the last case should be weaker and, therefore, these hydroxyl groups should be more acidic. The adsorption and dissociation of water was assumed to take place on pairs of surface acid-base centers, anion vacancy-O^{2-}, yielding an acidic OH and an anion vacancy placed between coordinatively-unsaturated La^{3+} and reduced metal ions (M^{2+}, M^{+}) (center a), and a basic OH^{-} on a coordinatively-unsaturated oxygen ion bonded to a La or M ion (center b). The fact that dissociation of water takes place even on the oxidic samples suggests that pumping at 773 K generates some anion vacancies on the surface, although presumably in low concentrations. The H_2-reduction treatment greatly enhances the formation of these vacancies and, consequently, the dissociation of water in the reduced perovskite samples.

Acid sites in perovskite-type oxides have been evaluated through adsorption of pyridine. Infrared spectra of pyridine adsorbed at 298 K on $LaMO_3$ (M = Cr, Mn, Fe, Co, Ni) oxides (26) are given in Fig. 8. Spectra corresponding to pyridine adsorbed on oxidized $LaMnO_3$ and $LaFeO_3$, shown in Fig. 8a, exhibit bands at 1595, 1490 and 1440 cm^{-1}, whose intensity remains constant after water adsorption at 298 K or at 673 K. For $LaMnO_3$, the appearance of a shoulder at 1540 cm^{-1} was noticed. A similar spectrum was obtained for $LaCrO_3$ (34) but, in this case, a decrease in intensity of the above-mentioned bands suggests that they are associated with Lewis acidity.

Pyridine adsorbed on reduced $LaMO_3$ oxides yielded (Fig. 8b) a spectrum similar to that of the unreduced samples, and also a weak Bronsted band at 1540-1545 cm^{-1} whose intensity increased upon water adsorption (Fig. 8b). The increase of Bronsted acidity after water adsorption on reduced perovskites can be explained on the basis of a mechanism of dissociative adsorption of water on pairs of surface acid-base centers (O^{2-} vacancy) as indicated above. The band at 1490-1495 cm^{-1} was assigned to pyridine adsorbed on coordinatively-unsaturated exposed metal cations which constitute Lewis-acid sites, by analogy with the same band found after pyridine adsorption on other catalysts. The presence in these oxides of two metal cations with largely different charge/radius ratio probably generates acid centers of markedly different strength. In summary, the weak intensity of IR bands of adsorbed pyridine and the removal of these bands upon outgassing at 673-873 K shows the rather low acidic character of these oxides (26).

In connection with catalytic studies of hydrocarbon conversions, the adsorption of hydrocarbons was carried out also on perovskite-type oxides. Ichimura et al. (4) studied the TPD spectra of $LaCoO_3$, $LaAlO_3$ and $LaFeO_3$ exposed to ethane and ethylene. In the case of $LaCoO_3$, ethane adsorption at 300 K gave rise to single peaks of C_2H_2 and C_2H_4, and then two peaks of C_2H_6 and one peak of CH_4 were observed at increasing temperature. Almost the same spectra were obtained from the $LaCoO_3$ sample exposed to ethylene, although the fraction of CH_4 formed was smaller. These results, together with XPS data, emphasize the contribution of cobalt ions in the perovskite structure as sites for C-C bond scission. The TPD spectra obtained after ethane and ethylene adsorption on $LaAlO_3$ and $LaFeO_3$ surfaces were, however, relatively simple: they gave only the peaks of the undissociated molecules, which indicated a first order kinetics of desorption to be obeyed.

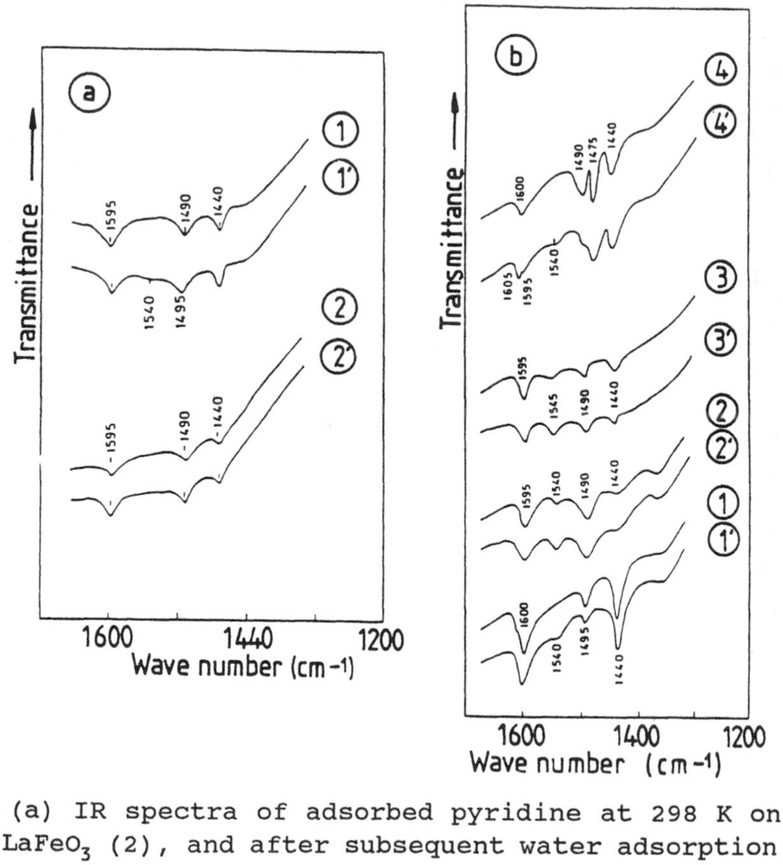

Fig. 8. (a) IR spectra of adsorbed pyridine at 298 K on $LaMnO_3$ (1) and $LaFeO_3$ (2), and after subsequent water adsorption (1' and 2'). (b) IR spectra of adsorbed pyridine at 298 K on reduced $LaMO_3$ samples (M = Cr (1), Mn (2), Fe (3), and Co (4)) and after subsequent water adsorption (1', 2', 3', 4'). Reproduced with kind permission of The Royal Society of Chemistry from J. Chem. Soc., Faraday Trans. I (Ref.26).

In many cases the adsorption of hydrocarbon molecules on perovskites yielded adsorbed oxidation products, viz. carbon dioxide. Thus, Min and Pei-yan (18) studied by IR the nature of species formed upon adsorption of ethane on $Nd_{0.7}Sr_{0.3}MnO_3$. They identified one adsorbed CO_2^- structure at room temperature and 573 K, and two adsorbed species, CO_2^- and CO_3^{2-}, at 723 K. Similarly, Kremenic et al. (39) by means of IR spectroscopy identified the species formed upon propylene and isubutene adsorption between 298 and 673 K on $LaCrO_3$ and $LaMnO_3$. A wide band centered at 2320-2340 cm^{-1}, corresponding to linearly-adsorbed CO_2, was observed. The CO_2 bands found for $LaMnO_3$ were more intense than those for $LaCrO_3$, likely because of the higher catalytic activity of the former oxide for total oxidation of hydrocarbons. In some cases, e.g., propene on $LaMnO_3$ at 673 K, wide bands centered at 1700 and 1520 cm^{-1} of carbonate species were also found.

V. Conclusion

From the analysis of the most recent works, it is evident that careful analysis of IR and TPD spectra of adsorbed molecules provide relevant information on the surface of perovskite-type oxides. The formal oxidation state of the cation in position B, the degree of coordinative unsaturation and the symmetry can easily be revealed from these spectra. In any case, it must be ensured that surface sites are not altered during the test of chemisorption. In this respect, the analysis of the gas-phase must be monitored in both IR and TPD experiments.

By virtue of the special advantage of a specific adsorbed molecule-surface site interaction the selection of an appropriate probe molecule is extremely useful. Attention must be paid to the fact that reliable qualitative information can be obtained from both techniques. However, quantitative data are only achieved with TPD. Finally, the strong dependence of the surface properties of perovskites on pretreatments is another important factor to be considered. Hence, proper comparison of the experimental results rigorously requires the same pretreatments.

References

1. Subba Rao, G. V., C. N. R. Rao, and J. R. Ferraro, Infrared

and electronic spectra of rare earth perovskites: orthochromites, -manganites and -ferrites, *Appl. Spectrosc.* , **24**: 436 (1970).

2. Couzi, M., and P.V. Huong, Spectres infrarouges des pérovskite de terres rares, LZO_3 (Z = Al, Cr, Fe, Co), *J. Chim. Phys.*, **69**: 1339 (1972).
3. Couzi, M., and P. V. Huong, Spectres infrarouge and Raman des pérovskites, *Ann. Chim.*, **9**: 19 (1974).
4. Ichimura, K., Y. Inoue, I. Kojima, E. Miyazaki, and I. Yasumori, Comparative studies of mixed oxide perovskite catalysts, $LaCoO_3$, $LaFeO_3$ and $LaAlO_3$ for hydrogenation of alkenes and hydrogenolysis of alkanes, *Proc. 7th Intern. Congr. Catal.*, Tokyo, 1980, Kodansha-Elsevier, vol. B, p. 44.
5. Watson, P. R., and G. A. Somorjai, The formation of oxygen-containing organic molecules by the hydrogenation of carbon monoxide using a lanthanum rhodate catalyst, *J. Catal.*, **74**: 282 (1982).
6. Somorjai, G. A., Molecular ingredients in heterogeneous catalysis, *Chem. Soc. Rev.*, **13**: 321 (1984).
7. Tejuca, L. G., Temperature-programmed desorption study of hydrogen adsorption on $LaMO_3$ oxides, *Thermochim. Acta*, **126**: 205 (1988).
8. Tejuca, L. G., A. T. Bell, and V. Cortés Corberán, TPD and IR spectroscopic studies of CO, CO_2 and H_2 adsorption on LaCrO3, *Appl. Surf. Sci.*, **37**: 353 (1989).
9. Tejuca, L. G., A. T. Bell, J. L. G. Fierro and J. M. D. Tascón, Temperature-programmed desorption study of the interactions of H_2, CO and CO_2 with $LaMnO_3$, *J. Chem. Soc., Faraday Trans. I*, **83**: 3149 (1987).
10. Cortés Corberán, V., L. G. Tejuca, and A. T. Bell, Surface reactivity of reduced $LaFeO_3$ as studied by TPD and IR spectroscopies of CO, CO_2 and H_2, *J. Mat. Sci.*, **24**: 4437 (1989).
11. Tejuca, L. G., A. T. Bell, J. L. G. Fierro, and M. A. Peña, Surface behavior of reduced $LaCoO_3$ as studied by TPD of CO, CO_2 and H_2 probes and by XPS, *Appl. Surf. Sci.*, **31**: 301 (1988).
12. Tejuca, L. G., and J. L. G Fierro, XPS and TPD probe techniques for the study of $LaNiO_3$ perovskite oxide, Thermochim. Acta, **147**: 361 (1989).
13. Yamazoe, N., Y. Teraoka, and T. Seiyama, TPD and XPS study on thermal behavior of adsorbed oxygen in $La_{1-x}Sr_xCoO_3$, *Chem. Lett.*, 1767 (1981).

14. Nakamura, T., M. Misono, and Y. Yoneda, Catalytic properties of perovskite-type mixed oxides $La_{1-x}Sr_xCoO_3$, *Bull. Chem. Soc. Jpn.*, **55**: 394 (1982).
15. Seiyama, T., N. Yamazoe, and K. Eguchi, Characterization and activity of some mixed metal oxide catalysts, *Ind. Eng. Chem., Prod. Res. Dev.*, **24**: 19 (1985).
16. Seiyama, T., Total oxidation of hydrocarbons on perovskite oxides, *This book, Chapter 10.*
17. Nitadori, T., and M. Misono, Catalytic properties of $La_{1-x}A_x'FeO_3$ (A' = Sr, Ca) and $La_{1-x}Ce_xCoO_3$, *J. Catal.*, **93**: 459 (1985).
18. Min, Y., and L. Pei-yan, Total oxidation reaction of ethane over $Nd_{0.7}Sr_{0.3}MnO_3$ catalyst, *Chihua Xuebao*, **7**: 287 (1986).
19. Imai, H., T. Tagawa, and N. Kamide, Oxidative coupling of methane over amorphous lanthanum aluminum oxides, *J. Catal.*, **106**: 394 (1987).
20. Imai, H., T. Tagawa, N. Kamide, and S. Wada, Preparation and characterization of lanthanum aluminum mixed oxide catalysts for oxidative coupling of methane, *Proc. 9th Intern. Congr. Catal.*, Calgary, 1988, p. 952.
21. Tagawa, T., and H. Imai, Mechanistic aspects of oxidative coupling of methane over $LaAlO_3$, *J. Chem. Soc., Faraday Trans. I*, **84**: 923 (1988).
22. Shamsi, A., and K. Zahir, Oxidative coupling of methane over perovskite-type oxides and correlation of T_{max} for oxygen desorption with C_2 selectivity, *Energy Fuels*, **3**: 727 (1989).
23. Peña, M. A., J. M. D. Tascón, and L. G. Tejuca, Surface interactions of NO with $LaFeO_3$, *Nouv. J. Chim.*, **9**: 591 (1985).
24. Peña, M. A., J. M. D. Tascón, J. L. G. Fierro, and L. G. Tejuca, A study of NO and CO interactions with $LaMnO_3$, *J. Colloid Interface Sci.*, **119**: 100 (1987).
25. Tascón, J. M. D., L. G. Tejuca, and C. H. Rochester, Surface interactions of NO and CO with $LaMO_3$ oxides, *J. Catal.*, **95**: 558 (1985).
26. Tejuca, L. G., C. H. Rochester, J. L. G. Fierro, and J. M. D. Tascón, Infrared spectroscopic study of the adsorption of pyridine, carbon monoxide, and carbon dioxide on the perovskite-type oxides $LaMO_3$, *J. Chem. Soc., Faraday Trans. I*, **80**: 1089 (1984).
27. Oliván, A. M. O., M. A. Peña, L. G. Tejuca, and J. M. D. Tascón, A comparative study of the interactions of NO and CO with $LaCrO_3$, *J. Mol. Catal.*, **45**: 355 (1988).

28. Voorhoeve, R. J. H., J. P. Remeika, and L. E. Trimble, Nitric oxide and perovskite-type catalysts: solid state and catalytic chemistry, *The Catalytic Chemistry of Nitrogen Oxides* (R. L. Klimisch and J. G. Larson, eds.), Plenum Press, New York, p. 215 (1975).
29. Voorhoeve, R. J. H., J. P. Remeika, L. E. Trimble, A. S. Cooper, F. J. Disalvo, and P. K. Gallagher, Perovskite-like $La_{1-x}K_xMnO_3$ and related compounds: solid state chemistry and the catalysis of the reduction of NO by CO and H_2, *J. Solid State Chem.*, **14**: 395 (1975).
30. Tascón J. M. D., and L. G. Tejuca, Adsorption of CO on the perovskite-type oxide $LaCoO_3$, *Z. Phys. Chem. NF*, **121**: 63 (1980).
31. Tejuca, L. G., J. L. G. Fierro, and J. M. D. Tascón, Structure and reactivity of perovskite-type oxides, *Adv. Catal., 36: 237 (1989).*
32. Kawai, T., K. Kunimori, T. Kondow, T. Onishi, and K. Tamaru, Ferroelectric effect in the catalytic reactions over a $BaTiO_3$ surface at its Curie point, *Z. Phys. Chem. NF*, **86**: 268 (1973).
33. Tascón, J. M. D., and L. G. Tejuca, Adsorption of CO_2 on the perovskite-type oxide $LaCoO_3$, *J. Chem. Soc. Faraday Trans. I*, **7** 591 (1981).
34. Martin, M. A., J. L. G. Fierro, and L. G. Tejuca, Surface interactions between CO_2 and $LaFeO_3$, *Z. Phys. Chem. NF*, **127**: 237 (1981).
35. Fierro, J. L. G., and L. G. Tejuca, Surface interactions of carbon dioxide and pyridine with $LaCrO_3$ perovskite-type oxide, *J. Chem. Tech. Biotech.*, **34A**: 29 (1984).
36. Martin, M. A., J. M. D. Tascón, J. L. G. Fierro, J. A. Pajares, and L. G. Tejuca, The Freundlich model of adsorption for calculation of specific surface areas, *J. Catal.*, **71**: 201 (1981).
37. Wan, L., H. Qing, Z. Wan-Jing, L. Bing-Xiung, and L. Guang-Lie, Improving the SO_2 resistance of perovskite-type oxidation catalysts, *Catalysis and Automotive Pollution Control* (A. *Crucq and A. Frennet, eds.)*, Elsevier, Amsterdam, p. 405 (1987).
38. Fierro, J. L. G., and L. G. Tejuca, Surface properties of $LaCrO_3$. Equilibrium and kinetics of O_2 adsorption, *J. Catal.*, **87**: 126 (1984).
39. Kremenic, G., J. M. Lopez Nieto, J. M. D. Tascón, and L. G. Tejuca, Chemisorption and catalysis on $LaMO_3$ oxides, *J. Chem. Soc., Faraday Trans. I*, **81**: 939 (1985).

Chapter 9

Composition and Structure of Perovskite Surfaces

J.L.G. Fierro

Instituto de Catálisis y Petroleoquímica, CSIC,
Serrano 119, 28006 Madrid, Spain

I. Introduction

The precise knowledge of the influence of solid state properties on catalytic activity of bulk catalysts is critical for the design of more efficient preparations. In the pursuit of these relations, preparation of catalyst systems which can provide reproducible data facilitating such generalizations becomes imperative (1). Therefore, the correlation of bulk properties with the rate of reaction occurring at the surface necessarily requires that the systems are isostructural. It is also expected that surface properties do not differ from that of the bulk, at least significantly, since the surface represents a discontinuity of the crystal.

These requirements are fulfilled to a great extent in perovskite-type compounds. As the most numerous and most interesting perovskite compounds are oxides, this review will refer only to the study of oxides. In the ABO_3 perovskites the A ions are in general catalytically inactive, and the active metal ions at the B position are placed at such relatively large distances (ca. 0.4 nm) from each other that a reactant molecule interacts only with a single site. For a single A or B center, there is still the possibility to substitute partly these centers of the ABO_3 structure yielding an ample diversity of compounds, e.g., $AB_{1-x}B'_xO_3$, $A_{1-x}A'_xO_3$ and doubly substituted $A_{1-x}A'_xB_{1-y}B'_yO_3$, as well as the possibility of having ordered systems with general formula $A_2BB'O_6$ (2). Before generalizations can be drawn, a very important factor to be considered is the surface composition of these compounds. As the symmetry and coordination of A^{n+} and B^{m+} (n+m = 6) ions are lost at the surface, they show a great tendency to be saturated by reaction with gas molecules. For instance, profile analysis of O 1s photoelectron spectra of many ABO_3 perovskite-type compounds exposed to moisture or reduced in a hydrogen flow indicated the appearance of more than one oxygen species (3-8), i.e. the lattice O^{2-} ions were, in general, accompanied by other less electron-rich oxygen species such as OH^- or O^{2-} and CO_3^{2-} structures. Moreover, very small amounts of the individual oxides AO_x and AO_y not completely incorporated in the perovskite structure during the synthesis may be still present at the surface. It is therefore expected that these differences in surface composition with respect to the stoichiometric one may significantly modify both the catalytic and adsorption properties of pure perovskites. In catalytic studies one wonders if the catalytic activity can be correlated with a single catalyst parameter, viz. piezoelectricity, ferroelectricity, electrical conductivity, magnetism. Parravano (9) observed that the rate of catalytic oxidation of CO is affected by ferroelectric transitions in $MNbO_3$ (M = Na, K) and $LaFeO_3$, which supports an electronic mechanism for this reaction. However, such anomalies in catalytic rates in the vicinity of ferromagnetic and ferroelectric Curie temperatures of some perovskite oxides cannot be generalized. To find correlations between catalytic and solid state properties, a precise knowledge of the mechanism of the selected test reaction is required. Voorhoeve et al. (10) brought forward new ideas in interpreting the kinds of correlations to be expected. According to these authors, two different correlations should be distinguished: (i) suprafacial processes in which the catalyst surface provides a set of electronic orbitals of proper symmetry and energy for the bonding of reactants and intermediates, and (ii) intrafacial reactions in which the catalyst participates as a reagent that is partly consumed and regenerated in a continuous cycle.

This chapter aims to give an overview of the influence of both composition and surface structure and their relation with catalytic properties of perovskite-type oxides. For the purpose of the present review only the ABO_3 oxides, which are the most numerous and most interesting compounds, will be considered. Some other halides, nitrides, hydrides and nitrides also crystallize with the perovskite structure (11), but the analysis is beyond the scope of the chapter.

II. Influence of Surface Composition on Catalytic Properties

The surface composition of perovskite-type oxides has been found to deviate in many cases from stoichiometry. This is not surprising since these compounds are synthetised at very high temperatures, usually above 900 K, starting with decomposition of A and B precursors. In the course of the complex solid-state transformations, isolated AO_x and BO_y oxides can be formed, which then react not completely to form the single ABO_3 phase. Other changes in the surface of these compounds induced either by exposure to moisture or by a reducing atmosphere have been already observed. These changes are examined below.

A. Surface Enrichment of One Component

Gysling et al. (12) and Monnier and Apai (13) have clearly indicated the need for careful attention to the preparation of lanthanum rhodate ($LaRhO_3$), which is to be evaluated as catalyst for CO hydrogenation and the usefulness of X-ray photoelectron spectroscopy for monitoring its surface. To illustrate this, the chemical changes brought about upon exposing the surface of $LaRhO_3$ to various environments have been revealed by XPS.

The core level spectrum of Rd 3d levels of the fresh material (Fig. 1, spectrum a) contains contributions of both Rh_2O_3 (impurity) and $LaRhO_3$ phases. The Rh_2O_3 impurity is observed to disappear upon reduction in hydrogen at low temperature, with the subsequent formation of metallic Rh^o characterized by the binding energy of the most intense Rh $3d_{5/2}$ peak at 307.1 eV (spectrum b). Reduction at 523 K in flowing hydrogen was insufficient to completely reduce the surface of $LaRhO_3$, but it was fully reduced at 573 K (spectrum c). Contrary to this behavior, syngas reaction conditions promote the stabilization of Rh^{3+} ions in the $LaRhO_3$ matrix as revealed by the appearance of two separate Rh $3d_{5/2}$ peaks at 308.6 eV (Rh^{3+}) and at 307.1 eV (Rh^o) when the catalyst was contacted at 573 K with a 2:1 = H_2:CO reactant mixture at atmospheric pressure (spectrum d). As the full Rh 3d spectrum can be obtained by summing the separate spectra for untreated $LaRhO_3$ and completely reduced $LaRhO_3$, the presence of intermediate Rh^+ species can be excluded. This finding differs from the results

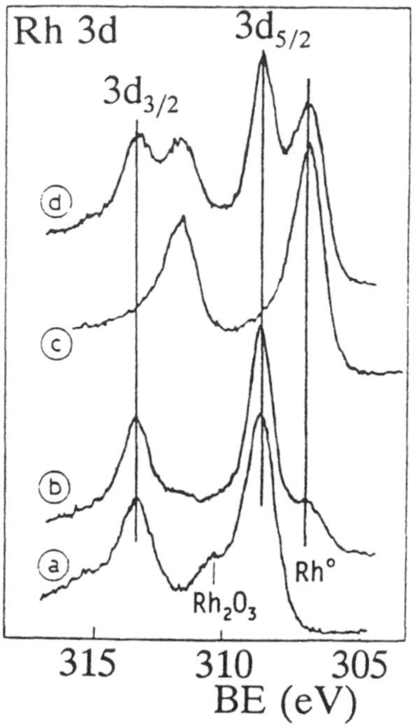

Fig. 1. (a), Rh 3d core level spectra of $LaRhO_3$ (containing trace impurity of Rh_2O_3). Reduction products on heating in pure hydrogen at: (b) 523 K for 0.5 h.; (c) 573 K for 0.5 h; (d) 573 K in a 2:1 H_2:CO mixture.

of Somorjai and co-workers (14,15), who studied the activity of clean and oxidized Rh foils and of $LaRhO_3$ at different temperatures, under H_2:CO mixtures, and thus at different levels of reduction. They observed that Rh_2O_3 produces large amounts of C_2 and C_3 hydrocarbons, and propionaldehyde when C_2H_4 was added to the H_2:CO feed, showing the carbonylation ability of rhodium oxide. However, C_2H_4 was quantitatively hydrogenated to C_2H_4 over Rh^o under similar conditions. Thus, higher oxidation states of Rh seem to be necessary to produce the oxygenated organic molecules. From X-ray photoelectron spectroscopy it has been reported that the active catalyst contains rhodium as Rh^+ and a small fraction as reduced Rh^o.

The origin of the above discrepancies may likely reside in the extent with which Rh_2O_3 impurity appears on the $LaRhO_3$ surface. The formation of pure monophasic $LaRhO_3$ requires an excess of La_2O_3 in the synthesis, since any excess present after the reaction can be readily removed with a warm, dilute acetic acid leach. It is also emphasized that surface enrichment of Rh_2O_3 on the $LaRhO_3$ may provide a source of isolated, simple Rh_2O_3 ensembles within a thermally stable perovskite crystal structure. In short, these results nicely illustrate how a careful control of sample surface preparation is needed if correlations between surface properties and catalytic behavior for the heterogeneous reaction are attempted.

Partly substituted $LaM_{1-x}Cu_xO_3$ (M = Mn, Ti) perovskites have been reported to exhibit high methanol selectivity in the hydrogenation of CO (16-18). It has also been concluded that copper is the catalytic agent for this reaction and that the catalytic sites involve Cu+ and/or Cuo in position B. It appears, therefore, that oxidized states of copper should play an important role in the formation of oxygenates from syngas over Cu-containing perovskites. However, the studies carried out so far, mostly of empirical character give little information on the effect of surface composition on the reactivity.

Rojas and Fierro (19) synthesized the family of $LaTi_{1-x}Cu_xO_3$ oxides and studied the chemical state and composition changes by XPS when subjected to various pretreatments. The binding energies (BE) for O 1s and the principal Cu 2p and Ti 2p peaks are summarized in Table 1. From this table, it appears that the BEs of Cu 2p electrons are very similar from sample to sample and somewhat lower than that of Cu2+ ions. Identification of the chemical state of copper is complicated since the BEs for Cu^o and Cu^+ are almost the same, appearing ca. 1.3 eV below that of Cu^{2+}. Cupric ions can easily be distinguished by the appearance of shake-up satellites by ca. 9.8 eV above the principal Cu 2p peaks. As can be seen, for substitution x = 1.0, and to a lesser extent for x = 0.6, shake-up satellites are observed. The presence of both Cu^{2+} and Cu^o in $LaCuO_{2.5}$ (x = 1) is expected because the reduction at 573 K only eliminates the CuO phase of the two constituent phases (CuO and La_2CuO_4) of $LaCuO_{2.5}$composition

$$CuO + La_2CuO_4\ (LaCuO_{2.5}) + H_2 \longrightarrow Cu^o + La_2CuO_4 + H_2O$$

while the La_2CuO_4 phase appears to be stable up to ca. 673 K. In order to monitor the changes undergone by copper upon H_2-reduction and use, the Cu/La atomic surface ratios have been calculated. In Fig. 2 the XPS Cu/La atomic ratios of $LaTi_{1-x}Cu_xO_3$ samples are plotted against the stoichiometric ones. It is observed that the Cu/La ratio increases with x, but deviates

Table 1. Binding Energies (eV) of Core Electrons in Spent Catalysts

Catalyst	O 1s	Cu $2p_{3/2}$	Ti $2p_{3/2}$
$LaTiO_3$	529.6	-	457.4
$LaTi_{0.8}Cu_{0.2}O_3$	529.2	932.8	457.3
$LaTi_{0.5}Cu_{0.50}3$	529.6	933.0	457.2
$LaTi_{0.4}Cu_{0.6}O_3$	529.9	933.1	457.3
$LaCuO_{2.5}$	530.4	932.5	-
Cu/La_2O_3	531.2	933.6	-

slightly from the stoichiometric composition, i.e., the surface Cu/La ratios become larger than x. Similarly, the XPS Cu/La ratio for the lanthana-supported copper catalyst is essentially the same as the x = 0.2 sample, though the latter catalyst has a lower bulk Cu content. All these results indicate that the surface of $LaTi_{1-x}Cu_xO_3$ catalysts becomes slightly Cu-enriched during reduction and use in reaction. However, for the Cu/La_2O_3 catalyst it seems that big Cu crystals build up on the top lanthana surface.

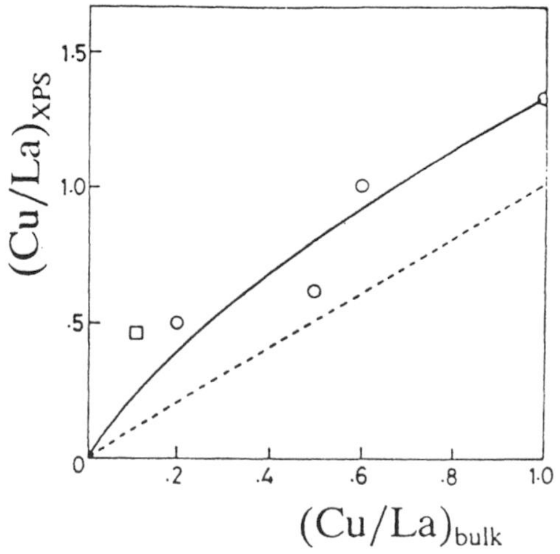

Fig. 2. Dependence of the bulk Cu/La atomic ratios on the surface XPS Cu/La ratios (O). The square refers to the Cu/La_2O_3 catalyst. The dashed line corresponds to the stoichiometric composition.

B. Nature of Surface Oxygen Species

The nature of the oxygen species on the surface of perovskites can be revealed by XPS analysis. To illustrate this, the O 1s and La 4d photoelectron spectra of $LaNiO_3$ subjected to a variety of pretreatments are given in Fig. 3A and 3B, respectively (20). The O 1s peaks show two maxima at 528.2-529.6 eV and at 530.9 eV. In addition, the peak at higher BE of the as-prepared (Fig. 3A, spectrum a) exhibits a shoulder towards higher BE which disappears upon outgassing (spectrum b) or reduction (spectra c-e). The intensity of the peak at higher BE undergoes a remarkable decrease for samples reduced at 473-673 K. The peak at lower BE is due to O^{2-} species, while that at higher BE seems to be due to other less electron-rich species of oxygen, i.e. OH^- (20, 21). The asymmetry of the peak at higher BE is assigned to molecular water adsorbed on the surface. Important changes are also observed for the La 4d peaks (Fig. 4B). Thus, the La $4d_{3/2}$/La $4d_{5/2}$ intensity ratio decreases markedly with increasing reduction temperature. It should be noted that this decrease runs parallel

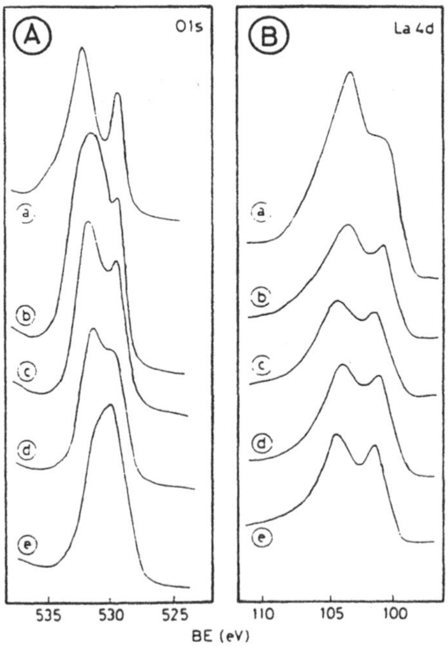

Fig. 3. O 1s (A) and La 4d (B) core level spectra of as- prepared $LaNiO_3$ (a), after outgassing under high vacuum for 2 h at 723 K (b) and after H_2-reduction at 473 K (c), 573 K (d) and 673 K (e).

with the decrease in the relative intensity of the higher BE O1s peak. These results are consistent with the formation of an oxyhydroxide LaO(OH) in the unreduced sample and its gradual thermal decomposition after reduction according to the reaction,

$$2\ LaO(OH) \longrightarrow La_2O_3 + H_2O$$

It must be stressed that this behavior for $LaNiO_3$ samples reduced under dynamic conditions is opposite to that observed for $LaCoO_3$ reduced under static conditions (21). In the latter case the intensity of the higher BE O 1s peak increases for increasing reduction temperature because the presence of the water produced in the reduction according to the above reaction shifts the equilibrium towards formation of LaO(OH).

III. Suprafacial Phenomena

A. Local Interactions

In ABO_3 perovskite-type oxides, the B^{3+} ions are placed in the center of an octahedron whose vertices are occupied by O^{2-} ions. Assuming that the most frequently exposed planes are those with the lower Miller index, on chemisorption of molecules the B^{3+} ion should recover the coordination of the bulk, changing from square pyramidal to octahedral in the (100) plane (Fig. 4a), from tetrahedral to octahedral in the (110) plane, and from trigonal to octahedral in the (111) plane. The change in the Crystal Field Stabilization Energy (CFSE) due to the change in coordination of B^{3+} ion usually shows two maxima when passing from d^0 to d^{10} (22). As can be seen below for some suprafacial reactions, the experimental results show a similar twin-peaked pattern, although this does not coincide with the theoretical CFSE, as was observed by Dowden et al. (23), Dowden and Wells (24) and Dixon et al. (25) in reactions involving hydrogen catalyzed by first-row transition metal oxides and by Boreskov et al. (26) in oxidation reactions catalyzed by a series of spinel-type oxides. This may be due to the appearance of surface defects, different in concentration and nature produced under conditions of adsorption and catalysis. In short, the similarity of the theoretical CFSE profile and the experimental observation shows the importance of localized interactions in these types of processes.

B. Oxygen Evolution

The catalytic performance of perovskite-type oxides for the electrolytic evolution of oxygen has been studied. This reaction has been chosen as a model reaction for kinetic studies that involve rate-determining step evaluation. Bockris and Otagawa (27, 28) carried out a systematic study on several substituted perovskites in an attempt to correlate their surface structure and electrocatalytic properties in the oxygen evolution from alkaline solutions laying particular attention on the bonding of oxygenates intermediates at the catalyst surface. These authors

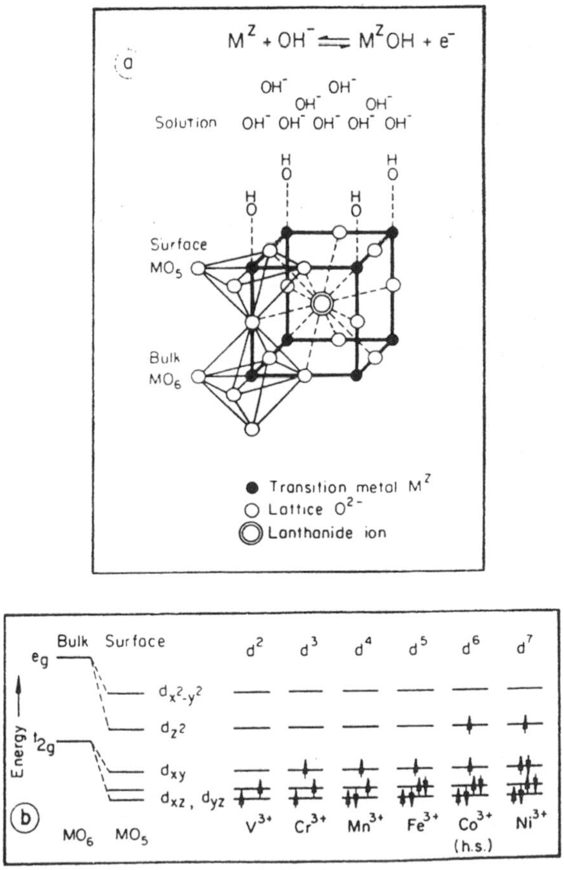

Fig. 4. (a) Schematic model for the active (001) plane of the AMO_3 perovskite in which transition metal M is the active component; (b) d-electron configuration of transition M^{3+} ions at the surface of perovskites.

show that the most likely mechanism involves a rate determining B-OH desorption step, where B is the transition element in the ABO_3 formula.

Based on molecular orbital (MO) theory, a model has already been proposed for the active (001) plane of the perovskite consisting of a MO_5 cluster to explain the chemical bonding between the surface B^{3+} ion and the OH^- group (Fig. 4a). According to this model, the 3d (e_g, t_{2g}) levels will split because of the lower symmetry at the surface. As shown by Wolfram et al. (29), MO's calculations for the $SrTiO_3$ perovskite revealed that the e_g levels of the BO_6 cluster (bulk) split into the d_{xy} and d_{z^2} levels, with the latter lying below the former at the surface, and the t_{2g} levels split into the doubly degenerated d_{xz} and d_{yz} states and the singlet d_{xy} state (Fig. 4b). If these calculations also hold for other perovskites of the type ABO_3, it is possible to assign the d-electron configuration of the B^{3+} ion at the surface. According to this diagram, the d_{z^2} orbital will be occupied by an electron only in Co (high spin configuration) and Ni. The presence of one unpaired electron in the d_z orbital in $ACoO_3$ (high spin configuration) and $ANiO_3$ perovskites may play an important role in achieving high reaction rates since they tend to occupy $\sigma*$ antibonding of the Bz-OH, resulting in a weaker bond. In addition, as the OH is a saturated ligand the back-bonding to B_z-OH would be negligible in these two perovskites, thus contributing to a weakening of the B_z-OH bond. These two characteristics support the most likely mechanism which involves B-OH desorption as the rate-determining step.

C. Oxidation of CO

Oxidation of CO has been widely studied as a test reaction for characterization of perovskites. Voorhoeve et al. (10), Shimizu (30) and Tascón and Tejuca (31) have shown a suggestive correlation between the activity for the oxidation of CO with oxygen and the electronic configuration of the B^{3+} cation. The catalytic activity of the reaction is represented in Fig. 5 as a function of the occupancy of the d levels of the transition B^{3+} ion. As it is known, the octahedral environment of B^{3+} ions splits the d-orbitals into two levels: the lower (t_{2g}) level contains orbitals that are repelled less by negative point charges (O^{2-}) than are the orbitals in the higher (e_g) level. According to the data in Fig. 5, it is seen that the maximum activity is attained in both cases for an occupation of the e_g levels of less than one electron, the t_{2g} levels being half-filled or totally filled.

The catalytic activity for the oxidation of CO on perovskites was found to depend markedly on their stoichiometry (4). A simple way of varying the oxidation state of the B^{3+} ion is by substitution of the A^{3+} ion by a different ion with an oxidation state lower than 3. This method has been largely used (32-35) to

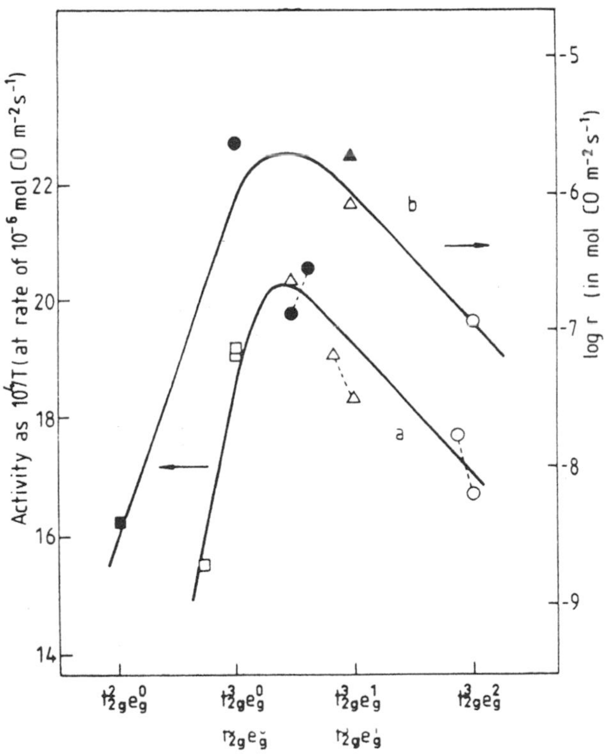

Fig. 5. Activities of $LaBO_3$ (B = 3d element) perovskites for CO oxidation in a 2:1 mixture of CO and O_2 (a) or in a 1:1 mixture of CO and O_2 at 500 K (b). The activities of vanadates (■), chromates (□), manganates (Δ), ferrates (O), cobaltates (●) and nickelates (▲) are plotted at the appropriate d-orbital occupation corresponding to the average valence of the B^{3+} ion. The activity is given either as the reciprocal of the temperature at which the activity is 1 μmole of CO per m^2 of catalyst and per second (a) or as the rate of mole CO converted per m^2 and unit time. (Readapted from refs. 1 and 31).

understand the role of the 3d-orbital occupancy in perovskite series. In the case of $(La,A')CoO_3$ oxides, the appearance of Co^{2+} ions by introduction of the Ce^{4+} ions in the A' position led to an increase in the rate of oxidation of CO, while the presence of Co^{4+} ions by introduction of Sr^{2+} in the A' position reduces the rate. This behavior was explained by assuming that the CO molecule is bonded to the cobalt ion as a carbonyl, as occurs on metals with donation of the carbon lone pair into the empty $3d_z$ orbital of this cobalt ion to form a σ-bond accompanied by back-donation of the t_{2g} electrons of Co^{3+} into the antibonding π-orbital of CO. It can be inferred that the binding of CO is optimal when both processes are posible. The correlation of the rate with the occupation of 3d orbitals is therefore consistent with the kinetic data which indicate that the rate of reaction varies as the partial pressure of CO (36).

As shown by Voorhoeve et al. (1), the values for the activation energy for the oxidation of CO are essentially the same for several substituted $(La,A')CoO_3$ (A' = Ce, Sr) perovskites, thus indicating that there is no significant variation of the activity among these cobaltates. A similar finding was reported by Le Coustumer et al. (37) on the combustion of methane, who observed that the intrinsic activity of Co^{3+} ions appears to be the same on different cobaltates. Consistently, it appears that the activity variations among the cobaltates should be associated with some parameter other than the spin state of cobalt ions. It was observed that the Arrhenius plots for the oxidation of CO on $LnCoO_3$ (Ln = La, Nd, Sm, Gd) perovskites exhibited abrupt changes around 470 K (38). In addition, the adsorption isobars showed a maximum or a minimum, corresponding to the temperatures at which changes were observed in the Arrhenius plots, showing that the changes in the mode of adsorption of reactants are responsible for the changes in the Arrhenius plots.

The above picture, which considers local interactions at the B^{3+} ions in perovskite surfaces, describes qualitatively the activity profiles for series of perovskites; however it must be stressed that the surface reaction between adsorbed CO and O_2 is stoichiometric and the presence of excess of either of the components results in lower values for the rate constants. This latter observation represents, therefore, another important factor to be considered if general conclusions or correlations between surface structure and catalytic activity are wished.

D. Oxidation of Hydrocarbons

The combustion of propylene and isobutene over a series of $LaBO_3$ (B = Cr, Mn, Fe, Co, Ni) perovskite-type oxides was found to show remarkable differences among these compounds (39). The catalytic activity for the total oxidation of propylene and isobutene at 573 K and for molar ratios $HC:O_2$ = 2 and 0.25 are shown in Fig. 6a. The rates for isobutene oxidation were found to be higher than those for propylene. In both cases two pronounced maxima for $LaMnO_3$ and $LaCoO_3$ were observed. This trend has also been found in plots of activity for N_2O decomposition and for oxygen isotope exchange (40) versus d-orbital occupancy in the same perovskite series. It must be stressed that the adsorption profile of oxygen at 298 K exhibits two maxima for $LaMnO_3$ and $LaCoO_3$ (Fig. 6b) that coincide with the maxima

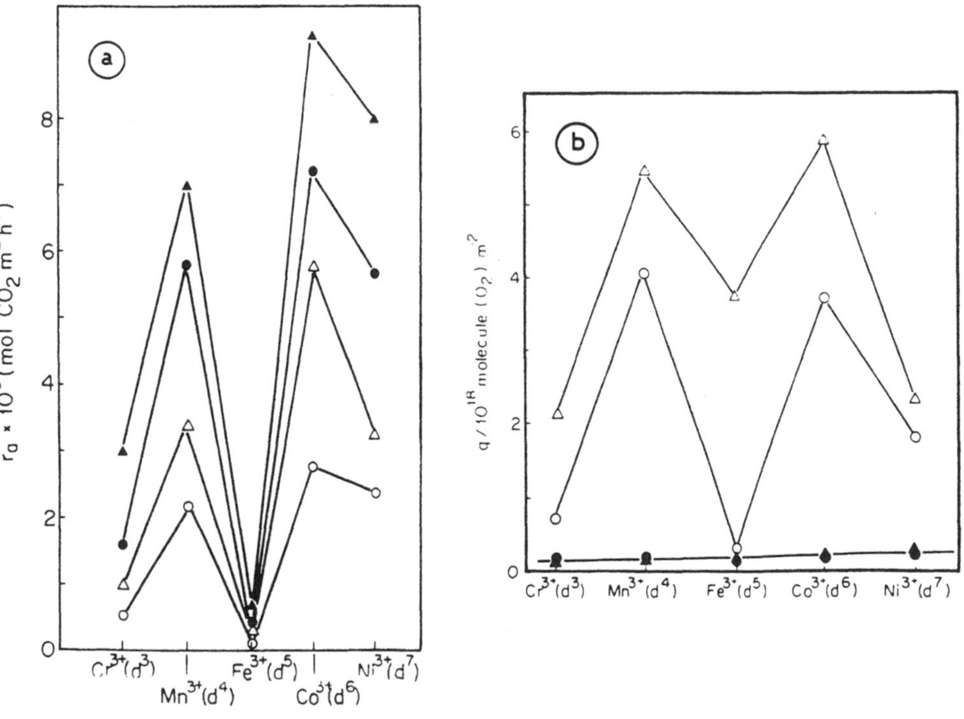

Fig. 6. (a) Activity profiles of $LaBO_3$ (B = Cr, Mn, Fe, Co, Ni) perovskites at 573 K in propylene (open symbols) and isobutene (filled symbols) oxidation. (b) Total (open symbols) and reversible (filled symbols) adsorption profiles of O_2 on the same perovskites at 298 K and 20 kPa. (Readapted from ref. 39).

observed by Iwamoto et al. (41) for the respective simple 3d-transition metal oxides and also for the catalytic activity of these perovskites for combustion of hydrocarbons (Fig. 6a).

The twin-peaked pattern observed for both oxygen adsorption and catalytic activity can be explained in terms of the change in CFSE as a result of the change in coordination of the B^{3+} ion. Although the position of the experimental peaks does not exactly match with the predictions of CFSE pattern because other effects such as surface defects may be operative, these results provide support to the ideas of Dowden and Wells (24) on the relationship between the local symmetry of the transition ion at the surface and its adsorption and catalytic properties. Furthermore, the close parallelism between catalytic activity for combustion of propylene and isobutene and the extent of oxygen adsorption indicates that both reactions occur through a suprafacial catalysis mechanism in which adsorbed oxygen is the dominant O-species participating in the oxidation of hydrocarbons.

IV. Intrafacial Processes

A. Reduction of NO

Reduction of NO with CO or H_2 has found to be an interesting example of an intrafacial reaction (10). Over transition metal oxides the reaction appears to be primarily related to the thermodynamic stability of the oxygen vacancies adjacent to the transition ion. As the reduction of NO over several perovskites leads to N_2 and N_2O products, the proposed reaction scheme includes participation of both dissociative and molecular adsorption of NO (10, 42, 43):

$$B - \square - B + NO \longrightarrow B - O - B + \tfrac{1}{2} N_2 \quad (1)$$

$$2\,e + NO \longrightarrow O^{2-} + N_{ads} \quad (2a)$$

$$2\,N_{ads} \longrightarrow N_2 \quad (2b)$$

$$N_{ads} + NO_{ads} \longrightarrow N_2O \quad (2c)$$

where □ is an oxygen vacancy. The influence of the reduction of the catalyst with the rate of conversion of NO was demonstrated by Sorenson et al. (44) who found a close parallelism between the onset of NO reduction and the onset of bulk reduction of $LaCoO_3$. This behavior was explained as being due to the formation of an oxygen deficient structure having the general formula $LaCo_{1-x}O_{3-y}$, where $0 < x < 0.08$ and $0 < y < 0.5$. From these observations, it is expected that perovskites that can easily release oxygen will be active for NO reduction. Voorhoeve et al. (1, 34, 43) demonstrated that manganites satisfy this requirement fairly well. In these perovskites the bond strength of the surface oxygen may be varied by introduction of La^{3+} vacancies that change the electronic configuration of the BO_6 octahedra, viz., by lowering the binding energy of oxygen in Mn-O and increasing the Mn^{4+} content. Thus, these authors (43) studied the rate of reduction of NO over stoichiometric $LaMnO_3$ and $La_{0.9}\square_{0.1}MnO_3$ perovskites, and found the latter to be more active than the former. To determine whether the large improvement in the rate was due to the Mn^{4+} ions or to the La vacancies, perovskites $(A_yA'_{1-y})Mn_{1-x}Mn^{4+}{}_xO_3$ (where A and A' represent one or more of the following entities: La, □'[cation vacancy], Bi, Pb, Sr, Na, K, Rb) containing almost the same proportion of Mn^{3+} and Mn^{4+} but different A-O binding energies, were tested for this reaction.

In the (001) plane of the perovskites the lattice oxygen is coordinated by two Mn nearest neighbors at a distance close to 0.2 nm and by two A positions:

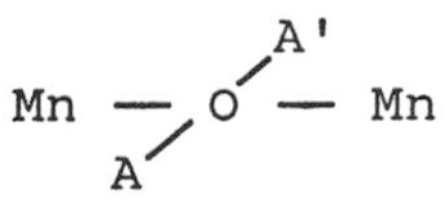

The heat of formation of an oxygen vacancy should be related to the sum of the A - O and A'- O binding energies, which are given by the expression:

$$\Delta(A - O) = \frac{\Delta H_f - m\,\Delta H_s - (n/2)\,D_o}{12\,m}$$

where ΔH_f, ΔH_s and D_o are the enthalpy of formation of 1 mole of A_mO_n, the enthalpy of sublimation of A, and the dissociation energy of oxygen, respectively. For $A_yA'_{1-y}M_{1-y}M_{1-x}Mn^{4+}O_3$ perovskites, this equation gave values for the binding energy of oxygen decreasing in the order: Bi, K > La,□' > La, Rb ≈ La, K ≈ La, Na > La, Pb > La, Sr. This trend is closely related to the rate of

NO reduction. These data clearly reflect the importance of the binding energy of surface oxygen, which determines the number of oxygen vacancies (active centers). The cation vacancies provide weakly bound oxygen to the surface and, hence, favor both the formation of oxygen vacancies and NO reduction.

B. Oxidation of Methanol

The oxidation of methanol on $LaCo_{1-x}Ni_xO_3$ and $LaCo_{1-x}Fe_xO_3$ perovskites has been found to be an intrafacial process (45). The change in activity of $LaCo_{1-x}Ni_xO_3$ compounds with increasing cobalt concentration was not large but contrasted with the large changes in electrical conductivity. Among the iron substituted compounds, the composition x = 0.01 showed the highest activity. In this case, the changes in catalytic activity with x are paralleled by changes in electrical properties. In general, an increase in conductivity is accompanied by an increase in activity. However, other pure compounds of the type La_2BO_4 (B = Ni, Cu) have quite high conductivity and are inactive. It seems, therefore, that the individual ionic species are more important that the collective properties in determining activity. In particular, it would seem that low spin states are essential for lowering the activation energy for the oxygen trnsfer reaction. In the $LaCo_{1-x}Ni_xO_3$ perovskites both Co^{3+} and Ni^{3+} ions may exist in the low spin state, so that although Co substitution may change the bulk electrical properties greatly, the catalyst activity is not much affected. In the $LaCo_{1-x}Fe_xO_3$, the Fe^{3+} ions exist only in the high spin state so that increasing Fe substitution decreases the concentration of low spin ions and hence the activity. The high spin ions which are screened less by its outer electrons act as active centers for the adsorption of alcohol while the low spin ions are involved in the oxygen transfer process of surface oxygen ions to the alcohol molecule.

V. Conclusion

As the perovskite structure can accommodate a wide range of ions and valence states, these systems lend themselves to chemical tailoring. By an appropriate formulation of the A and B elements many desirable properties can be tailored, including the valence of B^{n+} ions, the binding energy and the diffusion of oxygen in the lattice, the distance between the active sites and the collective properties of the solid. To find correlations between the solid state properties and reactivity one must be guided by the knowledge of the reaction mechanism. In the search of the parameters which may govern the catalyst activity, Voorhoeve et al. (10) distinguished among intrafacial processes, in which the catalyst participates as a reagent that is partly consumed and regenerated in a continuous cycle, and suprafacial processes in which the catalyst surface provides a set of electronic orbitals of proper symmetry and energy for the bonding of reactants and intermediates.

Although there is no doubt on the influence of the surface structure of perovskites on the catalytic behavior as evidenced by a few examples selected according to the ideas brought forward by Voorhoeve et al. (10), attention must be paid to the surface composition of these compounds, because impurities of the simple oxides, surface segregation of the active B element upon H_2-reduction, and presence of more than one surface oxygen species can markedly modify both the adsorption and catalytic properties. This is a very important factor to be considered before general conclusions about the type of surface process can be drawn.

References

1. Voorhoeve, R.J.H., Johnson, Jr., D.W., Remeika, J.P. and Gallagher, P.K., Perovskite oxides: materials science in catalysis, *Science* **195**: 827 (1977).
2. Viswanathan, B., Solid state and catalytic properties of rare earth orthocobaltites - New generation catalysts, *J. Sci. Ind. Res.* **43**: 151 (1984).
3. Ichimura, K., Inoue, Y. and Yasumori, I., Catalysis by mixed oxide perovskites. I. Hydrogenolysis of ethylene and ethane on $LaCoO_3$, *Bull. Chem. Soc. Jap.* **53**: 3044 (1980).
4. Yamazoe, N., Teraoka, Y. and Seiyama, T., TPD and XPS study on thermal behavior of adsorbed oxygen in $La_{1-x}Sr_xCoO_3$, *Chem. Lett.* 1767 (1981).
5. Fierro, J.L.G. and Tejuca, L.G., Non-stoichiometric surface behavior of $LaMO_3$ oxides as evidenced by XPS, *Appl. Surf. Sci.* **27**: 453 (1987).
6. Tejuca, L.G., Fierro, J.L.G. and Tascón, J.M.D., Structure and reactivity of perovskite-type oxides, *Adv. Catal.* **36**: 385 (1989).
7. Lombardo, E.A., Tanaka, K. and Toyoshima, I., XPS characte-

rization of reduced $LaCoO_3$ perovskite, *J. Catal.* **80:** 340 (1983).

8. Tabata, K., Matsumoto, I. and Kohiki, S., Effect of thermal treatment on catalytic properties of $La_{0.9}Ce_{0.1}CoO_3$, *J. Mat. Sci.* **22:** 1882 (1987).
9. Parravano, G., Ferroelectric transition and heterogeneous catalysis, *J. Chem. Phys.* **20:** 342 (1952).
10. Voorhoeve, R.J.H., Remeika, J.P. and Trimble, L.E., Defect chemistry and catalysis in oxidation and reduction over perovskite-type oxides, *Ann. NY Acad. Sci.* **272:** 2 (1976).
11. Galasso, F.S., *Structure, Properties and Preparation of Perovskite-Type Compounds*, Pergamon, Oxford, 1969.
12. Gysling, H.J., Monnier, J.R. and Apai, G., Synthesis, characterization and catalytic activity of $LaRhO_3$, *J. Catal.* **103:** 407 (1987).
13. Monnier, J.R. and Apai, G., Effect of oxidation states on the syngas activity of transition-metal oxide catalysts, *Preprints, Div. Petrol. Chem., Am. Chem. Soc.* **31:** 239 (1986).
14. Watson, P.R. and Somorjai, G.A., The hydrogenation of carbon monoxide over rhodium oxide surfaces, *J. Catal.* **72:** 347 (1981).
15. Watson, P.R. and Somorjai, G.A., The formation of oxygen-containing organic molecules by the hydrogenation of carbon monoxide using lanthanum rhodate catalyst, *J. Catal.* **74:** 282 (1982).
16. Broussard, J.A. and Wade, L.E., Catalytic conversion of syn gas with perovskites, *Preprints, Div. Fuel Chem., Am. Chem. Soc.* **31:** 75 (1986).
17. Brown-Bourzutschky, J.A., Homs, N. and Bell, A.T., Conversion of synthesis gas over $LaMn_{1-x}Cu_xO_3$ perovskites and related copper catalysts, *J. Catal.* **114:** 52 (1990).
18. Van Grieken, R., Peña, J.L., Lucas, A., Calleja, G., Rojas, M.L. and Fierro, J.L.G., Selective production of methanol from syngas over $LaTi_{1-x}Cu_xO_3$ mixed oxides, *Catal. Lett.* **8:** 335 (1991).
19. Rojas, M.L. and Fierro, Synthesis and characterization of $LaTi_{1-x}Cu_xO_3$ compounds, *J. Solid State Chem.* **89:** 299 (1990).
20. Tejuca, L.G. and Fierro, J.L.G., XPS and TPD probe techniques for the study of $LaNiO_3$ perovskite oxide, *Thermochim. Acta* **147:** 361 (1989).
21. Tejuca, L.G., Bell, A.T., Fierro, J.L.G. and Peña, M.A., Surface behaviour of reduced $LaCoO_3$ as studied by TPD of CO,

CO_2 and H_2 probes and by XPS, *Appl. Surf. Sci.* **31:** 301 (1988).
22. Clark, A., *The Theory of Adsorption and Catalysis*, Academic Press, New York, 1970, p. 360.
23. Dowden, D.A., Mackenzie, N. and Trapnell, B.M.W., The catalysis of H_2-D_2 exchange by oxides, *Proc. Roy. Soc. London, Ser. A* **237:** 245 (1956).
24. Dowden, D.A. and Wells, N., A crystal field interpretation of some activity patterns, in *Actes 2eme Congr. Inter. Catal.*, Technip, Paris, 1961, Vol. 2, p. 1499.
25. Dixon, G.M., Nicholls, G. and Steiner, H., Activity pattern in the disproportionation and dehydrogenation of cyclohexene to cyclohexane and benzene over the oxides of the first series of transition elements, in *Proceedings 3rd International Congress on Catalysis*, W.M.H. Sachtler, G.C.A. Schuit and P. Zwietering (Eds.), North-Holland, Amsterdam, 1965, Vol. 2, p. 981.
26. Boreskov, G.K., Catalytic activity of transition metal compounds in oxidation reactions, in *Proceedings 5th International Congress on Catalysis*, J.W . Hightower (Ed.), North-Holland, Amsterdam, 1973, Vol. 2, p. 981.
27. Bockris, J.O'M and Otagawa, T., Mechanism of oxygen evolution on perovskites, *J. Phys. Chem.* **87:** 2960 (1983).
28. Bockris, J.O'M and Otagawa, T., The electrocatalysis of oxygen evolution on perovskites, *J. Electrochem. Soc.* **131:** 290 (1984).
29. Wolfram, T., Hurst, R. and Morin, F.J., Cluster surface states for TiO_2, $SrTiO_3$ and $BaTiO_3$, *Phys. Rev. B* **15:** 1151 (1977).
30. Shimizu, T., Effect of electronic structure and tolerance factor on CO oxidation activity of perovskite oxides, *Chem. Lett.* 1 (1980).
31. Tascón, J.M.D. and Tejuca, L.G., Catalytic activity of perovskite-type oxides $LaMeO_3$, *React. Kinet. Catal. Lett.* **15:** 185 (1980).
32. Gallagher, P.K., Johnson, Jr., D.W. and Schrey, F., Studies of some supported perovskite oxidation catalysts, *Mat. Res. Bull.* **9:** 1345 (1974).
33. Nakamura, T., Misono, N. and Yoneda, Y., Reduction-Oxidation and catalytic properties of $La_{1-x}Sr_xCoO_3$, *J. Catal.* **83:** 151 (1983).
34. Gallagher, P.K., Johnson, Jr., D.W., Remeika, J.P., Schrey, F., Trimble, L.E., Vogel, E.M. and Voorhoeve, R.J.H., The

activity of $La_{0.7}Sr_{0.3}MnO_3$ without Pt and $La_{0.7}Sr_{0.3}MnO_3$ with varying Pt contents for the catalytic oxidation of CO, *Mat. Res. Bull.* **10:** 529 (1975).
35. George, S. and Viswanathan, B., Catalytic oxidation of CO on $La_{1-x}Sr_xCoO_3$ perovskite oxides, *React. Kinet. Catal. Lett.* **22:** 411 (1983).
36. Tascón, J.M.D., Fierro, J.L.G. and Tejuca, L.G., Kinetics and mechanism of CO oxidation on $LaCoO_3$, *Z. Phys. Chem. NF (Wiesbaden)* **124:** 249 (1981).
37. Le Coustumer, L.R., Bonnelle, J.P., Loriens, J. and Clerc, F., Catalytic activity of cobalt (3+) ions in octahedral sites for methane oxidation, *C. R. Acad. Sci. Paris, Ser. C* **285:** 49 (1977).
38. George, S. and Viswanathan, B., Carbon monoxide oxidation on $LnCoO_3$ perovskite oxides: effect of initial total pressure and gas composition, *Surface Technol.* **19:** 217 (1983).
39. Kremenic, G., López Nieto, J.M., Tascón, J.M.D. and Tejuca, L.G., Chemisorption and catalysis on $LaMO_3$ oxides, *J. Chem. Soc., Faraday Trans. I* **81:** 939 (1985).
40. Sazonov, L.A., Moskvina, Z.V. and Artamonov, E.V., Investigation of the catalytic properties of compounds of the $LnMeO_3$ type in the homomolecular exchange of oxygen, *Kinet. Catal.* **15:** 100 (1974).
41. Iwamoto, M., Yoda, Y., Yamazoe, N. and Seiyama, T., Study of metal oxide catalysts by temperature-programmed desorption. 4. Oxygen adsorption on various metal oxides, *J. Phys. Chem.* **82:** 2564 (1978).
42. Happel, J., Hnatow, M. and Bajars, L., in *Base Metal Oxide Catalysts*, Marcel Dekker, New York, 1977, p. 117.
43. Voorhoeve, R.J.H., Remeika, J.P., Trimble, L.E., Cooper, A.S., Disalvo, F.J. and Gallagher, P.K., Perovskite-like $La_{1-x}K_xMnO_3$ and related compounds: solid state chemistry and the catalysis of the reduction of NO by CO and H_2, *J. Solid State Chem.* **14:** 395 (1975).
44. Sorenson, S.C., Wronkiewicz, J.A., Sis, L.B. and Wirtz, G.P., Properties of $LaCoO_3$ as a catalyst in engine exhaust gases, *Ceram. Bull.* **53:** 446 (1974).
45. Ganguly, P., Catalytic properties of transition metal oxide perovskites, *Ind. J. Chem.* **15A:** 280 (1977).

Chapter 10

Total Oxidation of Hydrocarbons on Perovskite Oxides

T. Seiyama

Graduate School of Engineering Sciences, Kyushu University, Kasuga, Fukuoka 816, Japan

I. Introduction

Perovskite-type oxides containing transition metals are attracting great attention as catalysts for complete oxidation of hydrocarbons as well as electrochemical reduction of oxygen (1). As far as I know, the use of perovskite-type oxides as catalysts was first reported by Meadowcroft in 1970 (2) for the electrochemical reduction of oxygen. Soon after that, Voorhoeve et al. (3) reported the high catalytic activity of perovskite oxides for heterogeneous oxidation. These studies triggered many studies thereafter which are related to exhaust control catalysts and electrode catalysts.

In perovskite-type oxide, represented by ABO_3, the B site cation is surrounded octahedrally by oxygen, and the A site cation is located in the cavity made by these octahedra. In this oxide system, the replacement of A and/or B site cations by other metal cations often brings about the formation of lattice defects. Although such defects have been correlated with high catalytic activity, little is known of this relationship.

Perovskite-type oxides with A and/or B sites partly substituted desorb or adsorb a large amount of oxygen. The nature and reactivity of adsorbed oxygen may be greatly different from other oxygen which forms a rigid crystal lattice, because the adsorbed oxygen is more weakly bonded to metal cations than the normal lattice oxygen. Such weakly bonded oxygen is considered to be effective to complete combustion. Both the oxygen-sorptive properties of the compounds in relation with the defect structure and the role of absorbed oxygen and surface state in the catalytic activity will be mentioned.

II. General Aspects of Total Oxidation of Hydrocarbons on Various Perovskite Oxides

First of all, I will mention about a general aspect of unsubstituted perovskite oxide calysts (ABO_3). The activity of several perovskite oxides for catalytic oxidation of propylene with that of the component B oxides (BO_n) is compared in Fig. 1 (4). The activity is expressed in terms of temperature (T) at which the total oxidation of propylene takes place at a given rate. The plots below the straight line show that activity of B oxides is enhanced by forming a perovskite structure, while the

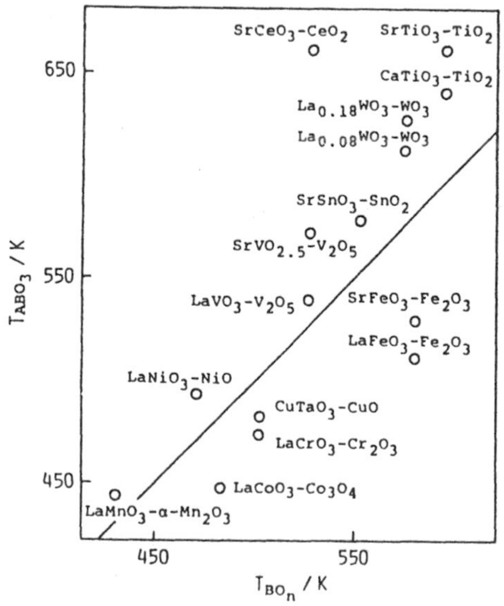

Fig. 1. Comparison of catalytic activities for propylene oxidation on perovskite-type oxides (ABO_3) and component B oxide (BO_n); T is the temperature at which the rate of propylene oxidation reaches $10^{-8.5}$ mol m^{-2} s^{-1}. From ref. 4.

negative effects are indicated by the plots above the straight line. As shown in the figure, some oxides such as $LaCoO_3$ and $LaFeO_3$ show positive effects. Roughly speaking, however, the plots are distributed near the straight line. This shows that the activity of unsubstituted perovskite oxides is determined mainly by component B oxides, and that the most active catalysts are those which contain Co and Mn. (More interesting effects come out when the perovskite oxide is partly substituted for A or B sites, which will be stated later.)

Nextly, catalytic activity for methane combustion is also investigated over various perovskite-type oxides of transition metals. In every case, carbon dioxide is the sole product but carbon monoxide or partial oxidation products are scarcely detected during oxidation. Catalytic activities, surface areas, and apparent activation energies of various La-based perovskite-type oxides are summarized in Table 1 (5). The oxidation activity is evaluated as the temperature at which 50 % conversion of methane oxidation is attained. Combustion of methane was promoted and the apparent activation energy is lowered by the presence of catalysts, as compared with noncatalytic thermal reaction. The activity of Pt/alumina catalyst is higher than the perovskite systems with a single B site cation. This is partly a result of

Table 1. Methane Oxidation Activity of Perovskite-Type Oxides, ABO_3, and Pt/Alumina Catalysts (From Ref. 5)

Catalyst	Surface Area (m^2/g)	$T_{50\%}$[a] (K)	E_a[b] (kcal/mol)
$LaCoO_3$	3.0	798	22.1
$LaMnO_3$	4.0	852	21.8
$LaFeO_3$	3.1	844	18.2
$LaCuO_3$	0.6	945	23.8
$LaNiO_3$	4.8	975	19.4
$LaCrO_3$	1.9	1053	28.8
Pt(1wt%)Al_2O_3	146.5	791	27.6
Thermal Data		1107	61.5

[a] Temperature at which conversion level is 50%
[b] Apparent activation energy

the larger surface areas of supported metal catalysts versus those of perovskite-type oxides. The activities of $LaCoO_3$, $LaMnO_3$, and $LaFeO_3$ catalysts were quite close to that of Pt/alumina catalyst.

We can see in these two total oxidations that the catalytic activity is mainly dependent on component B oxides, and the activity sequence is similar to those of single B oxides. The formation of perovskite-type oxides, ABO_3, from oxide A and B does not always give rise to the positive effect in promoting catalytic activity, but the negative effect is sometimes observed.

Further investigation is focused on the three active perovskite systems. Many metal cations are known as components of perovskite-type oxides. The oxidation activity is often improved by substituting foreign cations for the A and/or B site. Table 2 summarizes the activities for methane oxidation on A site

Table 2. Oxidation of Methane on $La_{1-x}A_xBO_3$[a] and Pt/Alumina Catalysts

$La_{1-x}A_xBO_3$ B site	A site	x	Surf. area (m^2/g)	E_a[b] (kcal/mol)	$T_{50\%}$[c] (K)
		0	3.0 (0.3)	22.1 (13.8)	798 (911)
	Sr	0.2	4.7 (0.5)	21.3 (15.8)	791 (925)
		0.4	3.6 (0.4)	19.0 (25.4)	843 (938)
Co	Ba	0.2	5.1 (0.4)	16.9 (16.1)	808 (945)
	Ca	0.2	2.0 (0.6)	18.1 (14.6)	879(1015)
	Ce	0.2	3.1 (0.9)	19.7 (15.5)	772 (935)
		0	4.6 (0.4)	21.3 (12.6)	852 (977)
Mn	Sr	0.2	8.6 (0.7)	19.7 (15.7)	783 (891)
		0.4	3.3 (0.3)	20.1 (18.7)	755 (856)
	Ca	0.2	6.7 (0.6)	18.9 (17.8)	814 (922)
Fe	Sr	0	3.1 (0.5)	18.2 (14.2)	844 (970)
		0.2	4.7 (0.4)	17.8 (15.4)	815 (937)
Pt(1wt%)/Al_2O_3			146.5 (38.5)	27.6 (38.5)	791 (851)

[a] Samples calcined at 1123 K, in parentheses samples calcined at 1473 K
[b] Apparent activation energy
[c] Temperature at which conversion level is 50 %

substituted perovskite-type oxides. The activities of perovskite-type oxides are strongly affected and Sr^{2+}-substituted lanthanum manganate, $La_{0.6}Sr_{0.4}MnO_3$, is the most active for methane oxidation.

Temperature dependencies of methane oxidation over La-Mn oxide system and Pt/alumina catalyst are shown in Fig. 2 (5). Though $LaMnO_3$ was less active than the Pt/alumina catalyst, the catalytic activity increased as Sr ions were substituted for La. Or the perovskite-type oxides, $La_{0.8}Sr_{0.2}MnO_3$ exhibited an almost similar level of activity to that of Pt/alumina in a temperature range between 623 and 823 K. However, the $La_{0.8}Sr_{0.2}MnO_3$ sample became less active than Pt/alumina above 873 K. Although the $T_{50\%}$ values of the two catalysts were almost the same, the slope of activity increase was more gradual at a high conversion level in the case of perovskite catalyst. Suppression of activity increase was more or less observed for every perovskite-type oxide at elevated temperatures. The difference in reaction pattern between the perovskite-type oxides and Pt/alumina at the high conversion level is further investigated in the latter section by kinetic analyses of the reaction rate.

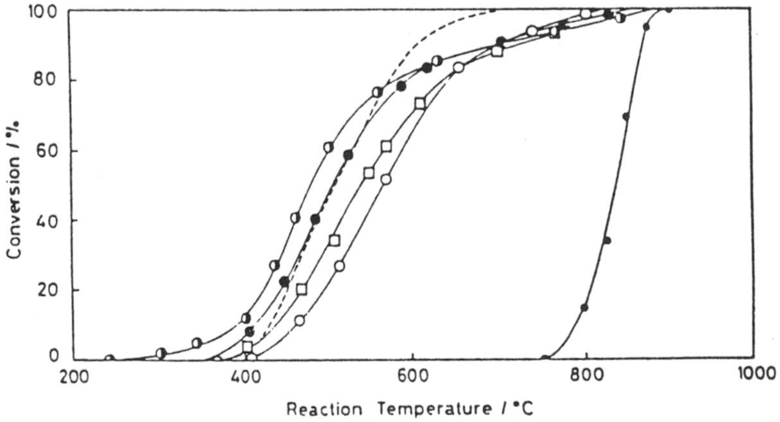

Fig. 2. Methane oxidation activity on $La_{1-x}A_xMnO_3$ catalysts: ○, $LaMnO_3$; ●, $La_{0.8}Sr_{0.2}MnO_3$; ◑, $La_{0.6}Sr_{0.4}MnO_3$; □, $La_{0.8}Ca_{0.2}$ MnO_3; ---, Pt(1wt%)/Al_2O_3; •, thermal data. Perovskite-type oxides and Pt/alumina were calcined at 1123 and 773 K, respectively, for 5 h. Reactant, 2 vol % methane in air; space velocity = 45.000-50.000 h^{-1}. From ref. 5.

III. Sorption and Desorption of Oxygen in Perovskite Oxides

As stated above, some perovskite-type oxides, in particular $LaCoO_3$ and $LaMnO_3$ families, are attracting great attention as catalysts for the complete oxidation of hydrocarbons as well as for electrode processes. The properties of these oxides can be easily modified by the substitution of A site cations, as in the case of $La_{1-x}Sr_xCoO_{3-\delta}$. In this oxide system, the partial or total substitution of Sr^{2+} for La^{3+} leads to the formation of Co^{4+} (or positive hole) and/or oxygen vacancies. Their high catalytic activities are considered to be related to such a defective structure.

It has been reported that on cooling in oxygen or heating in vacuo, the oxide absorbs or liberates a large amount of oxygen (6). Probably, this phenomenon is related to the defective structure and also to the oxidation activity of the oxide. However, little is known on the nature of the absorbed oxygen in this oxide system. So, we tried to collect information on the thermal behavior and nature of absorbed oxygen by using temperature programmed desorption (TPD) technique and thermogravimetry.

As a result, it is found that the effects of partial substitution are seen most obviously on the oxygen-sorptive properties (7,8). TPD chromatograms of oxygen from Sr-substituted $LaCoO_3$ are shown in Fig. 3. Oxygen was preadsorbed by cooling the samples from 1073 K to room temperature in an oxygen atmosphere. All chromatograms are characterized by the appearance of two desorption peaks, α and β. α is a broad plateau-like peak appearing below ca. 1073 K, while β is a very sharp one centered around 1093 K. Curve 1 for the unsubstituted sample shows only a small β peak, while curves 2 to 4 for Sr-substituted samples additionally show α peaks in the lower temperature region. The amounts of oxygen desorbed at α and β peaks are presented in Table 3. The amounts of both α and β oxygen increase with an increase in Sr substitution. The surface coverages of these oxygens which are presented in parentheses far exceed unity except for a case of α oxygen from Sr-unsubstituted oxide. This indicates that both α and β oxygen are not only adsorbed on the surface but they are also absorbed in the bulk. The situation is quite similar when La ions are substituted by other alkali or alkali earth metals. We have confirmed that in every case A site substitution increases the α and β desorption of oxygen.

The effect of A site substitution is considered by

Table 3. The Amounts of Oxygen Desorbed from $La_{1-x}Sr_xCoO_{3-\delta}$

Catalyst	Surface area m^2g^{-1}	Amount of O_2 desorbed,[a] mol O_2 g^{-1}	
		$\alpha(\Theta_\alpha)$	$\beta(\Theta_\beta)$
$LaCoO_3$	2.2	2.7(0.3)	29.4(3.4)
$La_{0.8}Sr_{0.2}CoO_{3-\delta}$	5.2	81.6(3.9)	72.0(3.5)
$La_{0.6}Sr_{0.4}CoO_{3-\delta}$	5.2	207.3(9.1)	85.5(3.8)
$SrCoO_{2.5+\delta}$	13.4	348.0(6.5)	172.0(3.2)

[a] Θ_α and Θ_β denote surface coverages (in unit of surface monolayer). From ref. 4.

substituting Sr for La as an example. The substitution of divalent Sr for trivalent La requires charge compensation, which is achieved either by the formation of tetravalent Co or positive holes as shown in formula 2. Formula 3 is a mixture of the above two. In the X-ray photoelectron spectrum of $La_{1-x}Sr_xCoO_3$, two distinct peaks, O_A and O_L appeared at BE = 530.2 - 531.4 eV and BE = 528.2 eV, respectively, in O 1s region. Obviously the O_A signal is usually much broader than the O_L. We consider that this broadness reflects the heterogeneity of absorbed oxygen: the adsorbed oxygen is not in a uniform state but in broadly dispersed states. The absorbed oxygen with lower BE is liberated more easily than that with higher BE.

$$La^{3+}{}_{1-x}Sr^{2+}{}_xCo^{3+}{}_{1-x}Co^{4+}{}_xO_3 \quad (1)$$
$$La^{3+}{}_{1-x}Sr^{2+}{}_xCo^{3+}O_{3-x/2}Vo_{x/2} \quad (2)$$
$$La^{3+}{}_{1-x}Sr^{2+}{}_xCo^{3+}{}_{1-x+2\delta}Co^{4+}{}_{x-2\delta}O_{3-\delta}Vo_\delta \quad (3)$$

Thus we have to know how the two types of oxygen desorption, α and β, are associated with such defect formation. The β desorption is observed for the substituted as well as unsubstituted samples and similar in shape for all samples. This suggests that the β desorption is more specific to the B cation, though it is also affected by A site substitution. This suggestion was investigated by using various B cations. Results of TPD chromatograms of La-based oxides where B sites are 3d transition metals from Cr to Ni indicate that the desorption of β oxygen is ascribable to partial reduction of B site cations to lower valencies accompanied with the evolution of equivalent lattice oxygen.

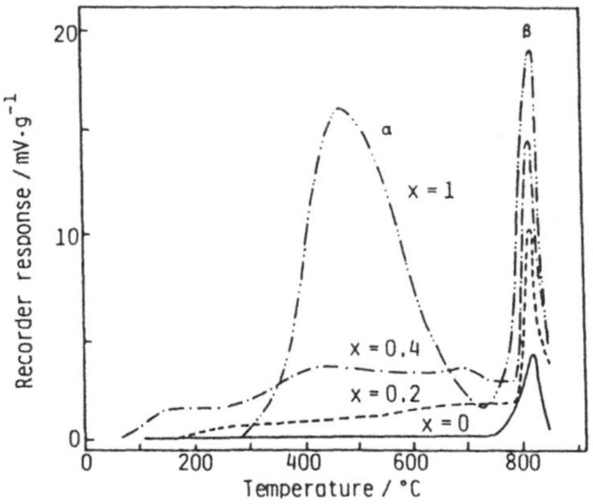

Fig. 3. TPD profiles of oxygen from $La_{1-x}Sr_xCoO_{3-\delta}$ after preadsorption of 100 Torr O_2 at 1073-300 K.

In order to elucidate the effect of A site substitution, Sr-substituted $LaCoO_3$, $LaFeO_3$, and $LaMnO_{3+\delta}$ are subjected to TPD experiments (Fig. 4) (10). As stated before, in the case of $LaCoO_3$, A site substitution increased α desorption. A similar phenomenon can be seen for Fe-perovskite. In the case of $LaMnO_{3+\delta}$, Sr substitution for 20% La simply decreases the β desorption, while that for 40% La brings about the appearance of a small α desorption accompanied by a further decrease of β desorption. In $LaCoO_3$ and $LaFeO_3$, the Sr substitution is expected to lead to the formation of oxygen vacancies. The observed promotion of α desorption by A site substitution suggests that α oxygen is accommodated in the oxygen vacancies and evolved in heating. On the other hand, $LaMnO_{3+\delta}$ is likely to accommodate excessive oxygen in the lattice. Since no interstitial site is available for excessive oxide ion, in the perovskite lattice, the charge of oxide ions is balanced by formation of Mn^{4+} and cationic defects at A and B sites. As stated before, the β oxygen desorption from this unsubstituted sample corresponds to the reduction of tetravalent Mn ions. In La-Mn perovskite, the partial substitution of Sr for La simply decreases these cation vacancies and Mn^{4+} concentration without formation of oxygen vacancies. This is why the 20% Sr substitution is not effective to promote α desorption. As the Sr substitution becomes higher, the formation of oxygen vacancies becomes easy and the appearance of α desorption occurs.

The nature of α and β oxygen described above was supported

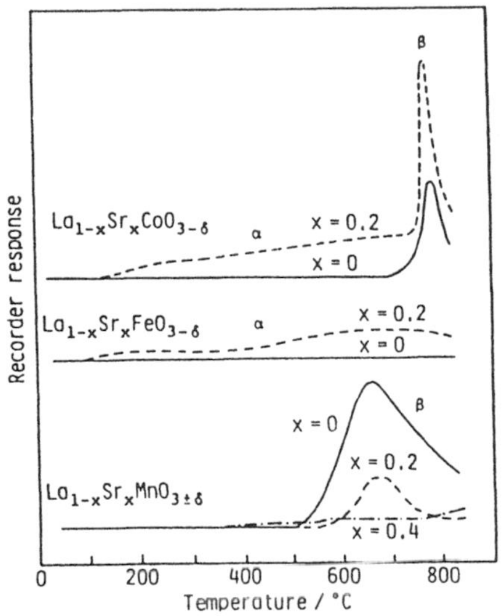

Fig. 4. TPD profiles of oxygen from $La_{1-x}Sr_xMO_{3+\delta}$ (x = 0, 0.2; M = Co, Fe, Mn) after preadsorption of 100 Torr O_2 at 1073-300 K.

by the determination of exact oxygen composition of $La_{1-x}Sr_x CoO_{3-\delta}$ (x=0, 0.2, 0.4, and 1) before and after the respective desorption peaks. The observed oxygen compositions and the corresponding oxidation states of Co ions estimated from the oxygen compositions are listed in Table 4 (9). After α desorption, Co ions take essentially normal oxidation states, namely trivalent with the disappearance of the tetravalent one in every case. This means that, at this stage, the electric charges resulting from Sr substitution are compensated solely by the formation of oxygen vacancies. Thus we understand that oxygen is dissociatively adsorbed and accommodated in these oxygen vacancies as α oxygen, resulting in the formation of Co^{4+} ions or positive holes. On the other hand, β desorption is accompanied by the partial reduction of trivalent Co ions to divalent. The amounts of α and β oxygen estimated from the above oxygen compositions agreed well with the results obtained from the TPD experiments.

In conclusion, α oxygen is accommodated in the oxygen vacancies formed by the partial substitution of A site cations, while β desorption is ascribable to the reduction of B cations to lower valencies. The uptake or α and β oxygen may be described by (4) and (5), respectively.

Table 4. Oxygen Compositions and Oxidation States of Co Ions in $La_{1-x}Sr_xCoO_{3-\delta}$

x	after O_2 sorption*			after α desorpt.			after β desorpt.		
	Co^{4+}	Co^{3+}	Co^{2+}	Co^{4+}	Co^{3+}	Co^{2+}	Co^{4+}	Co^{3+}	Co^{2+}
0	$LaCoO_{3.000}$			$LaCoO_{3.000}$			$LaCoO_{2.987}$		
	0	100%	0%	0%	100%	0%	0%	97.4%	2.6%
0.2	$La_{0.8}Sr_{0.2}CoO_{2.930}$			$La_{0.8}Sr_{0.2}CoO_{2.906}$			$La_{0.8}Sr_{0.2}CoO_{2.88}$		
	6%	94%	0%	1.2%	98.8%	0%	0%	96.4%	3.6%
0.4	$La_{0.6}Sr_{0.4}CoO_{2.895}$			$La_{0.6}Sr_{0.4}CoO_{2.796}$			$La_{0.6}Sr_{0.4}CoO_{2.75}$		
	19%	81%	0%	0%	99.2%	0.8%	0%	90%	10%
1	$SrCoO_{2.895}$			$SrCoO_{2.506}$			$SrCoO_{2.473}$		
	23.6%	76.4%	0%	1.2%	98.8%	0%	0%	94.6%	5.4%

Oxygen adsorption was carried out at 1073 K followed by cooling to room temperature; Co^{4+} may be replaced by "positive hole".

$$La_{1-x}Sr_xCo^{3+}O_{3-x/2}Vo_{x/2} + y/2O_2 \;{}^{\alpha}\; La_{1-x}Sr_xCo_{1-2y}Co_{2y}O_{3-x/2+y}Vo_{X/2-Y} \quad (4)$$

$$(\text{ or } V_{\ddot{o}} + 1/2O_2 \;{}^{\alpha}\; O_o^{x} + 2h^{\cdot} \;)$$

$$La_{1-x}Sr_xCo_{2z}Co_{1-2z}O_{3-x/2-z}Vo_{x/2+z} + z/2O_2 \;{}^{\beta}\; La_{1-x}Sr_xCo^{3+}O_{3-x/2}Vo_{x/2} \quad (5)$$

$$(\text{ or } 2Co_{Co} + V_{\ddot{o}} + 1/2O_2 \;{}^{\beta}\; 2Co_{Co} + Oo^{x} \;)$$

Fig. 5 shows the schematic model for such defective formations in $La_{1-x}Sr_xCoO_{3-\delta}$ and $LaMnO_{3+\delta}$ lattices. Both α and β oxygen species occupy the normal oxygen site in the perovskite lattice of the bulk. But the oxygen atoms in the vicinity of Co^{4+} are apt to desorb as α oxygen or react with molecules accompanied by a valence change of Co^{4+} to Co^{3+}. Such a desorption or a reaction

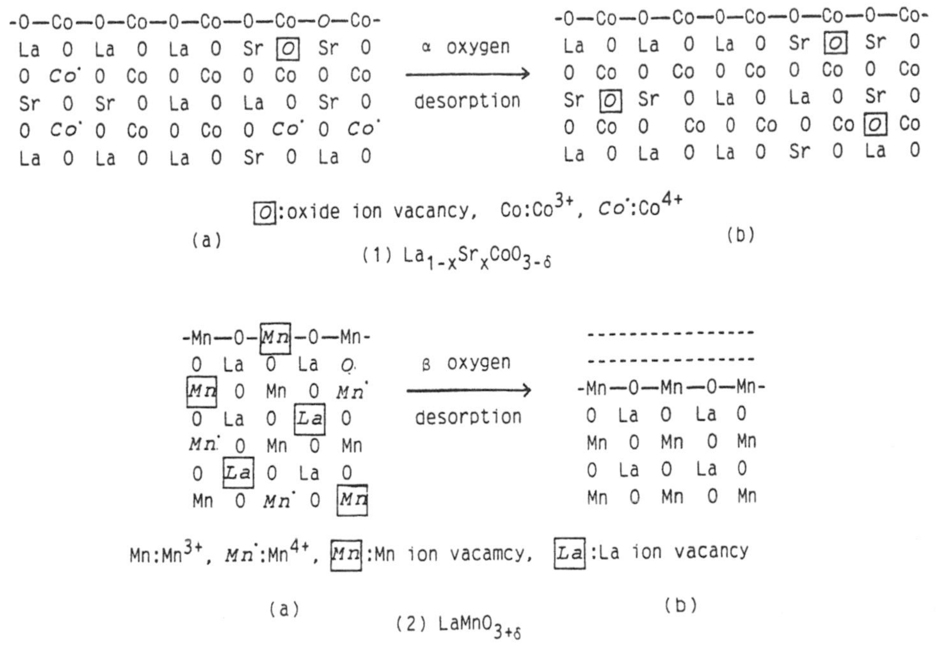

Fig. 5. Schematic models of oxygen desorption from $La_{1-x}Sr_xCoO_{3-\delta}$ (1) and $LaMnO_{3+\delta}$ (2).

scheme appears to be also applied to β oxygen.

In both Mn- and Co-based perovskite-type oxide systems, the catalytic oxidation is expected to be determined by formation of a high valence cation in the perovskite structure, i.e., tetravalent Co^{4+} or Mn^{4+}. Thus, the numbers of high valence cations more strongly influence methane oxidation activity than their oxygen sorptive properties.

IV. Hydrogen Oxidation by Sorbed Oxygen on Perovskite Oxides

The reactivity of absorbed oxygen was examined by pulse technique with hydrogen as a reducing reagent (7). Fig. 6 shows the TPD chromatograms of oxygen after various amounts of hydrogen were pulsed over the oxygen preadsorbed $La_{0.6}Sr_{0.4}Co\ O_{3-\delta}$ at 573 K. With increasing amounts of pulsed hydrogen, α oxygen is successively eliminated from the lower temperature region to higher temperature region, and finally β oxygen is also eliminated. This indicates that both α and β oxygen are reactive to hydrogen, but α oxygen is more active than β.

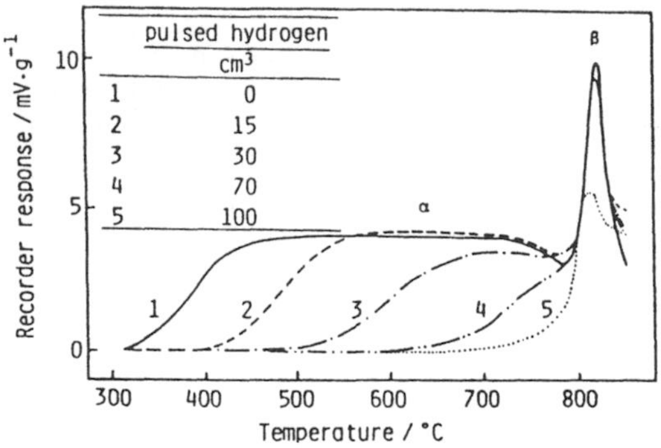

Fig. 6. TPD chromatograms of oxygen from $La_{0.6}Sr_{0.4}CoO_{3-\delta}$ after hydrogen pulse at 573 K. From ref. 4.

Experiments have been done to see how hydrogen uptake increases as hydrogen is repeatedly pulsed on various Sr substituted samples. The initial slopes of hydrogen uptake were taken as a measure of reactivity and plotted as a function of Sr concentration x in Fig. 7. Curve a is the rate of oxygen reduction vs. x thus obtained. It is estimated that a maximum reactivity is attained around the x value of 0.4. Curve c

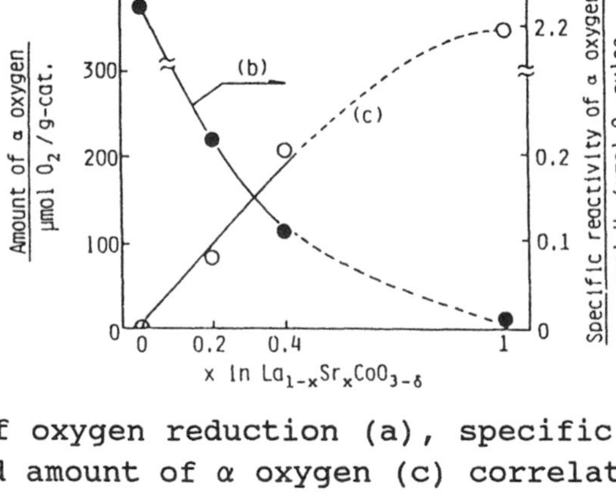

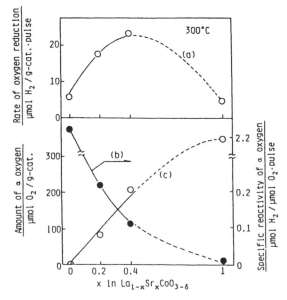

Fig. 7. Rate of oxygen reduction (a), specific reactivity of α oxygen (b), and amount of α oxygen (c) correlated with x.

represents the amount of α oxygen vs x, which increases with increasing x. If we assume that total reactivity of each sample (curve a) is given by the product of the amount of α oxygen and the specific reactivity of α oxygen, the specific reactivity of α oxygen is estimated to decrease with x as shown by curve b. In other words, total reactivity seems to result from these two factors, one increasing with x the other decreasing with x. The reactivity of α oxygen, and hence the reactivity of Co^{4+} or positive hole, are lowered as the number of α oxygen increases. α Oxygen appears to become thermodynamically stable with increasing x.

V. Detail of Total Oxidation of Hydrocarbons on Perovskite Oxides

The total reactivity for hydrogen oxidation, just discussed, seems to relate with the catalytic activity under oxidation conditions. The rate of n-butane oxidation over the samples with x = 0, 0.2, and 0.4 is shown in Fig. 8 (4). The highest catalytic activity is obtained for the sample of x = 0.4 followed by 0.2 and 0. This result shows the similar trend with the case of hydrogen oxidation. It is inferred that sorbed oxygen, especially α oxygen (or Co^{+4} in the oxide), is responsible for the high catalytic activities of the oxides. Thus we consider that the amount and reactivity of oxygen determines the catalytic activity. How to control sorbed oxygen seems to be important from the catalytic point of view.

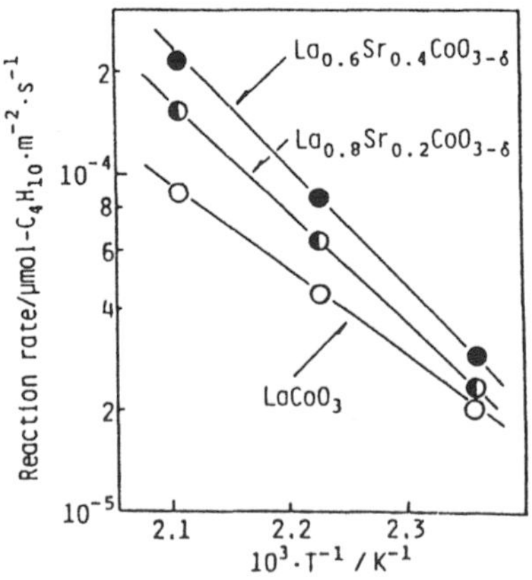

Fig. 8. Arrhenius plots of the rate of C_4H_{10} oxidation over $La_{1-x}Sr_xCoO_{3-\delta}$; W/F = 1.2 s x g cat/cm^3. From ref. 4.

On the other hand, the activity for total oxidation of methane is high over the $La_{1-x}Sr_xMnO_3$ system as mentioned in Section II. The activity in this case is determined by the occurence of a defective structure, such as Mn^{4+}. Furthermore, a thermal stability is one of the important properties for the catalysts of high temperature combustion. The activities for methane oxidation, examined in Section II were significantly lowered after calcination at 1472 K. Grain growth occurs at high temperatures, and, the surface areas of catalysts are significantly lowered after calcination at 1473 K and accompanies the decrease in catalytic activity.

Not only the activities were lowered, but the shapes of activity curve were changed by calcinig at 1473 K as compared with the samples prepared by calcination at 1123 K (Fig. 9). The activity of $LaMnO_3$ increased only gradually between 673 K and 1073 K. In the case of $La_{0.8}Sr_{0.2}MnO_3$, the increase in activity was strongly suppressed between 923 K and 1073 K, forming a plateau in a conversion curve at 60-80%. The activity again increased steeply above 1073 K and 100% conversion was attained at 1173 K.

The reaction kinetics of methane oxidation were examined by changing partial pressures of oxygen and methane (5). The results were expressed by the empirical rate equation

$$r = K\ P_{CH}{}^{m}\ P_{O}{}^{n} \tag{6}$$

where r is the oxidation rate of methane, and m and n are reaction orders for methane and oxygen partial pressure, respectively. Three catalysts in Table 3 showed the first order

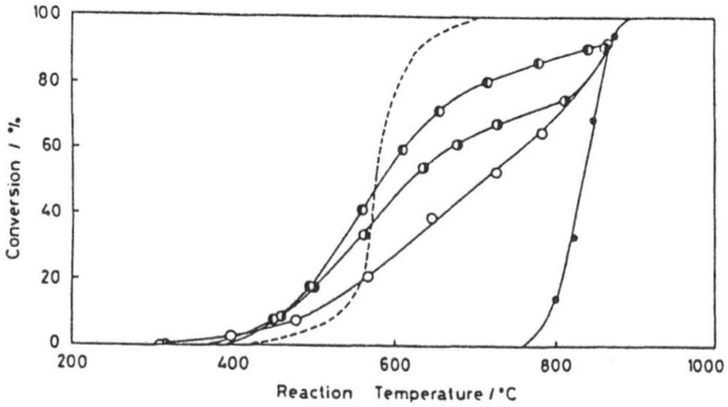

Fig. 9. Methane oxidation activity on $La_{1-x}Sr_xMnO_3$ catalysts: (○), $LaMnO_3$; (◐), $La_{0.6}Sr_{0.4}MnO_3$; (◑), $La_{0.8}Sr_{0.2}MnO_3$; (●), Thermal data; (---) Pt(1%)/Al_2O_3.

kinetics with respect to a partial pressure of methane. On the other hand, the exponents, n, for oxygen partial pressure were negative values for Pt/alumina, while those for two perovskite systems were positive values. The reaction order n for oxide systems tended to decrease as the reaction temperature rose.

The oxidation of hydrocarbon on noble metal catalysts generally proceeds by the surface reaction between adsorbed oxygen and methane. The kinetic data for Pt/alumina catalyst obeys the Langmuir-Hinshelwood type equation in which the surface reaction between adsorbed methane and adsorbed oxygen is the rate-determining step. Dissociatively adsorbed oxygen is assumed to be the active species.

$$r = \frac{k\, K_{CH}\, P_{CH}\, (K_O\, P_O)^{1/2}}{(1 + K_{CH}\, P_{CH} + (K_O\, P_O)^{1/2})^2} \qquad (7)$$

where K_O and K_{CH} are adsorption equilibrium constants of oxygen and methane, respectively, and k is a rate constant. The negative order kinetics of oxygen partial pressure in the empirical expression appears to be attributed to strong adsorption of oxygen on the metal surfaces, i.e., $(K_O\, P_O)^{1/2} >> K_{CH}\, P_{CH}$.

The oxidation kinetics on the perovskite-type oxides showed contrasting behavior with Pt/alumina. The reaction orders on the oxide catalysts were positive with respect to the partial pressure of oxygen. Although some perovskite-type oxides, such as $La_{0.8}Sr_{0.2}MnO_3$, were almost as active as Pt/alumina catalyst at low temperatures, these oxides became less active at high temperatures above 80% conversion (Section II).

As mentioned before, two kinds of oxygen species α and β with different bonding strength seem to coexist at the surface of perovskite-type oxides. The α oxygen, dissociatively adsorbed and accommodated in the oxygen vacancies, with the formation of neighboring high valent metal ions, is believed to be more active and reacts with hydrocarbons at lower temperatures than the original lattice oxygen. In this respect, the oxidation is understood essentially as a redox mechanism. At low temperatures, the oxidation rate largely depends on both P_{CH} and P_O. About half-order kinetics of P_O are characteristic of reaction with weakly bonded oxygen species which are in equilibrium with gaseous O_2. The reaction rate is expressed by the Rideal-Eley mechanism where adsorbed oxygen in the vacancy is assumed to react with gaseous methane.

$$r = \frac{k_a P_{CH} (K_O P_O)^{1/2}}{1 + (K_O P_O)^{1/2}} \tag{8}$$

However, the adsorption equilibrium constant, K_O, is expected to decrease with a rise in reaction temperature, as oxygen molecules desorb from the surface, and the reaction rate largely depends on the oxygen partial pressure. This expectation disagrees with the experimental fact that the reaction order of oxygen decreases with increasing temperature (Table 5).

To explain the observed temperature dependence and reaction kinetics, two oxygen species should be taken into account as active species. At low temperatures, the reaction is mainly operated by the adsorbed oxygen, when the activity of the lattice oxygen is negligibly low. With a rise in reaction temperature, not only α oxygen decreases but β oxygen, that is, normal lattice oxygen becomes reactive. As a result, the high temperature oxidation is dominantly operated by the lattice oxygen. In many oxide catalysts, the oxidation rate by the redox mechanism is zero order with respect to the oxygen partial pressure due to rapid incorporation of gaseous oxygen in the lattice.

$$r = k_1 P_{CH} \tag{9}$$

Table 5. Dependence of Methane Oxidation Rate on Partial Pressure of Methane and Oxygen ($r = KP_{CH4}^{m} PO_2^{n}$) (From Ref. 10)

Catalyst	Temp./K	m	n
Pt(1%)/Al_2O_3	723	0.9	-0.5
	923	1.1	-0.3
$La_{0.8}Sr_{0.2}MnO_3$	723	1.1	0.5
	823	0.9	0.3
	923	1.0	0.2
$La_{0.8}Sr_{0.2}CoO_3$	450	1.0	0.6
	550	1.0	0.4
	650	0.9	0.3

In perovskite-type oxides, the exponent, n, decreased at elevated temperatures. This phenomenon seems to correspond to the change in the active species from the adsorbed α oxygen to the lattice oxygen. The reaction by adsorbed oxygen greatly depended on the oxygen partial pressure, whereas that by lattice oxygen is zero order. The α oxygen contributes less to the reaction at elevated temperatures because of the decrease in content. Then, the lattice oxygen predominantly reacts with hydrocarbon molecules.

Based on the above consideration. The oxidation rates by the adsorbed oxygen and lattice oxygen are expressed by Eq. 8 and Eq. 9, respectively. The overall activity is given as a sum of two parallel reactions.

$$r = \frac{K_2 P_{CH4} (K_{O2} P_{O2})^{1/2}}{1 + (K_{O2} P_{O2})^{1/2}} + K_1 P_{CH4} \qquad (10)$$

At high temperatures, the reaction rate of the Rideal-Eley mechanism (Eq. 8) can be simply expressed as $r= K_a P_{CH4} (K_{O2} P_{O2})^{1/2}$, if the oxygen coverage is small ($1 >> (K_0\ P_0)^{1/2}$. Then Eq. 10 will be

$$r = K_a P_{CH4} (K_{O2} P_{O2})^{1/2} + K_1 P_{CH4} \qquad (11)$$

Contribution of each oxygen species can be separately estimated if the overall rate is plotted vs. $P_0{}^{1/2}$, since the reaction rate for adsorbed oxygen can increase in $P_0{}^{1/2}$, whereas that for lattice oxygen is zero order with respect to P_0 . Straight plots were obtained in Fig. 10 as expected at three temperatures. The first term in Eq. 11, which represents the contribution of adsorbed oxygen, is expressed by hatched areas in Fig. 10. It was proposed by Voorhoeve et al. (3) that oxidation on perovskite-type oxides are classified into two types, i.e., intrafacial reactions, where lattice oxygen becomes an active species, and suprafacial reactions, in which adsorbed or surface oxygen is an active species. The suprafacial reactions are believed to proceed at relatively low temperatures, though the intrafacial one is operative at high temperatures. In the present case, the reaction kinetics at low temperature are classified to the suprafacial one. In the whole temperature region, the α oxygen behaves like adsorbed oxygen in the rate equation. As the lattice oxygen largely contribute to the overall reaction at elevated temperatures, the kinetics are explained as an intrafacial

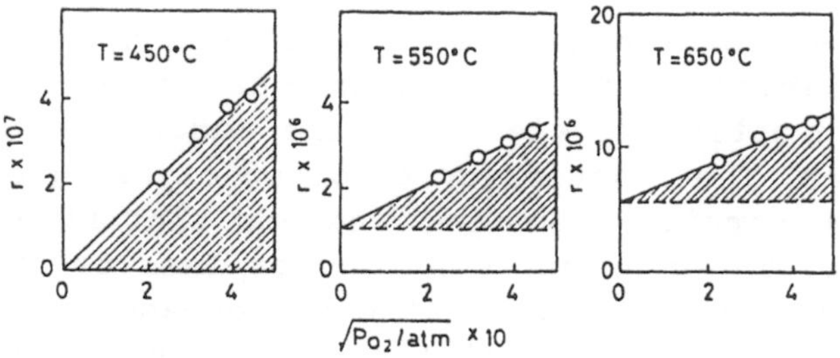

Fig. 10. Plots of methane oxidation rate, r, on $La_{0.8}Sr_{0.2}MnO_3$ vs. the oxygen pressure. Hatched areas denote contributions of adsorbed oxygen. From ref. 5.

reaction.

So far, we have been concerned with oxides which contain a single kind of B site cation. We refer to a possibility of modifying the properties of perovskite oxides with a combination of adequate B cations. For example, Fig. 11 shows TPD chromatograms of oxygen from complex oxides where the B site is occupied by Fe and Co ions at various compositions. By the combination of Fe and Co, the amounts of oxygen desorbed are increased, especially in the lower temperature region.

In coincidence with this change, the samples with mixed B site cations have been found to have catalytic activities for H_2O_2 decomposition in the liquid phase which are superior to the compounds with a single kind of B cation. In the methane oxidation reaction, the catalytic activity of the compound with

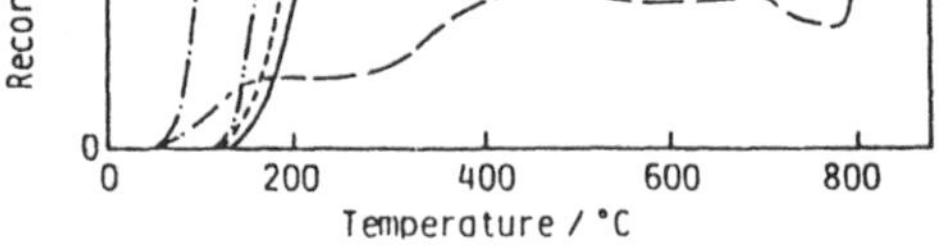

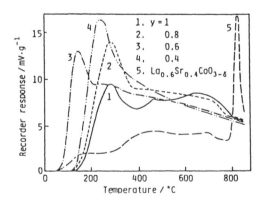

Fig. 11. TPD chromatograms of oxygen from $La_{0.2}Sr_{0.8}Fe_yCo_{1-y}O_{3-\delta}$ after exposure to 100 Torr oxygen at 1073 K followed by cooling to room temperature. From ref. 4.

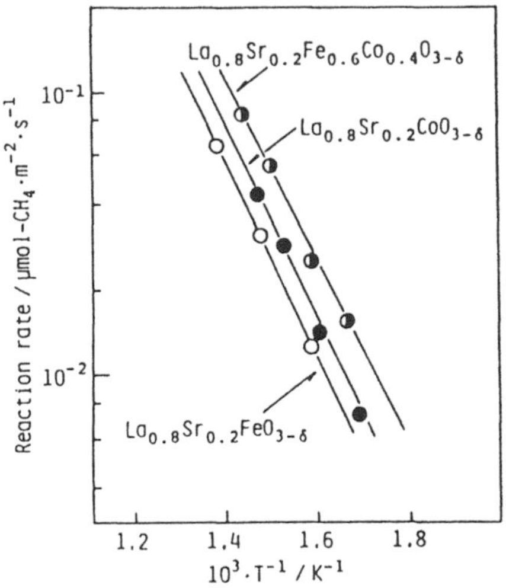

Fig. 12. Arrhenius plots of the rate of methane oxidation over $La_{0.8}Sr_{0.2}Fe_yCo_{1-y}O_{3-\delta}$. From ref. 4.

mixed B site cations is also higher than that of the compounds with mono B cation (Fig. 12) (11). It has been shown by many workers that the perovskite-type oxide which has only Co at the B site is one of the most active oxidation catalysts, while one which has only Fe is less active. Accordingly, the results obtained here suggest a possibility of finding more excellent catalysts by selecting a proper combination of B site cations.

I would briefly refer to the surface structure of perovskite-type oxides. According to our XPS measurements, the surface of Sr substituted $LaCoO_3$ is subjected to compositonal changes during high-temperature calcination. These changes can lead to a deactivation of catalytic properties of the compounds.

As mentioned so far, Co- and Mn-based perovskite-type oxides are very active for oxidation of hydrocarbons at low temperatures. For oxidation at elevated temperatures, e. g. above 1273 K, the magnetoplumbite-type oxides, such as $La_{1-x}Sr_xAl_{11}MnO_{19-\delta}$, are more suitable as catalysts because of their high thermal stability than the perovskite-type oxides (12). Cation-substitution and redox properties of oxides are also important in the magnetoplumbite-type oxides.

References

1. Voorhoeve, R. J. H., *Advanced Materials in Catalysis*, Academic Press, New York, 1977, p.129.

2. Meadowcroft, D. B., Low-cost oxygen electrode material, *Nature (London)* **226**: 847 (1970).
3. Voorhoeve, R. J. H., J.P. Remeika, P.E. Freeland and B.T. Mathias, Rare-earth oxides of manganese and cobalt rival platinum for the treatment of carbon monoxide in auto exhaust, *Science* **177**: 353 (1972).
4. Seiyama, T., N. Yamazoe and K. Eguchi, Characterization and activity of some mixed metal oxide catalysts, *Ind. Eng. Chem., Prod. Res. Dev.* **24**: 19 (1985).
5. Arai, H., T. Yamada, K. Eguchi and T. Seiyama, Catalytic combustion of methane over various perovskite-type oxides, *Appl. Catal.* **26**: 265 (1986).
6. Nakamura, T., N. Misono and Y. Yoneda, Reduction-oxidation and catalytic properties of perovskite-type mixed oxide catalysts ($La_{1-x}Sr_xCoO_3$), *Chem. Lett.* 1589 (1981).
7. Teraoka, Y., S. Furukawa, N. Yamazoe and T. Seiyama, Oxygen-sorptive and catalytic properties of defect perovskite-type $La_{1-x}Sr_xCoO_{3-\delta}$, *Nippon Kagaku Kaishi*, 1529 (1986).
8. Seiyama, T., Characterization of heterogeneous oxidation catalysts, *Oxidat. Commun.* **2**: 239 (1982).
9. Yamazoe, N., Y. Teraoka and T. Seiyama, TPD and XPS study on thermal behavior of absorbed oxygen in $La_{1-x}Sr_xCoO_3$, *Chem. Lett.* 1767 (1981).
10. Teraoka, Y., M. Yoshimatsu, N. Yamazoe and T. Seiyama, Oxygen-adsorption properties and defect structure of perovskite-type oxides, *Chem. Lett.*, 893 (1985).
11. Teraoka, Y., H. Zhang and N. Yamazoe, N., Oxygen-adsorption properties of defect perovskite-type $La_{1-x}Sr_xCo_{1-y}Fe_yO_{3-\delta}$, *Chem. Lett.*, 1367 (1985).
12. Machida, M., H. Kawasaki, K. Eguchi and H. Arai, Surface areas and catalytic activities of Mn-substituted hexaaluminates with various cation compositions in the mirror plane, *Chem. Lett.*, 1461 (1988).

Chapter 11

Hydrogenation and Hydrogenolysis of Hydrocarbons on Perovskite Oxides

K. Ichimura[1], Y. Inoue[2] and I. Yasumori[3]

[1] Department of Chemistry, Faculty of Science, Kumamoto University, Kurokami, Kumamoto 860, Japan

[2] Department of Chemistry, Nagaoka University of Technology, Nagaoka, Niigata 940-21, Japan

[3] Department of Applied Chemistry, Faculty of Engineering, Kanagawa University, Rokkakubashi, Yokohama 221, Japan

I. Introduction

A mixed oxide is one of a group of potentially important catalysts in heterogeneous catalysis, because different catalytic behavior would be expected to emerge with a combination of their components. For the fundamental research on the catalytic properties of the mixed oxides, a perovskite-type oxide with a chemical formula of ABO_3 has been used as a suitable catalyst because of their well-defined and stable structure. For some perovskite-type oxides, their electronic structures have been pointed out to be similar to those of transition metals on the basis of theoretical calculations of valence band structures (1-4). Thus, one of the catalytic features of perovskite-type oxides is to be active for hydrogen-involving reactions such as the equilibration between hydrogen and deuterium, the hydrogenation and isomerization of olefins, and in particular the hydrogenolysis of hydrocarbons (5-11).

Since Pedersen and Libby (5) first reported that $ACoO_3$ (A = lanthanide elements) compounds are more active than some transition metals for the hydrogenolysis of cis-2-butene, many studies have been done on $LaCoO_3$. Yasumori and his coworkers showed that $LaCoO_3$ has a catalytic function of promoting the hydrogenolysis of various alkanes from ethane to neopentane and producing predominantly methane in all cases (12-15), while the stability of $LaCoO_3$ oxide during ethylene hydrogenolysis was discussed by Crespin and Hall (16), and Lombardo et al. (17-18). In addition, the catalytic behavior of Co-substituted perovskite oxides was studied by Yasumori et al. (15), who found variable catalysis in hydrogen involving reactions when B ions are properly changed.

This review mainly deals with the hydrogenation of alkenes (C_2-C_4) and hydrogenolysis of alkanes (C_2-C_5) on $LaCoO_3$, different catalytic behavior of $LaAlO_3$ and $LaFeO_3$, and the structures of active sites of the perovskite oxides.

II. Kinetic Behavior of Hydrogenation of Alkenes and Hydrogenolysis of Alkanes on $LaCoO_3$

The hydrogenation and hydrogenolysis of C_2-C_5 hydrocarbons proceed according to the following reaction scheme (except for the butene isomers):

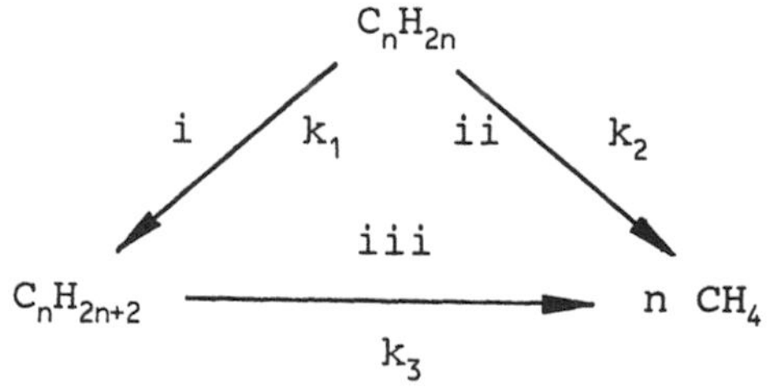

(1)

Processes i, ii and iii represent the hydrogenation of alkenes, the direct hydrogenolysis of alkenes, and hydrogenolysis of alkanes, respectively. By assuming that the reaction order with respect to the partial pressure of the respective hydrocarbons is unity, the following rate equation can be derived for the consumption of hydrocarbons under the conditions of an excess of hydrogen:

$$\frac{d\ P_{alkene}}{dt} = -\ (k_1 + k_2)\ P_{alkene} \qquad (2)$$

$$\frac{d\ P_{alkane}}{dt} = k_1\ P_{alkene} - k_3\ P_{alkane} \quad (3)$$

$$\frac{d\ P_{methane}}{dt} = n\ (k_2\ P_{alkene} + k_3\ P_{alkane} \quad (4)$$

where k_i denotes the rate constant for process i in the scheme (1).

On a $LaCoO_3$ oxide, the reaction of ethylene (n = 2) with hydrogen leads to the production of merely ethane below 420 K, that is, the process i of hydrogenation proceeds predominantly, whereas above this reaction temperature the hydrogenolysis of ethylene to form methane occurs, in addition to the hydrogenation. Figure 1 shows changes in the production of methane and ethane at 573 K as a function of the reaction time. This behavior is represented by consecutive production of methane and ethane. The rate of ethylene and ethane consumption are

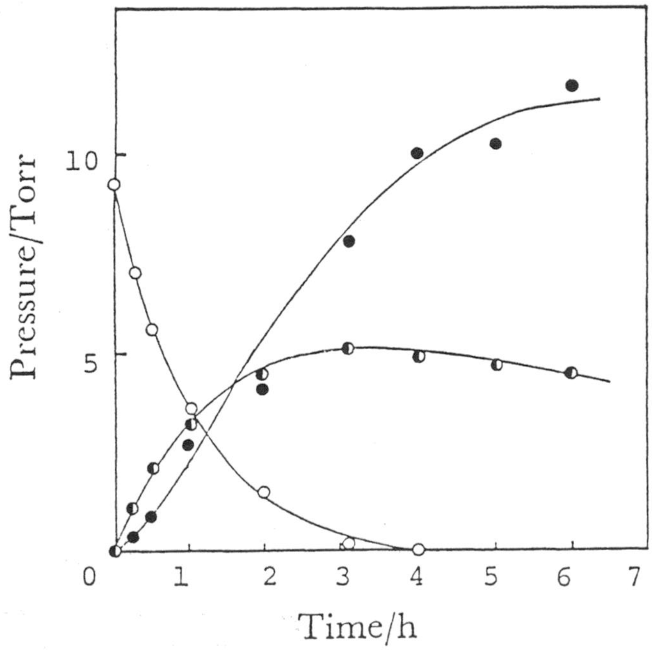

Fig. 1. Reaction of ethylene with hydrogen at 573 K (12): (○), C_2H_4; (◐), C_2H_6; (●), CH_4.

expressed in the pseudo-first order reaction as given by Eqs. (2), (3) and (4) with n = 2. The apparent rate constants, k_1, k_2 and k_3 are determined by applying the non-linear least square method to the experimental data; the calculated values are k_1 = 0.15 h^{-1}, k_2 = 0.014 h^{-1} and k_3 = 0.12 h^{-1}.

This analytical result indicates that the main reaction of ethylene hydrogenolysis proceeds via processes i and iii, that is, the direct conversion of ethylene to methane gives a minor contribution to the overall reaction. The solid curves in Fig. 1, which represent the change in the partial pressures with time by using the kinetic parameters determined above, show good agreement. A rate constant for the direct hydrogenolysis of ethane under similar reaction conditions has been already obtained by Ichimura et al. (12), who found that the value of k_3 is equivalent to the rate constant for ethane hydrogenolysis. The Arrhenius plot for ethane hydrogenolysis is compared in Fig. 2 with that for ethylene hydrogenation.

The hydrogenolysis of C_3-C_5 alkanes such as propane, butane, isobutane, pentane, isopentane and neopentane on $LaCoO_3$ produces only methane as a product. Neither the corresponding alkanes nor the fragmented hydrocarbons are formed in the gas phase. The rates of hydrocarbon consumption fell within the range of magnitude 0.5-1.5 x 10^{13} molecule $s^{-1}cm^{-2}$, except for the case of

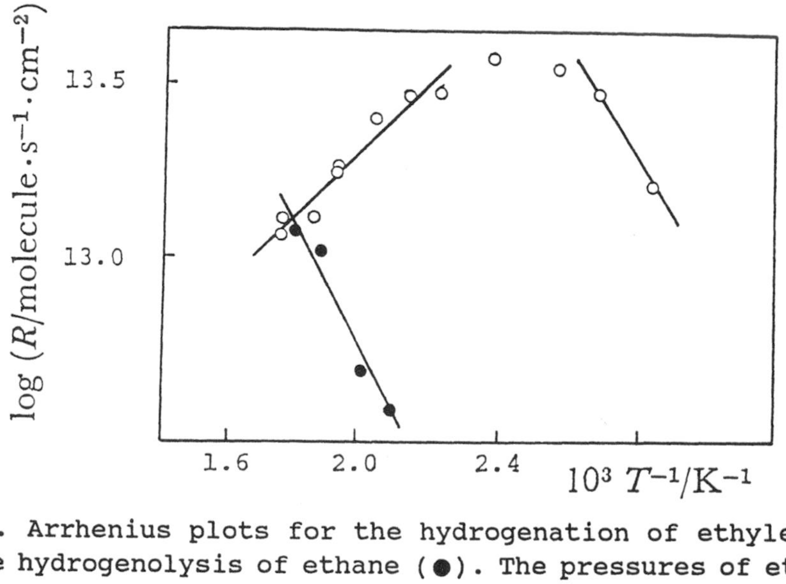

Fig. 2. Arrhenius plots for the hydrogenation of ethylene (○) and the hydrogenolysis of ethane (●). The pressures of ethylene, ethane and hydrogen were 10, 10 and 100 Torr, respectively. (reproduced with kind permission of Chem. Soc. Japan, Ref. 12).

Table 1. Kinetic Parameters for the Hydrogenolysis of C_2-C_5 Alkanes on $LaCoO_3$ (14)

Reaction			Order[a] m	n	E_a[b] kJ/mol	Rate[c] molec/s cm^2
$C_2H_6 + H_2$	$\longrightarrow$	2 CH_4	1.0	-0.5	35	1.0 x 10^{13}
$C_3H_8 + 2\ H_2$	$\longrightarrow$	3 CH_4	1.0	0.0	120	1.5
$nC_4H_{10} + 3\ H_2$	$\longrightarrow$	4 CH_4	1.0	1.0	32	7.2
$iC_4H_{10} + 3\ H_2$	$\longrightarrow$	4 CH_4	1.0	1.0	71	0.5
$nC_5H_{12} + 4\ H_2$	$\longrightarrow$	5 CH_4	1.0	2.0	55	1.5
$iC_5H_{12} + 4\ H_2$	$\longrightarrow$	5 CH_4	1.0	2.0	38	1.0
$C(CH_3)_4 + 4\ H_2$	$\longrightarrow$	5 CH_4	1.0	2.0	102	0.9

[a] Estimated error ± 0.1 according to the rate equation: Rate = $k_e \cdot P_{alkane}^m \cdot P_{hydrogen}^n$(573 K); [b] estimated error ± 3 kJ/mol; [c] pressures of hydrogen and alkane 100 and 10 Torr, respectively; [d] See ref. 1.

butane, which gives a higher rate of 7 x 10^{13} molecule $s^{-1}cm^{-2}$. The activation energy and other kinetic parameters are summarized in Table 1. The reaction order with respect to the hydrocarbon pressure are unity for all reactions, while the hydrogen order increases with the number of carbon atoms involved in the reactant hydrocarbons, that is, 0.0, 1.0 and 2.0 in the sequence $C_3H_8 \longrightarrow C_4H_{10} \longrightarrow C_5H_{12}$. The activation energy is remarkably high for propane and neopentane, but low for ethane butane, pentane and isopentane hydrogenolysis. An intermediate value is obtained for isobutane hydrogenolysis.

In the reaction of alkenes with hydrogen, methane and the corresponding alkanes were produced. For example, the reaction of propene with hydrogen produces a considerable amount of methane, together with propane at the initial stage of reaction.

The reaction of butenes with hydrogen undergoes rapid isomerization, followed by the formation of butane and then methane. For all reactions, ratios of the produced butene isomers remained almost unchanged until the disappearance of the reactant butenes; the cis-2-butene/trans-2-butene, cis-2-butene/1-butene and trans-2-butene/1-butene ratios are 1.0, 1.8 and 1.8, respectively. Since there is no remarked difference in the observed rates of the hydrogenation of butenes, the average of these rates could be employed as that of butene hydrogenation. The kinetic analysis by the non-linear least square method gives the value of k_1 = 0.040, k_2 = 0.003 and k_3 = 0.95 h^{-1}, indicating that the fraction of the direct formation of methane from butene is about 7 %. As it is demonstrated in Tables 1 and 2, there also exists a good agreement in the kinetic parameters and reaction rates between process iii and butane hydrogenolysis. For the reaction with propene, the values analyzed by the above mentioned

Table 2. Kinetic Parameters for the Hydrogenolysis of Propene and Butenes (14)

Reaction	Path[a]	Rate[b] (molec/s cm^2)	Reac. order[c] m	n^d	E_a^e (kJ/mole)
$C_3H_6 + H_2$	I	0.18×10^{13}	1	1.0	-33
	II	0.96×10^{13}	1	1.0	52
	III	1.30×10^{13}	1	0.0	119
$C_4H_8 + H_2$	I	0.31×10^{13}	1	1.0	-33
	II	~ 0.03×10^{13}	1	-	-
	III	7.40×10^{13}	1	1.0	32

[a] See text; [b] Under standard reaction conditions: Reaction temperature = 573 K; $P^o_{hydrogen}$ = 100 Torr; P^o_{alkene} = 10 Torr; and P^o_{alkane} = 10 Torr; [c] Rate = $k'_e\ P^m_{hydrocarbon}\ P^n_{hydrogen}$; [d] Estimated error ± 0.1; [e] Estimated error ± 4.

method are $k_1 = 0.023$, $k_2 = 0.123$ and $k_3 = 0.167$ h^{-1}, which indicates that the fraction of the direct hydrogenolysis via process ii is about 84 %.

The hydrogenolysis of alkanes has mostly been studied for transition-metal catalysts (19); the kinetic behavior is characterized by negative orders with respect to hydrogen pressure and by a wide variety of fragmented hydrocarbons, except for the selective formation of ethane in the case of nickel catalysts (20,21). The proposed mechanism involves the dissociative adsorption of alkanes, liberating a hydrogen atom on the surface, followed by the step-by-step scission of carbon-carbon bonds. The most striking features in the hydrogenolysis on $LaCoO_3$ are not only the high reaction order with respect to hydrogen pressure but also the selective formation of methane; neither the other fragmented nor hydrogenated hydrocarbons are observed in the gas phase. The deuterium distributions observed in the reaction of alkanes with deuterium had a common feature that large amounts of methane (CHD_3 and CD_4) are involved even from the initial stage of hydrogenolysis, whereas the fractions of the deuterium-exchanged alkanes are negligibly small. These results indicate that the adsorption of alkanes is irreversible, but the surface species thus produced undergo a rapid hydrogen exchange.

The following mechanisms are proposed as the most plausible pathway for the hydrogenolysis of C_3-C_5 alkanes:

$$H_2 \rightleftharpoons 2\ H(a) \tag{5}$$

$$H_2 \rightleftharpoons 2\ H(a^*) \tag{6}$$

$$CnH_{2n+2} + 2(n-2)H(a) \rightarrow 2\ CH_3(a^*) + (n-2)[H\text{--}CH_2(a)\text{--}H] \tag{7}$$

$$CH_3(a^*) \rightleftharpoons CH_2(a^*) + H(a) \tag{8}$$

$$[H\text{--}CH_2(a)\text{--}H] \rightleftharpoons CH_3(a) + H(a) \tag{9}$$

$$CH_3(a) \rightleftharpoons CH_2(a) + H(a) \tag{10}$$

$$CH_3(a^*) + H(a) \rightarrow CH_4 \tag{11}$$

$$[H\text{--}CH_2(a)\text{--}H] \rightarrow CH_4 \tag{12}$$

where symbols (a*) and (a) represent the adsorbed species on the Co(III) ion and those on the other sites. The surface Co(III) favors the adsorption of the methyl group rather than the methylene group. This is not unexpected, since a similar selective adsorption occurs on the Cr(III) ion in α-Cr_2O_3 (22). The formation of highly deuterium-exchanged methane is due to the rapid hydrogen exchange between monocarbon species, such as methyl and carbene, as shown in Steps (8), (9) and (10). The presence of surface carbene as an intermediate is possible, since this species is confirmed in the adsorption of acetylene on the nickel (110) surface (23).

From the deuterium distributions in methane and the reactants, Step (7) is proposed to be rate-determining. Provided that the requisite number of the site on the trivalent cobalt ion is two and a negligibly small fraction of the sites is occupied by hydrocarbon species, the following rate equation is given:

$$R = k_7 \cdot P_{alkane} \cdot \Theta_V^{*2} \cdot \Theta_H^{2(n-2)}$$

where Θ_V^* denotes the fraction of the vacant Co(III) site and Θ_H the fraction of the La(III) and O(II) sites occupied by hydrogen atoms. Thus, this reaction can be written as:

$$R = k_7 \cdot P_{alkane} \cdot \frac{(K_5 \cdot P_{hydrogen})^{n-2}}{(\sqrt{1+K_6 \cdot P_{hydrogen}})^2 (1+\sqrt{K_5 \cdot P_{hydrogen}})^{2(n-2)}}$$

Under the conditions of a strong hydrogen adsorption on cobalt ion sites and a weak adsorption on the other sites, the equation is simplified as :

$$R = k_7 \cdot (K_5^{n-2}/K_6) \cdot P_{alkane} \cdot P_{hydrogen}^{n-3} \qquad (13)$$

For C_3–C_5 alkanes, this equation represents the observed pressure dependences of the rates. For ethane hydrogenolysis, the rate-determining Step (7) is described as:

$$C_2H_6 + H(a) \longrightarrow CH_3(a^*) + CH_3(a^*){-}{-}{-}H(a) \qquad (14)$$

By the same procedure mentioned above, the rate equation of ethane hydrogenolysis is given by:

$$R_{C2} = k_{14} \cdot (\sqrt{K_5}/K_6) \cdot P_{ethane} \cdot P_{hydrogen}^{-0.5}$$

III. Comparative Studies of $LaCoO_3$, $LaFeO_3$ and $LaAlO_3$

The principal difference in the catalytic properties of $LaCoO_3$ and the other two perovskites, $LaFeO_3$ and $LaAlO_3$, is that only $LaCoO_3$ catalyses the hydrogenolysis of alkanes. Fig. 3 shows the Arrhenius plots of ethylene hydrogenation on these three perovskite oxides. The striking common feature is that the reaction exhibits a maximum rate at a given temperature, which depends on the nature of each catalyst: ca. 420 K for $LaCoO_3$ and $LaFeO_3$, and 490 K for $LaAlO_3$. Such behavior has been observed for the same reaction on nickel (24,25). The kinetic parameters of the reaction in the high and low temperature regions are summarized in Table 3. The distributions of deuterium atoms in the reactants and products when deuterium was used instead of hydrogen are given in Table 4. Highly deuterium exchanged ethanes can be seen for the reaction in both temperature regions, whereas there is a difference in the deuterium distributions among gaseous hydrogen isotopes; an equilibrium is nearly established in the high temperature region and much smaller amounts of HD and H_2 are produced in the lower temperature region. These isotope distributions indicate that the hydrogenation proceeds via an ethyl radical as intermediate and the rate determining step varies from the adsorption of hydrogen in the low temperature

Table 3. Kinetic Parameters for Ethylene Hydrogenation (15)

Catalyst	Surface Area (m^2/g)	Reaction Orders[a] m	n		log V^o [b]	E_a (kJ/mol)
$LaFeO_3$	0.40	0	1	(333 K)	13.2	92
		1	1	(573 K)	(420 K)	-62
$LaAlO_3$	7.9	0	1	(353 K)	11.6	7
		1	1	(573 K)	(490 K)	-85
$LaCoO_3$[a]	0.33	0	1	(353 K)	13.6	34
		1	1	(573 K)	(420 K)	-19

[a] $V = k.P_{ethylene}^{m}.P_{hydrogen}^{n}$; [b] The maximum initial rate (molecule. $s^{-1}.cm^{-2}$); [c] Taken from ref. 12.

Table 4. Deuterium Distribution in C_2H_4 + H_2 Reaction System

High temp. region	Ethylene			Ethane			Hydrogen		
	$LaFeO_3^a$ (523K)	$LaAlO_3^a$ (573K)	$LaCoO_3^b$	$LaFeO_3$	$LaAlO_3$	$LaCoO_3$	$LaFeO_3$	$LaAlO_3$	$LaCoO_3$
D_0	11	19	31.9	0	5	3.3	2	0.3	0.2
D_1	31	30	28.5	62	12	10.6	24	9.5	8.4
D_2	33	38	38.5	15	20	14.3	74	90.2	91.4
D_3	20	12	0	11	24	19.5			
D_4	5	1	1.1	8	21	24.9			
D_5				4	11	16.4			
D_6				0	7	11.1			
Low temp. region	$LaFeO_3$ (333K)	$LaAlO_3$ (353K)	$LaCoO_3$	$LaFeO_3$	$LaAlO_3$	$LaCoO_3$	$LaFeO_3$	$LaAlO_3$	$LaCoO_3$
D_0	61	32	19.5	45	4	11	0.7	0.5	1
D_1	31	38	42.8	33	26	25	2.8	3.1	4
D_2	4	24	28.8	15	32	32	96.5	96.4	95
D_3	3	5	8.1	5	20	22			
D_4	1	1	0.8	2	14	9			
D_5				0	3	2			
D_6				0	1	0			

[a] Conversion = 16 %; [b] Conversion = 24 % (573 K), 42 % (353 K), Taken from ref. 12.

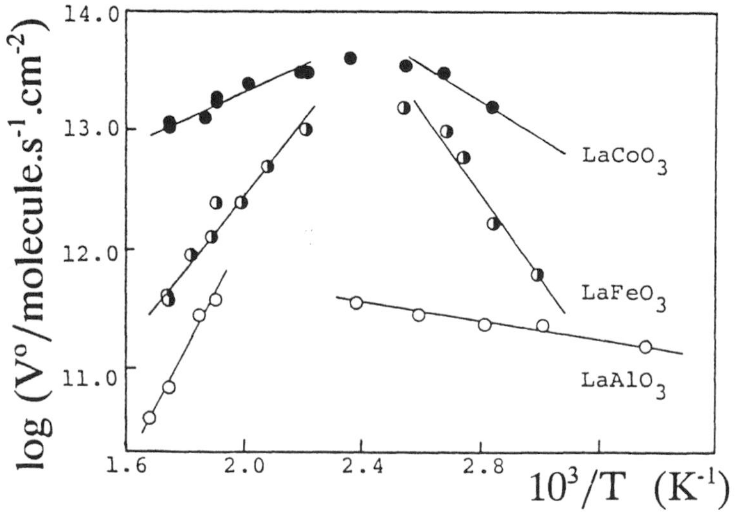

Fig. 3. Arrhenius plots of ethylene hydrogenation. From ref. 15.

region to the hydrogenation of ethyl radical in the high temperatures. It is rational to consider that the different temperature dependence described above is associated with this transfer in the rate determining step. The value of the activation energy differed markedly each other for the respective reactions on the three perovskites, depending upon the nature of the component cations.

Figure 4a shows the thermal desorption spectra of hydrocarbons from $LaCoO_3$ exposed to ethane at 300 K; single peaks of C_2H_2 and C_2H_4 appear at ca. 380 K and two peaks of C_2H_6 at ca. 380 and 540 K. Methane is observed above 450 K. Almost the same thermal desorption spectra are obtained from the surface exposed to C_2H_4, although the fraction of methane formed is smaller. The thermal desorption spectra of hydrogen from $LaCoO_3$ showed two peaks, referred to as α-peak appearing below 340 K and β-peak above this temperature. When the surface retaining β-hydrogen was exposed to C_2H_6 at 298 K for 10 min, observed changes in the thermal desorption spectra were the complete disappearance of C_2H_2 and C_2H_4 peaks and the considerable enhancement of C_2H_6 and CH_4 peaks without any shift in their positions (Fig. 4b). The consumption of the β-hydrogen located at the lower temperature

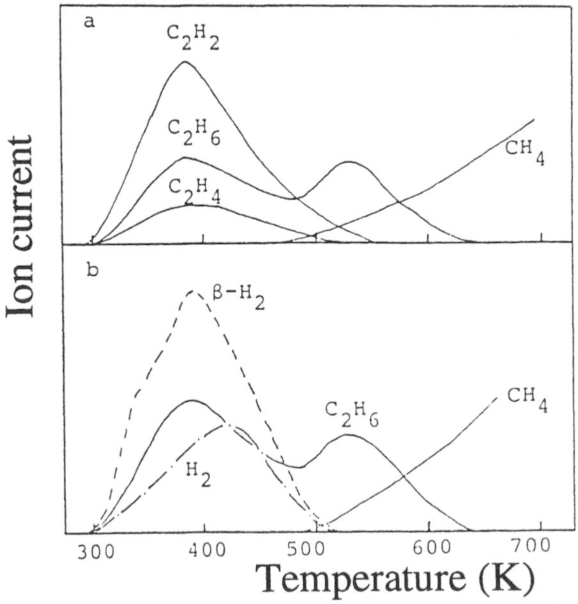

Fig. 4. Thermal desorption spectra of adsorbed C_2H_6 on $LaCoO_3$ (top) and hydrogen-preadsorbed $LaCoO_3$ (bottom). From Ref. 15.

side of β occurred as a consequence of contact with C_2H_6, thus suggesting the species to be responsible for the hydrogenolysis and also for the hydrogenation of C_2H_4 and C_2H_2.

Hydrogen was desorbed from the $LaFeO_3$ and $LaAlO_3$ surface, and their thermal desorption spectra consisted of α and β peaks similar to those observed for $LaCoO_3$. However, the thermal desorption spectra of C_2H_4 and C_2H_6 were relatively simple; only the original molecules are observed. The hydrogen-preadsorbed $LaFeO_3$ and $LaAlO_3$ surfaces convert the subsequently adsorbed ethylene only to ethane. On these surfaces, β-hydrogen is also consumed to a large extent.

Figure 5 compares the X-ray photoelectron spectra in the valence band region of the three perovskite catalysts. The characteristic feature is that $LaCoO_3$ is different from the other two oxides in giving rise to a sharp peak at 0.3 eV below the

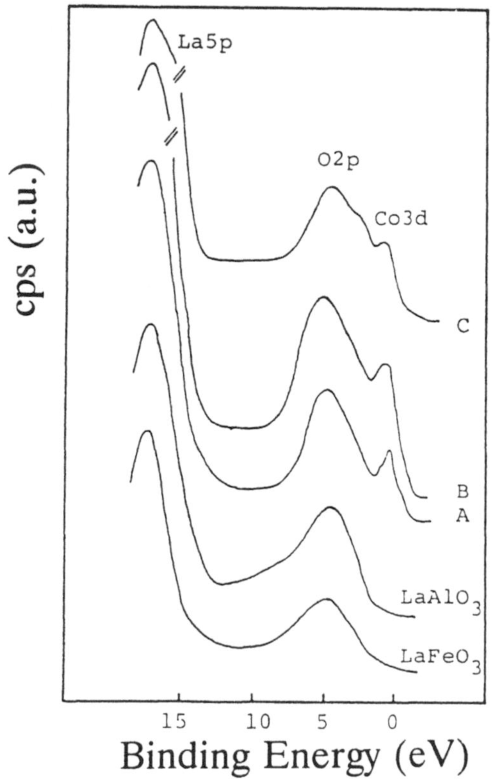

Fig. 5. X-ray photoelectron spectra in the valence band region (15). A, fresh $LaCoO_3$; B, 50%-deactivated $LaCoO_3$; C, 100%-deactivated $LaCoO_3$.

Fermi level, which is assigned to Co 3d level. The X-ray photoelectron spectra of core levels of $LaCoO_3$ surfaces showed interesting electronic states in relation to the deactivation of the catalytic activity. With the O 1s spectra for $LaCoO_3$ surfaces evacuated at 300 K there were two kinds of oxygen species; the peak of the lower binding energy is associated with the lattice oxygen and shifted to a higher energy side with decreasing activity until it becomes similar to that of cobalt oxides (Fig. 6). The O 1s peak of higher binding energy diminishes in intensity to the level shown by the dotted lines upon evacuation at 770 K. From the comparison with spectra of Cu_2O, CuO and NiO obtained by Robert et al. (26), this peak is assigned to adsorbed oxygen.

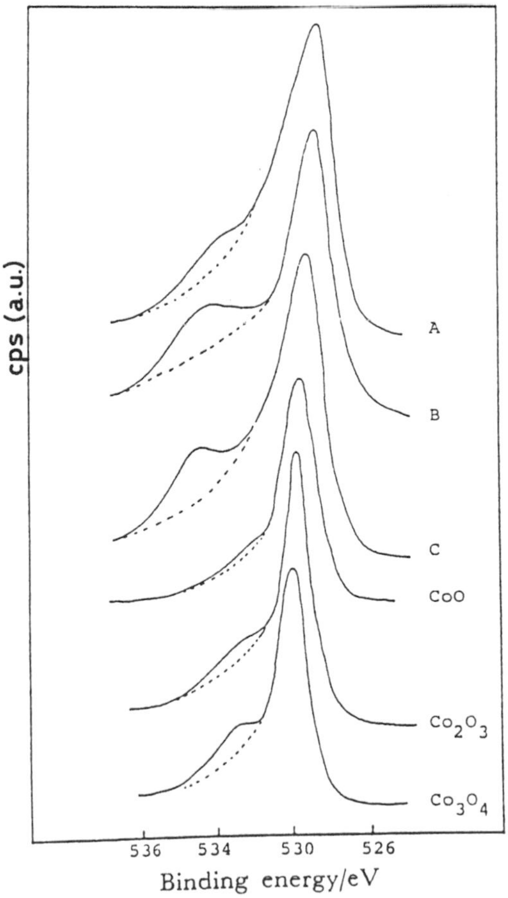

Fig. 6. X-ray photoelectron spectra in O 1s region (12). Spectra A, B and C: see caption in Fig. 5.

As shown in Fig. 7, the values of spin-orbit splitting and of full width at half maximum (FWHM) of Co 2p peaks are respectively 15.3 and 3.2 eV for the fresh $LaCoO_3$ (A), and 15.9 and 4.7 eV for the completely deactivated one (C). The value of the splitting for the (A) is similar to that for Co_2O_3, 15.2 eV. The values for the (C) are almost the same as those for CoO, 16.0 and 4.7 eV. These differences can be used to distinguish the oxidation states of cobalt (27-29), and lead to the conclusion that the reduction from a trivalent ion Co(III) to a bivalent ion Co(II) occurs, resulting in the structural change and the degradation of catalytic activity. On the other hand, no significant change of La 3d spectra is observed for all the catalysts in comparison with that for La_2O_3.

IV. Active Sites of Perovskite Oxides

Results of the thermal desorption and X-ray photoelectron spectroscopic studies showed the role of trivalent cobalt ion for the C-C bond scission in the catalytic hydrogenolysis of alkanes. From the measurements of adsorption of hydrogen, ethylene and carbon monoxide at room temperature, the number of adsorption sites on $LaCoO_3$ is evaluated at 4×10^{14} cm^{-2} which corresponds to about 40 % of the total surface ions. These results suggest that most of the exposed surface is available for the reaction and the (110) and (102) lattice planes are likely to be exposed preferably at the surface because of their higher thermodynamic stability than other planes. The (110) plane is illustrated in Fig. 8. For hydrogenolysis, the geometric arrangement of the (110) plane substantiates the co-functional role in which the Co(III) ion ruptures the C-C bond, and the nearby oxygen ions provide hydrogen adatoms to the resulting monocarbon species. These synergetic effects by the component ions are pronounced in the hydrogenolysis of alkanes with longer carbon chains. The distance between the adjacent Co(III) in the (110) plane is 0.542 nm. The carbon-atom chain of butane or higher alkanes is long and flexible enough to interact with two Co(III) in a bridge form, as shown in Fig. 9. Although the La(III) and oxygen ion fail to adsorb alkanes dissociatively both ions are likely to accommodate the methyl and carbene groups formed as a result of the bond rupture. Thus, such a bridged adsorption makes the nearby La(III) and O(II) ions effective in supplying hydrogen atoms to the decomposed species, thus lowering the activation energy.

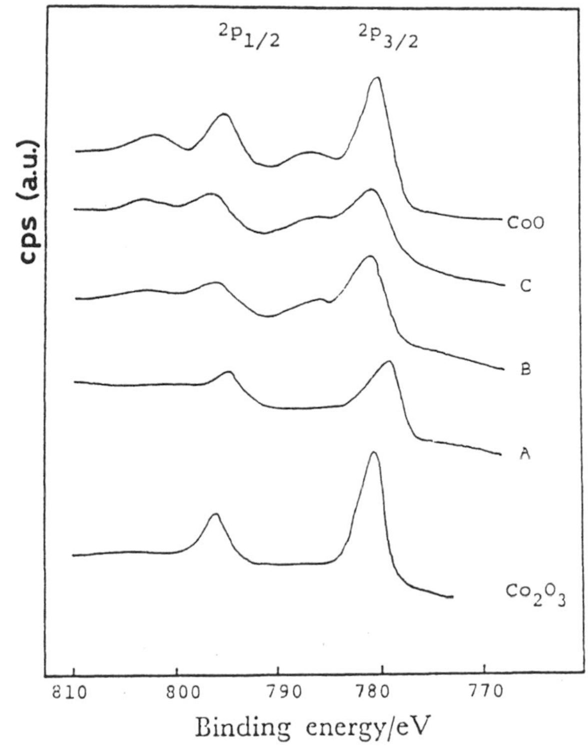

Fig. 7. X-ray photoelectron spectra in Co 2p region (12). Spectra A, B and C: see caption in Fig. 5.

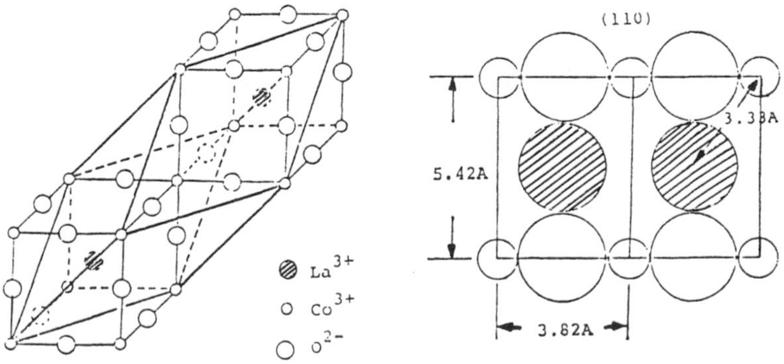

Fig. 8. Structure and geometry of $LaCoO_3$ (12). $LaCoO_3$ has a perovskite structure D^5_{3d}-$R_{\bar{3}m}$, but the (110) plane is drawn as a cubic structure for simplicity.

It is known that $LaCoO_3$ and $LaFeO_3$ are stable at 1273 K even under the extremely low pressure of oxygen, 1×10^{-4} and 1×10^{-14} Torr, respectively (30). These facts are indicative of the stability of the rigid framework of the perovskite structure. However, there has been a controversy as to the stability of $LaCoO_3$ oxide when it is exposed to a reductive atmosphere at high or even moderate temperatures and is employed to the catalytic hydrogenation of alkenes. Crespin and Hall investigated the surface states of perovskite-type oxides, $BaTiO_3$, $SrTiO_3$ and $LaCoO_3$ during reduction and found that Co(III) in $LaCoO_3$ were easily reduced to Co(II) by heating in hydrogen at 673 K, and the compound is then decomposed to Co(0) and La_2O_3 at 773 K (16). They also showed that the reduced mixture recovers its original perovskite structure by reoxidation at 673 K. They described that those phenomena are due to the structure containing vacancies and to the resultant high mobility of ions. By using $LaCoO_3$ prepared by the same procedures, Lombardo and his coworkers reported changes in the catalytic activity for ethylene hydrogenation with an increase in the degree of reduction of the oxide; a maximum activity at 253 K is observed on the $LaCoO_{3-x}$ catalyst at the $x = 0.7$ composition for the first oxidation-reduction cycle and at $x = 1.2$ for the second cycle (17). The catalyst prepared by Lombardo et al. is more reducible than that prepared by Crespin and Hall, since their catalyst is reduced to some extent at 353 K (18). Lombardo et al. have proposed that the active site of $LaCoO_3$ is finely dispersed zero-valent cobalt particles on the La_2O_3 surface.

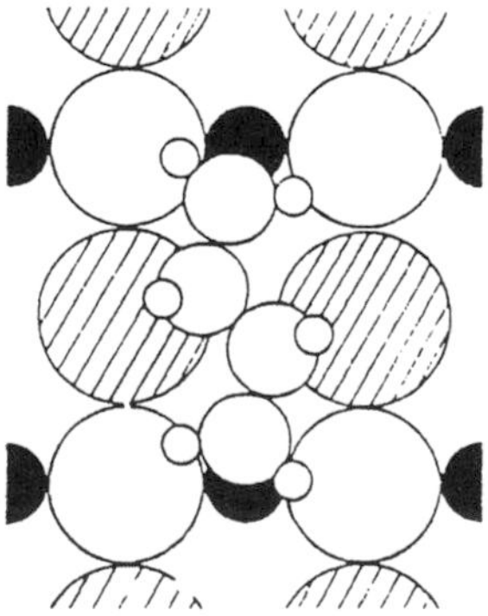

Fig. 9. Schematic representation of butane configuration on the (110) plane of $LaCoO_3$ (14). ◍, La(III); ○, O(II); ●, Co(III); ○, carbon atom; ○, hydrogen atom.

It is not possible to compare the behavior of their catalyst for ethane hydrogenolysis with that described in the section I, since they have examined only the hydrogenation of ethylene at temperatures below 323 K. As for the hydrogenolysis of ethane over $LaCoO_3$ studied by Yasumori et al., it is to be noted that the values of kinetic parameters such as the reaction order with respect to hydrogen pressure and the activation energy were quite different from those for the reaction on metallic cobalt dispersed on oxide supports (13,19,31). The stability of the catalyst in a reductive atmosphere may be influenced by the density of vacancies involved in the oxides, associated with the different conditions of the oxide preparation (13).

A CNDO-MO calculation was performed for ethane adsorbed on a linear atomic cluster, O-M-O-M-O (M = Co, Fe and Al) which represents a part of the (110) plane of perovskite (15). The adsorption energy of ethane and the change in C-C bond energy are estimated by changing the distance between the axes of ethane and the cluster laid in parallel. In the case of [Co-O] and [Fe-O] clusters, the energetically stable adsorption takes place when the center of the molecule is located on the metal ion, whereas for [Al-O] cluster the favorable position of admolecule shifted to the top of the central oxygen atom. The percentage decreases in the C-C bond energy induced by adsorption at equilibrium distances are estimated at 3.7, 1.7 and 6.4 % for [Al-O], [Fe-O] and [Co-O] clusters, respectively; the largest value for the [Co-O] cluster reflects the high capability of $LaCoO_3$ to rupture the C-C bond, compared to the other two perovskites. A more precise calculation using a larger three-dimensional cluster of perovskite will reveal the exact feature of adsorption and reactivities of alkanes and alkenes with hydrogen on these catalyst surfaces.

References

1. Goodenough, J.B. and P.M. Raccah, Complex versus band formation in perovskite oxides, *J. Appl. Phys.*, **36:** 1031 (1965).
2. Mattheiss, L.F., Energy Bands for $KNiF_3$, $SrTiO_3$, $KMoO_3$ and $KTaO_3$, *Phys. Rev. B*, **6:** 4718 (1972).
3. Bhide, V.G., D.S. Rajaria, G. Rama Rao and C.N.R. Rao, Mössbauer studies of the high spin-low spin equilibria and delocalized-collective electron transition in $LaCoO_3$, *Phys.*

Rev. B, **6(3):** 1023 (1972).

4. Kraut, E.A., T. Wolfram and W. Hall, Electrostatic potentials for semi-infinite and lamellar cubic lattices containing several different kinds of ions per unit cell, *Phys. Rev. B*, **6(4):** 1499 (1972)
5. Pedersen, L.A. and W.F. Libby, Unseparated rare earth cobalt oxides as auto exhaust catalysts, *Science*, **175:** 1355 (1972).
6. Voorhoeve, R.J.H., J.P. Remeika, P.E. Freeland and B.T. Matthias, Rare earth oxides of manganese and cobalt rival platinum for the treatment of carbon monoxide in auto exhaust, *Science*, **177:** 353 (1972).
7. Voorhoeve, R.J.H., J.P. Remeika, L.E. Trimble, A.S. Cooper, F.J. Disalvo and P.K. Gallagher, Perovskite-like $La_{1-x}K_xMnO_3$ and related compounds: solid state chemistry and the catalysis of the reduction of NO by CO and H_2, *J. Solid State Chem.*, **14:** 395 (1975).
8. Shimizu, T., T. Nishida and Y. Morikawa, *Nippon Kagaku-Kaishi*, 1936 (1974).
9. Yao, Y.F.Y., The oxidation of hydrocarbons and CO over metal oxides, *J. Catal.*, **36:** 266 (1975).
10. Ishiyama, S., *M. Thesis*, Tokyo Institute of Technology (1975).
11. Voorhoeve, R.J.H., G.K.N. Patel, L.E. Trimble, R.J. Kerl and P.K. Gallagher, HCN from the reduction of NO over platinum, palladium, ruthenium, monel and perovskite catalysts, *J. Catal.*, **45:** 297 (1972).
12. Ichimura, K., Y. Inoue and I. Yasumori, Catalysis by mixed oxide perovskites. I. Hydrogenolysis of ethylene and ethane on $LaCoO_3$, *Bull. Chem. Soc. Japan*, **53:** 3044 (1980).
13. Ichimura, K., Y. Inoue and I. Yasumori, Active State of the $LaCoO_3$ catalyst. A reply to the comment "Hydrogenolysis and Hydrogenation of ethylene on $LaCoO_3$", *Bull. Chem. Soc. Japan*, **55:** 2313 (1982).
14. Ichimura, K., Y. Inoue and I. Yasumori, Catalysis by mixed oxide perovskites. II. The hydrogenolysis of C_3-C_5 hydrocarbons on $LaCoO_3$, *Bull. Chem. Soc. Japan*, **54:** 1787 (1981).
15. Ichimura, K., Y. Inoue, I. Kojima, E. Miyazaki and I. Yasumori, Comparative studies of mixed oxide perovskite catalysts, $LaCoO_3$, $LaFeO_3$ and $LaAlO_3$ for hydrogenolysis of alkanes, in *New Horizonts in Catalysis*, T. Seiyama and K. Tanabe (Eds.), Kodansha-Elsevier, 1281 (1981).

16. Crespin, M. and W.K. Hall, The surface chemistry of some perovskite oxides, *J. Catal.*, **68:** 359 (1981).
17. Petunchi, J.O., M.A. Ulla, J.A. Marcos and E.A. Lombardo, Characterization of hydrogenation active sites on $LaCoO_3$ perovskite, *J. Catal.*, **70:** 356 (1981).
18. Ulla, M.A. and E.A. Lombardo, Hydrogenolysis and hydrogenation of ethylene on $LaCoO_3$, *Bull. Chem. Soc. Japan*, **55:** 2311 (1982).
19. Sinfelt, J.H., Specificity in catalytic hydrogenolysis by metals, *Adv. Catal.*, **23:** 91 (1973); Catalytic hydrogenolysis over supported metals, *Catal. Rev.-Sci. Eng.*, **3:** 175 (1969).
20. Matsumoto, H., Y. Sato and Y. Yoneda, Contrast between nickel and platinum catalysts in hydrogenolysis of saturated hydrocarbons, *J. Catal.*, **19:** 101 (1970).
21. C.G. Myers and G. W. Munns, Jr., Platinum hydrocracking of pentanes, hexanes and heptanes, *Ind. Eng. Chem.*, **50:** 1727 (1958).
22. Pass, G., A.B. Littlewood and R.L. Burwell, Jr., Reactions between hydrocarbons and deuterium on chromium oxide gel. II. Isotope exchange of alkanes, *J. Am. Chem. Soc.*, **82:** 6281 (1960).
23. Demuth, J.E., The reaction of acethylene with Ni(110) surfaces at room temperature, *Surf. Sci.*, **93:** 127 (1980).
24. Rideal, E.K., The hydrogenation of ethylene in contact with nickel, *J. Chem. Soc.*, 309 (1922).
25. Sato, S. and K. Miyahara, Hydrogenation of ethylene on metallic catalysts. Part 4. Pressure dependence of optimum temperature on evaporated nickel film, *J. Res. Inst. Catal. (Hokkaido Univ.)*, **13:** 10 (1965).
26. Robert, T., M. Bartel and G. Offergelt, Characterization of oxygen species adsorbed on copper and nickel oxides by X-ray photoelectron spectroscopy, *Surf. Sci.*, **33:** 123 (1972).
27. Frost, D.C., A. Ishitani and G.A. McDowell, X-ray photoelectron spectroscopy of copper compounds, *Mol. Phys.*, **24:** 861 (1972).
28. Briggs, D. and V.A. Gibson, Direct observation of multiplet splitting in 2p photoelectron peaks of cobalt complexes, *Chem. Phys. Lett.*, **25:** 493 (1974).
29. Okamoto, Y., N. Nakano, T. Imanaka and S. Teranishi, X-ray photoelectron spectroscopic studies of catalysts. Supported cobalt catalysts, *Bull. Chem. Soc. Japan*, **48:** 1163 (1975).
30. Nakamura, T., G. Petzow and L.J. Lauckler, Stability of

perovskite phase $LaBO_3$ (B = V, Cr, Mn, Fe, Co, Ni), *Mat. Res. Bull.*, **14:** 649 (1979).

31. Bond, G.C., *Catalysis by metals*, Academic Press, London and New York (1962), p. 396.

Chapter 12

Hydrogenation of Carbon Oxides over Perovskite-Type Oxides

J.L.G. Fierro

Instituto de Catálisis y Petroleoquímica, CSIC,
Serrano 119, 28006 Madrid, Spain

I. Introduction

Transition metals, metal oxides and metal-containing mixed oxides have been extensively used for Fischer-Tropsch hydrocarbon synthesis (1-9) and their ability to yield oxygenated compounds. During the last two decades, an extensive search has been carried out in examining the mechanism of catalytic hydrogenation of CO and CO_2 which produced hydrocarbons and oxygenate compounds. The catalytic models for hydrocarbon synthesis have already been reviewed by Vannice (10), which gives detailed information and discusses the data which were obtained during their development. As a broad product distribution is obtained during the FT synthesis, the mechanisms are extremely complicated and not clearly understood; therefore this topic is beyond the objective of this chapter and hence no further attention will be paid.

Many investigations were devoted to understanding which of the two carbon oxides is the most efficient starting molecule for the production of alcohols. In Klier's review (11) three basic mechanisms of adsorbed CO hydrogenation were discussed: carbon-down, oxygen-down and side-on hydrogenation. The latter pathway to methanol is not impeded by large thermodynamic barriers and therefore feasible, but adsorbed formaldehyde intermediate has never been observed. However, it was shown that for at least one catalyst (Rh/TiO_2), a second C-O bond was not formed during CO hydrogenation, which indicates that methanol synthesis does not require formate intermediate. Klier concluded that Pt, Pd, Ir and Cu will behave similarly to the Rh/TiO_2 catalyst, and will not favor the formate route. The alternative oxygen-down hydrogenation involves formate. For the CO_2 hydrogenation the mechanism seems to be different. Erdöhelyi et al. (12) suggested that CO_2 hydrogenation proceeds via formate, as an intermediate, which yields mostly methane but not methanol. Moreover, formate species have also been detected during the CO hydrogenation on CeO_2 but formyl species are involved in methanol synthesis over Pd/CeO_2 surfaces (13). Thus, there are strong implications as to the importance of the nature and oxidation state of the active ingredient and type of catalyst in controlling the product distribution in CO or CO_2 hydrogenations.

To better understand the relationship between the oxidation state of the active ingredient and the catalytic activity for CO or CO_2 hydrogenations, perovskite-type oxides ABO_3, with reducible B ions under typical reaction conditions, are excellent model compounds for that purpose. Accordingly, perovskites containing noble metals in position B or copper-substituted perovskites of the type $LaM_{1-x}Cu_xO_3$ (M = Mn, Ti) will be considered for the hydrogenation of CO. Finally, this chapter also summarizes results on CO_2 hydrogenation over unsubstituted and partially substituted cobaltites.

II. The Different Synthesis Processes from Carbon Oxides

The thermodynamics for the production of hydrocarbons from carbon oxides has been calculated and discussed in detail (14). With the exception of a few compounds such as acetylene and formaldehyde, the ΔG° values at 298 K are negative; therefore it is theoretically possible to produce a large variety of compounds at reasonable reaction temperatures. Some of the typical and most desirable reactions of the hydrogenation of CO are summarized in Table 1. Other products such as acids, ketones, aldehydes and aromatics can also be produced by similar reactions. From Table 1, it is clear that the stoichiometric feed ratio H_2/CO is 3 for the methanation reaction whereas for the synthesis of high molecular weight hydrocarbons and alcohols, the H_2/CO usage ratio approaches a limiting value of 2.

A number of complicating reactions can occur concurrently with the synthesis of hydrocarbons and alcohols. Since water is

Table 1. Reactions Which Can Occur During Hydrogenation of CO

Desirable	Methanation:	$3\ H_2 + CO$	$\longrightarrow$	CH_4
	Hydrocarbons:	$2n\ H_2 + nCO$	$\longrightarrow$	$C_nH_{2n}\ (C_nH_{2n+2})$
	Alcohols:	$2n\ H_2 + nCO$	$\longrightarrow$	$C_nH_{2n+1}OH$
Undesirable	Water Gas Shift:	$CO + H_2O$	$\longrightarrow$	$CO_2 + H_2$
	Bouduard Reaction:	$2\ CO$	$\longrightarrow$	$C + CO_2$
	Coke Formation:	$CO + H_2$	$\longrightarrow$	$C + H_2O$

a primary product in most of the synthesis reactions, the water gas shift reaction can occur between this water and CO from the feed, which changes the O-containing by-products from water to carbon dioxide and alters the stoichiometry of the CO and H_2 in the feed stream. Carbon deposits can be formed at the catalyst surface by the direct disproportionation of CO and also by the direct reaction of CO with H_2.

The production of different compounds can be achieved today by a careful design of the catalyst and an appropriate control of the temperature and pressure. In general, metal catalysts favor the production of paraffins, whereas metal oxides and doped metal oxides are required for the production of alcohols. This later feature is fullfilled by perovskites since they combine a bifunctional character, i.e., the B^{n+} cation reducible during on-stream on which hydrogen is dissociated, and the A^{m+} cation possessing basic properties on which carbon oxides are adsorbed.

A. Noble Metal Perovskites

Supported rhodium appear to be one of the most promising catalysts for the hydrogenation of carbon oxides. However it leads to very different product distributions. To understand the origin of these differences among nominally similar catalysts model rhodium perovskites have been used. Somorjai et al. (15,16) have studied the activity of $LaRhO_3$ at different temperatures under H_2:CO mixtures, and thus at different levels of reduction. The product distribution was highly temperature dependent and can be accounted for by a mechanism that involves non-dissociative adsorption of CO to form methanol and a dissociative mechanism for other products. The change of selectivity with temperature is due to competing processes of hydrogenation and carbonylation and variable concentrations of molecularly and dissociatively adsorbed CO and hydrogen on the surface. Thus, at temperatures below 500 K methanol production through hydrogenation of chemisorbed CO predominates. However above 620 K CO adsorbs dissociatively; therefore methanol production is severely depressed while methane is the main product. In the middle temperatures dissociative adsorption of CO occurs, yielding CH_x species that can then undergo either CO insertion to produce oxygenates or hydrogenation to methane. They also conclude that the active catalyst contain Rh^+ species and a small fraction of the metal (Rh^0).

Gysling et al. (17) studied the Fischer-Tropsch activities for $LaRhO_3$ at reaction temperatures from 493 to 623 K. The Schulz-Flory plot for the formation of oxygenates is displayed in Fig. 1a. At reaction temperatures of 573 K the formation rates of C_2 oxygenates were found to be greater than for methanol indicating that the mechanism is different in the two cases. This implies that the methanol is formed by the reduction of adsorbed CO by a mechanism not containing intermediates common to those by which higher oxygenates are formed. The higher oxygenates are apparently formed by a CO insertion into surface alkyl fragments, as already observed by Ichikawa and Fukushima (18) and Watson and Somorjai (15). At variance with the early reported data, they concluded that the rhodium species active in the hydrogenation of CO seems to be Rh^0, in agreement with the TPR profiles and X-ray photoelectron spectroscopic analysis. The comparison of these data with the Fischer-Tropsch distributions for a conventional Rh/SiO_2 catalyst over almost the same temperature interval (Fig. 1b) reveals the same catalytic behavior. It can be noted that the two distributions in Fig. 1 are almost indistinguishable, suggesting the same type of catalytically active rhodium species for both catalysts. Since the Rh present on the Rh/SiO_2 catalysts is considered to be Rh^0 under such reaction conditions (19,20), it is consistent that the active rhodium present in $LaRhO_3$ is also Rh^0.

The kinetic modeling of the CO insertion mechanism for linear oxygenate formation from $CO+H_2$ has been developed by Baetzold and Monnier (21) who observed that by lowering the activation energy for CO dissociation, the rate of alkyl chain growth increased dramatically. In such a situation it was possible to precisely reproduce the experimentally obtained Schulz-Flory distributions for both linear oxygenates and hydrocarbons over a Ru/SiO_2

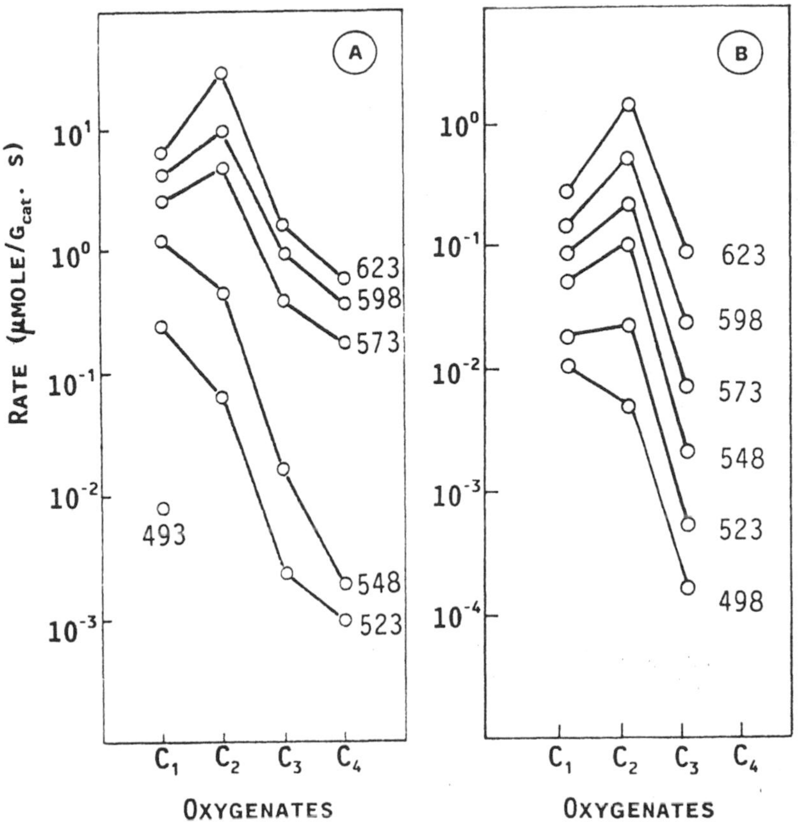

Fig. 1. Schultz-Flory plots for the production of oxygenates from ($CO+H_2$) over $LaRhO_3$ (a) and 1.5% Rh/SiO_2 (b) catalysts at an overall pressure of 800 psig and H_2:CO = 1:1 molar ratio. From Ref. 17.

catalyst. For Rh catalysts the rate of chain growth is more critical than for Ru homologues at a given temperature. Consequently, for similar rates of alkyl chain growth, higher reaction temperatures are required for Rh catalysts.

To test the conclusion that no Rh^+ species are involved in the $CO+H_2$ reaction, Gysling et al. (17) examined and evaluated the $LaRhO_3$ in the hydroformylation of ethylene in which Rh^+ is pressumably the species responsible for the selective formation of aldehyde. Upon H_2-reduction at temperatures 473-598 K the selectivity to propionaldehyde was almost constant, implying that Rh^+ species is not formed at any temperature. Thus, the results for ethylene hydroformylation over $LaRhO_3$ as a function of H_2-

pretreatment temperature corroborate the hypothesis that Rh^{o} is the only catalytically active rhodium species present on the surface, since hydrogen pretreatment at 573 K of a similar catalyst gave in the photoelectron spectrum a binding energy for the Rh $3d_{3/2}$ peak close to that of Rh^{o} species.

Perovskites of barium with platinum group metals, $BaMO_3$ (M = Rh, Ru, Ir, Pt), were also evaluated as catalysts for the ($CO+H_2$) conversion (2). These materials prepared by low temperature calcination (< 670 K) were amorphous; however they became pure crystalline phases upon recalcination at higher temperatures. High oxygenate selectivities, almost exclusively methanol, were obtained with $BaRhO_3$ and $BaPtO_3$ (maximum, 62% and 54%, respectively). On the contrary, the $BaRuO_3$ perovskite yielded the lowest oxygenate selectivity. All these perovskites were pure crystalline phases prior to catalyst testing. The bulk oxidation states of the platinum metals were the expected M^{4+}. However all of these materials, except $BaRuO_3$, contained surface platinum metal cations in lower oxidation states instead of, or in addition to the M^{4+} state. Both crystalline and amorphous barium platinum metal perovskites are unstable to the synthesis gas and its reaction products. The platinum metals were found to be reduced to the elemental state, and the barium was converted to $BaCO_3$. Therefore, it appears that these materials after an induction period are transformed to the zero valent platinum metal crystallites on the surface of $BaCO_3$.

B. Copper-Based Perovskites

The crystalline structure of perovskite oxides with its capability to accomodate many different cations (1,22-27) affords an interesting frame of reference to test the substitution of B cations by copper in the perovskite on the catalytic behavior for the hydrogenation of CO. This idea was first developed by Broussard and Wade (2) and then by Brown-Bourzutcshky et al. (3) who noted that substitution of Mn by Cu in $LaMnO_3$ shifted the product distribution from 100 % hydrocarbons to methanol and small amounts of C_2^+ oxygenates. Later on, Rojas et al. (4,5) synthesized copper-containing perovskites of the type $LaM_{1-x}Cu_xO_3$ (M = Mn, Ti) with the aim of understanding the effect of partial replacement of Mn or Ti by copper on the activity and product distribution of the hydrogenation of CO.

Activity data for CO hydrogenation as a function of the substitution degree (x) are summarized in Table 2. It can be

Table 2. Activity and Product Distribution for CO Hydrogenation over $LaTi_{1-x}Cu_xO_3$ Perovskite-Type Oxides

		Selectivity (%)					
Catalyst	x_{CO}(%)	MeOH	CO_2	CH_4	C_2H_6	DME	C_2^+OH
$LaTiO_3$	0.6	31	31	29	-	9	-
$LaTi_{0.8}Cu_{0.2}O_3$	12.7	39	34	9	6	10	2
$LaTi_{0.5}Cu_{0.5}O_3$	22.4	78	13	5	1	1	2
$LaTi_{0.4}Cu_{0.5}O_3$	21.0	83	10	4	1	1	2
$LaCuO_{2.5}$	21.3	73	15	7	2	1	1
Cu/La_2O_3	7.7	18	38	25	10	9	-

Reactions conditions were: T = 573 K; H_2/CO = 3; overall pressure = 5 MPa.

observed that the conversion level of CO strongly depends on the Cu substitution, attaining the highest values in the composition range $0.5 \leq x \leq 1.0$ and the lowest for x = 0. The product distribution was also influenced by copper substitution. Methane, ethane, dimethyl-ether and alcohols were the principal organic products. Table 3 also summarizes the product distribution observed at a fixed set of reaction conditions. The unsubstituted $LaTiO_3$ produced methanol, methane and carbon dioxide with selectivities by about 30 %, and a substantially lower proportion of dimethyl-ether. Upon incorporation of Cu, the alcohols selectivity increased whereas that of methanation decreased. Interestingly, for x = 0.5 and 0.6 selectivity to methanol ranged from 78 to 83 %, while that of CO_2 varied from 13 to 10 %. The product distribution obtained with the reference lanthana-supported Cu catalyst gave quite a dissimilar result to that of the perovskites x = 0.5 and 0.6. In particular, methanol selectivity decreased markedly whereas CO_2, CH_4, C_2H_6 and DME followed an opposite trend.

The drastic increase of CO hydrogenation activity with increasing Cu substitution is similar to that reported by Broussard and Wade (2) for a $LaM_{0.5}Cu_{0.5}O_3$ (M = Mn, Ti) perovskite and by Brown-Bourzutschky et al. (3) for the $LaMn_{1-x}Cu_xO_3$ catalyst series. The methanol synthesis rate was observed to pass through a maximum at substitution x = 0.6, for which the methanol synthesis rate was 9.7×10^{-2} g/h m^2, which is similar to the one found by Sheffer and King (7) for a 1.2 atom % K promoted unsupported Cu catalyst using similar reaction parameters. All these findings indicate clearly the importance of copper for both activity and selectivity toward methanol.

In order to assess the chemical state of copper, the catalysts after on-stream have been examined by X-ray photoelectron spectroscopy. Figure 2 shows the copper core level spectra of various $LaTi_{1-x}Cu_xO_3$ catalysts used in reaction. From this figure, it results that the binding energy (BE) values of Cu 2p electrons are very similar from catalyst to catalyst and

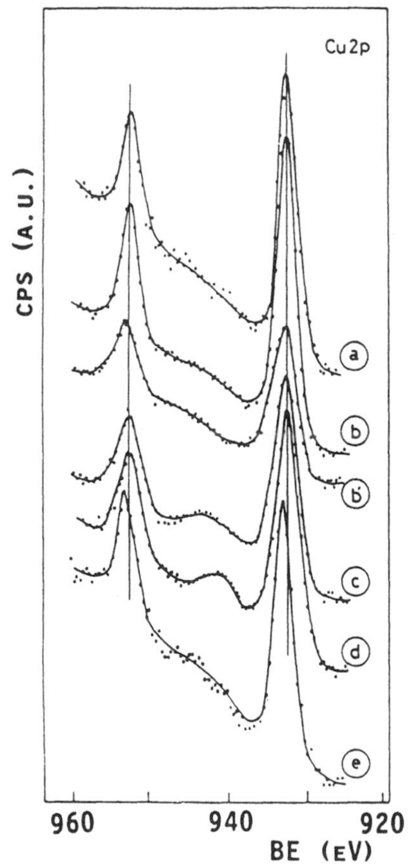

Fig. 2. Copper 2p core level spectra of various $LaTi_{1-x}Cu_xO_3$ catalysts, prereduced at 573 K and used in the hydrogenation of CO. (a), x = 0.2; (b), x = 0.5; (c) x = 0.6; (d) x = 1.0; (e), Cu/La_2O_3. For comparative purpose the $LaMn_{0.5}Cu_{0.5}O_3$ (b') catalyst is also included.

somewhat lower than that of Cu^{2+} ions. Cupric ions can be distinguished by the appearance of a shake-up satellite line by ca. 10 eV above the principal Cu 2p peaks. As can be noted, for x = 1.0, and to a lesser extent for x = 0.6, a satellite peak is observed. The presence of Cu^{2+} ions, together with Cu^{o}, is expected since for these compositions the two CuO and La_2CuO_4 phases have been detected by XRD, but the reduction in hydrogen at 573 K only eliminates the CuO phase (4,6), according to the equation:

$$CuO + La_2CuO_4 + H_2 \longrightarrow Cu^{o} + La_2CuO_4 + H_2O$$

while the La_2CuO_4 one appears to be stable up to 673 K.

The identification of the reduced copper species, i.e., Cu^{o} and/or Cu^{+}, becomes extremely difficult, if not imposible, by XPS data alone since the BE for Cu^{o} and Cu^{+} are almost the same, appearing at ca. 1.3 eV below that of Cu^{2+} ions. This is feasible only through observation of the L_3VV X-ray induced Auger parameter of copper. The modified Auger parameter α_A, defined by the equation: $\alpha_A = h\nu + KE_{LMM} - KE\ Cu\ 2p_{3/2}$, is generally used. K_{LMM} and KE $Cu2p_{3/2}$ are the kinetic energies of the L_3VV X-ray induced Auger emitted electrons and the Cu $2p_{3/2}$ photoemitted electrons, respectively. The X-ray induced Auger spectrum for a representative $LaMn_{0.5}Cu_{0.5}O_3$ catalyst prereduced and used in CO hydrogenation at 573 K is shown in Fig. 4. This catalyst was

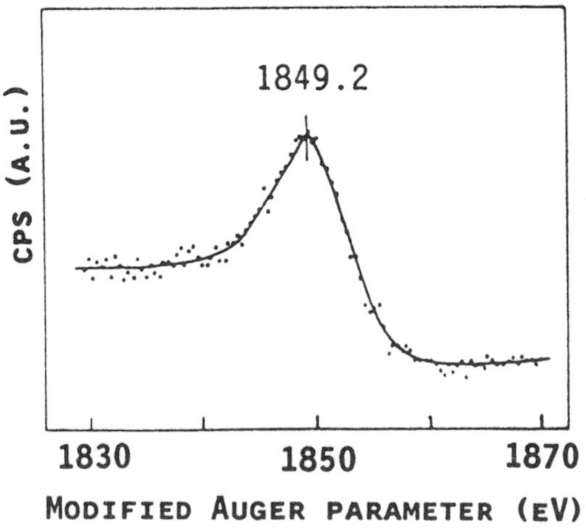

Fig. 3. L_3VV X-ray induced Auger parameter for the $LaMn_{0.5}Cu_{0.5}O_3$ catalyst prereduced and used in CO hydrogenation at 573 K.

found to have the Cu L_3VV peak at 1849.2 eV, which agrees with that at 1848.8 eV observed by Sheffer and King (7) for a prereduced and synthesis gas-exposed unsupported $CuK_{0.36}$ catalyst. Hence, as no other peak was observed in Fig. 3 at energies as high as 1851.0 eV, where both Cu^0 and Cu^{2+} species are expected to appear, it can be concluded that Cu^+ is the dominant copper species in the spent catalyst.

On the basis of this spectroscopic analysis, and considering that substitutions of the element M by copper in the range x = 0.5-1.0 in $LaM_{1-x}Cu_xO_3$ compounds result in active and selective catalysts toward methanol synthesis it can be inferred that Cu^+ locations should be responsible for methanol synthesis. Methanol synthesis activity has been correlated with the presence of Cu^+ ions in Cu/ZnO (8) and Cu/Cr_2O_3 (9) catalysts. In both cases, the host oxides stabilizes Cu^+ ions by substitutional or interstitial occupancy in the ZnO lattice in the former case, and by solid-state reaction in the latter. The mechanism by which the perovskite lattice stabilizes Cu^+ ions can be derived from the analysis of temperature-programmed reduction profiles. These profiles for the $LaM_{1-x}Cu_xO_3$ (M = Mn, Ti) (4,5) catalyst series demonstrated that Cu^{2+} ions placed in the perovskite lattice come to be reduced at temperatures much higher than in massive CuO. Consequently, the extent of reduction is inhibited and the sintering of the resulting reduced phase is retarded with respect to the reduced bulk CuO or conventional supported catalysts.

The quite different catalytic behavior of $LaTi_{1-x}Cu_xO_3$ and the reference Cu/La_2O_3 catalyst suggests that under the reaction conditions the structure and probably the chemical state of copper are different. As large Cu^0 particles have been found in the Cu/La_2O_3 catalyst it seems that formation of hydrocarbons very likely occurs on metallic Cu. Since CO does not dissociates over Cu^0 (28), reaction probably proceeds via hydrogenation of CO to form H_xCO intermediates, which then undergo dissociation to yield H_yC fragments, thus constituting the starting precursor of hydrocarbons. Even though the dominant Cu species in the spent lanthana-supported copper catalyst is metallic Cu, in line with the high proportion of hydrocarbons in the product distribution, the presence of a small part of ionic copper on the surface of Cu crystals, or in close contact with them, cannot be excluded since it seems an important requirement for the methanol synthesis.

IV. Hydrogenation of Carbon Dioxide

Although the hydrogenation of CO_2 is a potential source of higher valued organic compounds this has not been as extensively studied as the hydrogenation of CO over perovskite-type compounds. The main reason for this lies doubtless in the fact that the hydrogenation of carbon dioxide usually yields methane, which is a highly hydrogen-consuming process. Accordingly, the literature related CO_2 hydrogenation is very scarce. One of the few papers where the reaction is analyzed in detail has been reported by Lombardo and co-workers (29). They studied the catalytic behavior for the carbon dioxide hydrogenation of the $La_{1-x}M_xCoO_3$ (M = Sr, Th) systems at different degrees of reduction. The reaction rates and the product distributions at two degrees of reduction are summarized in Table 3. For the unsubstituted $LaCoO_3$ the initial reaction rate of the reduced perovskite fluctuated with the extent of reduction, but it did not reach a maximum.

The production of C_2^+ products decreased at higher reduction degrees, whereas the CH_4 selectivity followed an opposite trend.

Table 3. Catalytic Activity[a] and Selectivity for the CO_2 Hydrogenation over $La_{1-x}M_xCoO_3$ (M = Sr, Th) Perovskites

Catalyst	Extent Red.[b]	Rate[c]	S_{CH4}[d]	C_2^+
$LaCoO_3$	1.2	0.010	0.52	0.13
	3.0	0.018	0.67	0.02
$La_{0.8}Th_{0.2}CoO_3$	1.1	0.130	0.89	0.02
	2.9	0.120	0.82	0.02
$La_{0.6}Sr_{0.4}CoO_3$	1.3	0.040	0.91	0.03
	3.3	0.004	0.32	0.30

[a] Reaction temperature = 553 K; total pressure = 160 Torr in a batch gas recirculating system; $H_2:CO_2$ = 4:1 molar.
[b] Reduction of Co^{n+} corresponds to n electrons per formula.
[c] mmol of CO_2 converted per min and square meter of cobalt.
[d] The C_2^+ fraction included hydrocarbons containing more than one carbon atom (measured at constant CO_2 conversion of 0.20). Adapted from ref. 29.

The $La_{0.8}Th_{0.2}CoO_3$ sample was, however, much more active and selective to methane than its unsubstituted $LaCoO_3$ homologue. The extent of reduction did not significantly affect the initial rate, and the C_2^+ was very low. The other substituted perovskite ($La_{0.6}Sr_{0.4}CoO_3$) had two completely different behaviors. The unreduced, and the H_2-treated sample at temperatures below 570 K, yielded a product distribution close to the Th-substituted specimen. However, at higher reduction degrees it showed a behavior completely different from those of the other oxides, i.e., the methane selectivity decreased and the the C_2^+ products increased with the extent of reduction. Particularly, the fully reduced catalyst gave the highest selectivity to C_{2+} products.

It has already been observed that the surface Co/La ratio in the reduced catalysts correlate with the rate of methanation and the C_2^+ production. This points out that the metallic cobalt is the active catalyst component in the hydrogenation reaction of carbon dioxide. It is widely accepted that CO_2 is dissociatively adsorbed in the presence of hydrogen over most Group VIII metals (29-32) according to the equation:

$$CO_{2(g)} \rightleftharpoons CO_* + O_*$$

Alternatively, a formate group has been detected by infrared spectroscopy (33-35). This species could in turn be the starting point of another sort of intermediates for CO production according to the equation:

$$CO_{2(g)} + 2\ H_* \rightleftharpoons COOH_* + H_* \rightleftharpoons CO_* + H_2O_*$$

It is also well known (36-39) that an essential step leading to hydrocarbon synthesis is the dissociation of CO_* to produce an active surface carbon C_* which would be the key to achieve a high catalytic activity:

$$CO_* + 2\ H_* \rightleftharpoons C_* + H_2O_*$$

To get an insight on how the specific activity of the metallic sites is affected by the surrounding matrix, a comparison with a conventional silica-supported cobalt catalyst is needed. Weatherbee and Bartholomew (40) have reported turnover frequencies (TOF) for methane over Co/SiO_2 catalysts which are substantially lower than that obtained by Lombardo et al. (29) under similar reaction conditions. This led to the conclusion that metallic cobalt present on both Th- and Sr-substituted perovskites has a higher specific methanation activity than Co on silica. Such a matrix effect may arise from the fact that the perovskite permits a higher dispersion of cobalt, and therefore the very small metal particles would then be affected by the basic properties of the perovskite components. Finally, it is inferred that such a structure can be essential to obtain a high selectivity toward the formation of larger hydrocarbons (C_2^+), which can even be enhanced by a suitable hydrogen pretreatment.

V. Conclusion

In the CO hydrogenation, the stabilization of transition metal ions by incorporating them into perovskite-type structures (ABO_3, A = Lanthanide, B = transition element) enables a high selectivity to oxygenate products. Particularly, noble metals and copper in position B provide relatively simple and well-defined model catalysts, in which the in situ surface characterization easily contributes to understand the nature of the active centers. Although there is not a general explanation for the whole data of the literature, one must be sure that no isolated phases of the individual oxides (A_2O_3 and B_2O_3) remain at the surface of the perovskite after the synthesis. Finally, the unsubstituted and partly substituted $ACoO_3$ perovskites, when reduced, show activity in the CO_2 hydrogenation. In this latter case, the dominant reaction is methanation with a minor chain growth to C_2^+ hydrocarbons, but in no case oxygenates were found.

References

1. Tejuca, L.G., J.L.G. Fierro and J.M.D. Tascón, Structure and reactivity of perovskite-type oxides, *Adv. Catal.*, **36:** 237 (1989).
2. Broussard, J.A. and L.E. Wade, Catalytic conversion of syn gas with perovskites, *Preprints, Am. Chem. Soc., Div. Fuel Chem.*, **31:** 75 (1986).
3. Brown-Bourzutschky, J.A., N. Homs and A.T. Bell, Conversion of synthesis gas over $LaMn_{1-x}Cu_xO_3$ perovskites and related copper catalysts, *J. Catal.*, **124:** 52 (1990).
4. Rojas, M.L., J.L.G. Fierro, L.G. Tejuca and A.T. Bell, Preparation and characterization of $LaMn_{1-x}Cu_xO_3$ perovskite

oxides, *J. Catal.*, **124:** 41 (1990).
5. Rojas, M.L. and J.L.G. Fierro, Synthesis and characterization of $LaTi_{1-x}Cu_xO_3$ compounds, *J. Solid State Chem.*, **89:** 299 (1990).
6. Van Grieken, R., J.L. Peña, A. Lucas, G. Calleja, M.L. Rojas and J.L.G. Fierro, Selective production of methanol from syngas over $LaTi_{1-x}Cu_xO_3$ mixed oxides, *Catal. Lett.*, **8:** 335 (1990).
7. Sheffer, G.R. and T.S. King, Potassium's promotional effect of unsupported copper catalysts for methanol synthesis, *J. Catal.*, **115:** 376 (1989).
8. Herman, R.G., K. Klier, G.W. Simmons, B.P. Finn, J.B. Bulko and T.P. Kobylinski, Catalytic synthesis of methanol from CO/H_2. I. Phase composition, electronic properties and activities of the $Cu/ZnO/M_2O_3$ catalysts, *J. Catal.*, **56:** 407 (1979).
9. Monnier, J.R., M.J., Hanrahan and G.R. Apai, A study of the catalytically active copper species in the synthesis of methanol over Cu-Cr oxide, *J. Catal.*, **92:** 119 (1985).
10. Vannice, M.A., The catalytic synthesis of hydrocarbons from carbon monoxide and hydrogen, *Catal. Rev.-Sci. Eng.*, **14:** 153 (1976).
11. Klier, K., Methanol synthesis, *Adv. Catal.*, **31:** 243 (1982).
12. Erdöhelyi, A., M. Pasztor and F. Solymosi, Catalytic hydrogenation of CO_2 over supported palladium, *J. Catal.*, **98:** 166 (1986).
13. Diagne, C., H. Idriss, I. Pepin, J.P. Hindermann and A. Kiennemann, Temperature-programmed desorption studies on Pd/CeO_2 after methanol and formic acid adsorption and carbon monoxide-hydrogen reaction, *Appl. Catal.*, **50:** 43 (1989).
14. Anderson, R.B., in *Catalysis, P.H. Emmett (Ed.)*, Reinhold, New York, 1956, Vol. 4.
15. Watson, P.R. and G.A. Somorjai, The formation of oxygen-containing organic molecules by the hydrogenation of carbon monoxide using a lanthanum rhodate catalyst, *J. Catal.*, **74:** 282 (1982).
16. Somorjai, G.A. and S.M. Davis, The surface science of heterogeneous catalysis, *CHEMTECH*, **13:** 502 (1983).
17. Gysling, H.J., J.R. Monnier and G. Apai, Synthesis, characterization and catalytic activity of $LaRhO_3$, *J. Catal.*, **103:** 407 (1987).
18. Ichikawa, A. and T. Fukushima, Mechanism of syngas

conversion into C_2-oxygenates such as ethanol catalysed on a SiO_2-supported Rh-Ti catalyst, *J. Chem. Soc., Chem. Commun.*, 321 (1985).
19. Foley, H.C., S.J. DeCanio, K.D. Tau, K.J. Chao, J.H. Onuferko, C. Dybowsky and B.C. Gates, Surface organometallic chemistry: reactivity of silica bound rhodium alkyl complexes and the genesis of highly dispersed supported rhodium catalysts, *J. Am. Chem. Soc.*, **105:** 3074 (1983).
20. Jackson, S.D., J. Chem. Soc., Interaction of carbon monoxide with rhodium catalysts. Studies of adsorption and thermal desorption, *J. Chem. Soc., Faraday Trans. I*, **81:** 2225 (1985).
21. Baetzold, R.C. and J.R. Monnier, Kinetic model of hydrocarbon and oxygenate formation, *J. Phys. Chem.*, **90:** 2944 (1986).
22. Galasso, F.S., *Structure, Properties and Preparation of Perovskite-Type Compounds*, Pergamon Press, Oxford, 1969.
23. Happel, J., M. Hnatow and L. Bajars, *Base Metal Oxide Catalysts*, Marcel Dekker Inc., 1977.
24. Voorhoeve, R.J.H., Perovskite-related oxides as oxidation-reduction catalysts, in *Advanced Materials in Catalysis*, *J.J. Burton and R.L. Garten (Eds.)*, Academic Press, 1977.
25. Libby, W.F., Promising catalysts for auto exhaust, *Science*, **171:** 499 (1971).
26. Viswanathan, B., Solid state and catalytic properties of rare earth orthocobaltites - New generation catalysts, *J. Sci. Ind. Res.*, **43:** 151 (1984).
27. Fierro, J.L.G., Structure and composition of perovskite surface in relation to adsorption and catalytic properties, *Catal. Today*, **8:** 153 (1990).
28. Broden, G., T.N. Rhodin, C. Brucker, R. Brendow and Z. Hurych, Synchrotron radiation study of chemisortive bonding of CO on transition metals - Polarization effect on Ir(100), *Surf. Sci.*, **59:** 593 (1976).
29. Falconer, J.L. and Zagler, A.E., Adsorption and methanation of carbon dioxide on a nickel/silica catalyst, *J. Catal.*, **62:** 280 (1980).
30. Rofer-De Poorter, C.K., A comprehensive mechanism for the Fischer-Tropsch synthesis, *Chem. Rev.*, **81:** 447 (1981).
31. Izuka, T., Y. Tanaka and K. Tanabe, Hydrogenation of CO and CO_2 over rhodium catalysts supported on various metal

oxides, *J. Catal.*, **76:** 1 (1981).
32. Henderson, M.A. and S.D. Worley, An infrared study of the dissociation of carbon dioxide over supported rhodium catalysts, *Surf. Sci.*, **149:** L1 (1985).
33. Amenomiya, Y., Active sites of solid acid catalysts. III. Infrared study of the water gas conversion reaction on alumina, *J. Catal.*, **57:** 64 (1979).
34. Solymosi, F., T. Bansagi and A. Erdohelyi, Infrared study of the reaction of adsorbed formate ion with H_2 on supported rhodium catalysts, *J. Catal.*, **72:** 166 (1981).
35. Schild, C., A. Wokaun and A. Baiker, On the mechanism of CO and CO_2 hydrogenation reactions on zirconia-supported catalysts: a diffuse reflectance FTIR study. Part II. Surface species on copper/zirconia catalysts: implications for methanol synthesis selectivity, *J. Molec. Catal.*, **63:** 243 (1990).
36. Vance, K.C. and C.H. Bartholomew, Hydrogenation of carbon dioxide on Group VIII metals. III. Effects of support on activity selectivity and adsorption properties of nickel, *Appl. Catal.*, **7:** 169 (1983).
37. Vannice, M.A., Catalytic activation of carbon monoxide on metal surfaces, in *Catalysis-Science and Technology, J.R. Anderson and M. Boudart (Eds.)*, Springer-Verlag, Heidelberg, 1982, Vol. 3, Chap. 3, p. 144.
38. Weatherbee, G.D. and C.H. Bartholomew, Hydrogenation of CO_2 on Group VIII metals. II. Kinetics and mechanism of CO_2 hydrogenation on nickel, *J. Catal.*, **77:** 460 (1982).
39. Kort, J.G.E. and A.T. Bell, Synthesis of hydrocarbons from CO and H_2 over silica-supported Ru: reaction rate measurements and infrared spectra of adsorbed species, *J. Catal.*, **58:** 170 (1979).
40. Weatherbee, G.D. and C.H. Bartholomew, Hydrogenation of CO_2 on Group VIII metals. IV. Specific activities and selectivities of silica-supported Co, Fe, and Ru, *J. Catal.*, **87:** 352 (1984).

Chapter 13

CO Oxidation and NO Reduction on Perovskite Oxides

B. Viswanathan

Department of Chemistry, Indian Institute of Technology,
Madras 600 036, India

I. Introduction

The search for substitutes of noble metal supported exhaust control catalysts has led to the study of a number of perovskite oxides (1). The mechanisms of redox processes on these surfaces are topotactic in nature, thus accounting for the reversible loss and uptake of bulk oxygen or for the concomitant creation and annihilation of vacancies rendering these systems as attractive oxidation catalysts. The variable oxygen stoichiometry of perovskite structure is responsible for the exotic behavior of these materials as a new class of high T_c superconducting oxides (2). The extreme ease of removing oxygen from the framework but still retaining the basic ingredients of the perovskite structure and the possibility of deriving different structural frameworks from ideal perovskite structure (e.g. insertion of rock salt layer between perovskite layers results in a K_2NiF_4 structure, while insertion of double $[Bi_2O_2]$ layers between perovskite layers results in an Aurivillius phase) and the possibility of substitution of either or both A and B site cations by foreign metal ions make them function as chemical chameleons (3) with a wide variety of solid state and catalytic properties. Among the various redox catalytic processes promoted by them the oxidation of CO and the reduction of NO seem to be interesting in view of their relevance for auto-exhaust emission control. The present chapter attempts to consider the generalities that have been evolved in these studies, which will be useful for the formulation of auto-exhaust control catalysts based on perovskites in the future.

II. Oxidation of CO

This is one of the various oxidation reactions which are studied extensively on a variety of perovskite type oxides. Earlier studies reported on this reaction have been periodically reviewed by Voorhoeve (1), Balasubramanian et al. (4), Viswanathan (5) and Tejuca (6).

The catalytic oxidation of CO has been studied on a number of perovskite oxides with a view to correlate the observed activity with the electronic state of the transition metal ions (7) or the defect chemistry of perovskites (8). Shimizu (9) has pointed out the need for correlating the activity with a tolerance factor (10) (defined in terms of radius ratio of A,B and oxide ions which should satisfy the condition $0.75 < (r_A + r_0)/\sqrt{2}\ (r_B + r_0) < 1.00$), a fundamental parameter deciding the crystallization of the system in the perovskite phase as well as with the binding energy of oxide ions (defined in terms of the enthalpy of formation of oxide A_mO_n (ΔH_f), sublimation energy of metal (ΔH_s) and the dissociation energy of oxygen D_o by the following relation (11): $\Delta_{A\text{-}O} = [\Delta H_f - m\Delta H_s - (n/2)D_o]/12m$.

The data generated by Voorhoeve et al. (7) on the oxidation of CO by $LaBO_3$ type of oxides have been analysed and the correlations obtained with the tolerance factor and the binding energy of oxygen are shown in Fig. 1. It is seen that activity expressed as the temperature at which the rate constant is 4 x 10^{-7} moles m^{-2} min^{-1} decreases linearly with respect to the ionic radius of the B ions, while it increases with the increase of the tolerance factor. However, the binding energy increases with the decrease of the tolerance factor. It can be rationalized that the increase of the tolerance factor may promote oxygen vacancies at the surface which could cause considerable effect on the activity for CO oxidation. These correlations can be considered to imply: (i) the active sites involved in the oxidation of CO consists of the B-O-B clusters; this is in agreement with the conclusions drawn by Tascón and Tejuca (12-14), and (ii) the B-O-B bond interactions depend upon the size of the B ion, the lower binding energy for the surface oxygen species is favorable for the oxidation of CO. This might imply the participation of surface

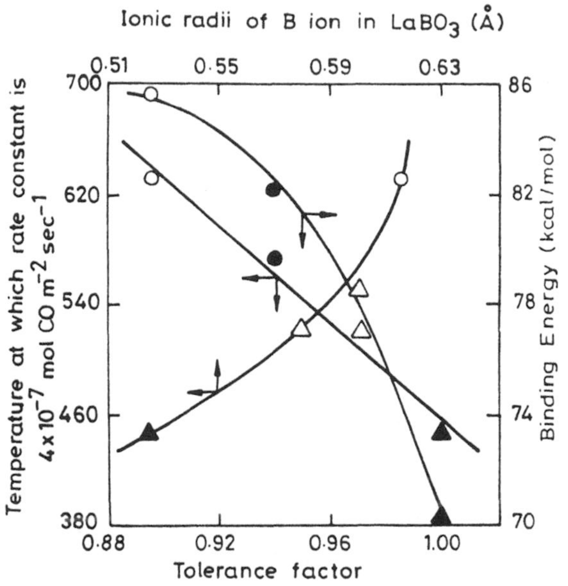

Fig. 1. Relationship between tolerance factor or ionic radius of B ions with either activity for CO oxidation or with binding energy of oxygen for $LaBO_3$ type perovskites: (○), Cr; (●), Fe; (△), Mn; (▲), Co.

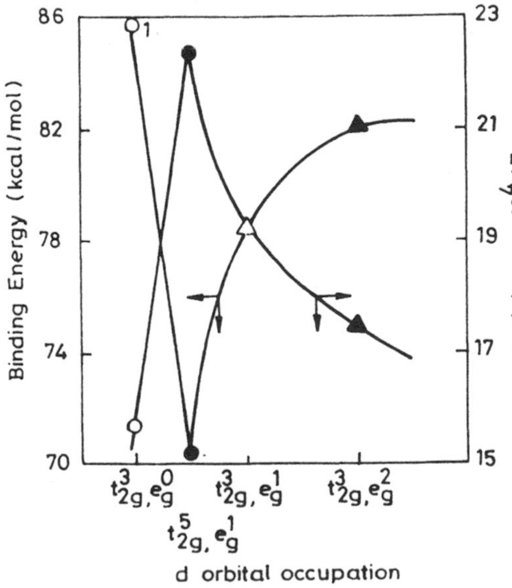

Fig. 2. Relationship of binding energy of oxygen or activity for CO oxidation with d electron configuration of the B ions in $LaBO_3$ type perovskites: (○), Cr; (●), Co; (△), Mn; (▲), Fe.

oxygen species in the oxidation of CO by a redox mechanism.

In Fig. 2, the binding energy of $LaBO_3$ perovskites and the catalytic activity are plotted as a function of the electronic occupation in the t_{2g} and e_g levels of the transition metal ion according to Goodenough's classification (15). It is evident that the decrease in binding energy corresponds with an increase in catalytic activity thereby supporting the conclusion that the active surface species involves the B-O-B clusters and the activity is controlled by the binding energy of oxygen with the B ions.

In Fig. 3, the variation of activity and the binding energy (16) is shown as a function of the tolerance factor for the series of solid solutions $Sm_{0.5}A_{0.5}CoO_3$ (A = Ca, Sr, Ba). In this case also, it is observed that the activity increases as the tolerance factor increases. Though the activity variation among the series of solid solutions is considerable, the net change in binding energy is quite small (1-2 kcal/mol).

In order to evaluate the role of rare earth ions, the relationships between the tolerance factor with either activity or binding energy have been examined (17) for a number of $LnMnO_3$ oxides and the results obtained are shown in Fig. 4. It is seen that the activity increases as the tolerance factor increases, but there is not much variation in binding energy with variation of tolerance factor. These results imply that the rare earth ion essentially plays a role of modifying the B-O bond only and does

Table 1. Catalytic Activity and Spin State of Cobalt in Rare Earth Cobaltites for the Oxidation of CO (19)

Catalyst	Cobalt states	Surface area (m^2/g)	T (K)	Conversion (% per g)
$LaCoO_3$ (R)	Co^{3+}, Co^{III} Co^{IV}, Co^{II}	1.0	463	1.6
$PrCoO_3$ (O)	-do-	1.2	452	3.2
$NdCoO_3$ (O)	-do-	1.3	428	10.0
$GdCoO_3$	-do-	1.4	468	1.6
HoCoO3 (O)	Co^{3+}, Co^{III} (1:1)	1.3	427	4.9

R - Rhombohedral, O - Orthorhombic

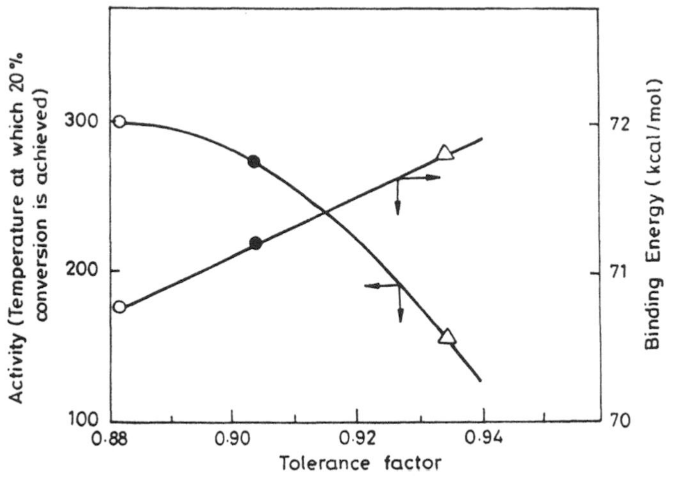

Fig. 3. Relationship between tolerance factor with either binding energy of oxygen or activity of CO oxidation on a series of solid solutions $Sm_{0.5}A_{0.5}CoO_3$ (A = Ca (○), Sr (●), Ba (△)).

not become directly involved in the actual reaction. Correlations between the sum of the Ln-O and Co-O binding energies or reducibility and the catalytic activity in CO oxidation have been proposed (6,18).

Some data on CO oxidation on a series of rare earth cobaltites as reported by Om Prakash et al. (19) are given in Table 1. Magnetic susceptibility and Mössbauer studies conducted by them showed that the ratio of the concentration of high spin Co^{3+} to that of other states is unity in $NdCoO_3$ and $HoCoO_3$, while in the other cobaltites it is significantly higher than unity. They argued that it is possible that the Co^{3+} sites facilitate the adsorption of CO while the other states, particularly low spin Co(III) provide sufficient conductivity to favor the process: $CO + O^{2-} \rightarrow CO_2 + 2e$.

This explanation, however, needs further examination with the following points in mind. (i) The spin state equilibrium and transition between spin states are strictly possible only with Co(III) ions at the octahedral coordination of the oxide ions, since only in this case, $\Delta_{cf} \approx \Delta_{ex}$. (ii) Above 200 K in many of the rare earth cobaltites, the concentration of Co^{3+} ions either decreases due to electron transfer, leading to the formation of Co(II) and Co^{4+} or remains almost constant as in case of $HoCoO_3$. If Co^{3+} ions act as centers for CO adsorption then an anomalous variation in the Arrhenius plot is expected around 200 K. (iii) As the number of active [Co^{3+} as well as Co(III)] sites decreases with an increase in temperature in the case of $LaCoO_3$ and related cobaltites, a decrease in activity is expected with an increase in temperature. (iv) The values of activation energy and frequency factor for the oxidation of CO on a series of rare earth transition metal perovskites are given in Table 2 (7). In the series of compounds $LaMnO_3$, $LaFeO_3$ and $LaCrO_3$ exhibit localized behavior, whereas $LaCoO_3$ exhibits spin state equilibrium. In spite of this basic difference, the values of activation energy are almost equal and no anomalous behavior is exhibited by $LaCoO_3$. (v) It is presumed that the reaction $CO + O_2 \rightarrow CO_2 + 2e$ is favored by the presence of Co(III) states, which provide sufficient conductivity for the reaction. The activation energy for conduction in the temperature region of catalytic activity is highest (~0.3 eV) for $NdCoO_3$ and $HoCoO_3$ (high active systems) compared to other cobaltites (~0.1 eV). A set of experimental data collected from the literature is presented in Table 3. It appears that the activity data cannot be rationalized only on the magnitude and type of electrical conductivity. It is clear from these values that no such simple correlation between electrical conductivity and activity is possible in the case of these perovskite oxides, though a common feature appears to be

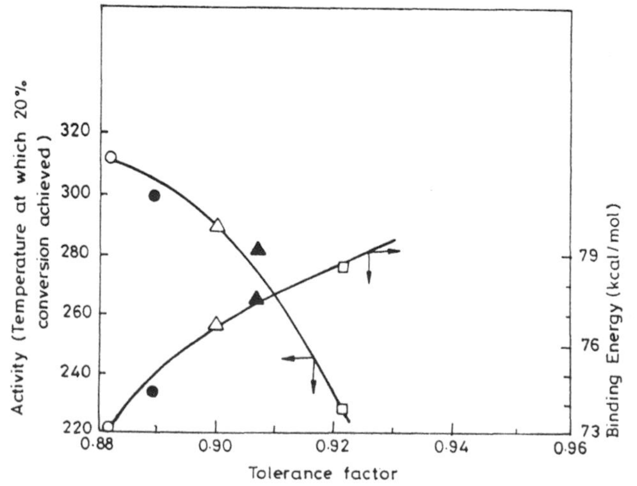

Fig. 4. Relationship between tolerance factor with either binding energy of oxygen or activity for CO oxidation on a series of $AMnO_3$ manganites (A = La (○), Pr (●), Nd (△), Sm (▲), Eu (□).

Table 2. Values of Kinetic Parameters for the Oxidation of CO on $LaBO_3$ Perovskites (B - First Row Transition Metal)

Catalyst	Activation energy (kcal/mol)	Frequency factor
$LaCrO_3$	27.4	5.4×10^5
$LaMnO_3$	22.1	9.5×10^2
$LaFeO_3$	26.5	6.0×10^3
$LaCoO_3$	21.2	6.9×10^3
$La_{0.9}Sr_{0.1}CoO_3$	11.5	
$La_{0.9}\ Ce_{0.1}CoO_3$	18.6	

that the activity increases when the Fermi level moves towards the center of the lowest e_g level, i.e. at the surface d_{z^2} antibonding level. This indicates that d_{z^2} level has the right energy and symmetry for the adsorption of CO (20) which could be the rate-determining step for CO oxidation.

The coordinatively unsaturated BO_x species on the surface of cubic perovskites possess orbitals (usually d_{xy} and d_{yz}) of suitable symmetry for interaction with antibonding states of CO (21). In the isostructural series of titanates of Ca, Sr and Ba (20) the mixing of cationic d orbitals with anionic p orbitals increases in the order Ca < Sr < Ba. Values of activation energy in kcal/mol for the oxidation of CO on these systems are in the same order Ca(19.2) < Sr (26.3) < Ba(32) (20). The reactivity variations between systems containing transition metal ions in high and low spin states could also be visualized in terms of the orbital symmetry considerations. Systems with high spin ions like in $LaCrO_3$ and $LaFeO_3$ are poor catalysts (22) because they do not possess empty d states of suitable symmetry for backbonding interaction with the reacting CO molecule.

Though the influence of the spin state equilibrium of the B ions on the observed catalytic activity has been postulated (19,23), experimental support for this interesting possibility is insufficient at this stage especially in view of the reports attributing the observed catalytic behavior to the defective structure of the perovskites (24). The values of the activation energy for the oxidation of CO on some of the rare earth

Table 3. Electrical Properties of CO Oxidation Catalysts (7)

Catalyst	ρ^a	Semic.[b]	T^c	Configuration[d]
$LaCrO_3$	0.01	p	623	t^3e^0
$La_{0.9}Ce_{0.1}CrO_3$	0.001	p	577	$t^3e^0t^3e^0$
$La_{0.85}Sr_{0.15}CoO_3$	-	p	476	
$La_{0.9}Sr_{0.1}CoO_3$	800	p	465	$t^6e^0+t^5e^0(ls)+t^4e^1(ls)$
$LaCoO_3$	180	p	447	$t^6e^0+t^4e^2(ls)$
$La_{0.9}Ce_{0.1}CoO_3$	70	p	431	$t^6e^0+t^6e^1(ls)+t^4e^2+$ $+t^5e^2(ls)$
$LaMnO_3$	0.17	p	521	t^3e^1
$LaFeO_3$	low	p	573	t^3e^2
(ls) = low spin				

[a] Electrical conductivity at 573 K ($Ohm^{-1}cm^{-1}$); [b] Type of semiconductor; [c] Temperature (K) at which rate is 4 x 10^{-7}; [d] Ground state electron configuration.

cobaltites are plotted against magnetic moments in Fig. 5 (25). Richter et al. (26) have postulated that the high spin state of cobalt ions at the surface may be favorable for the strong chemisorption of oxygen which accounts for the enhanced activity. This is not in conformity with the proposals of Kojima et al. (27) that the high spin states at the surface can still activate CO and thus account for the enhanced activity as e_g states are still unoccupied.

The values of activation energy for the oxidation of CO are plotted against the oxygen non-stoichiometry parameter of $LaCoO_3$ systems in Fig.6. The activation energy value decreases with an increase in oxygen non-stoichiometry implying that the activity is related to the ease of oxygen removal either from the adsorbed state or from the lattice (28,29).

In Fig. 7 the values of activation energy for the oxidation of CO are shown as a function of rare earth ion for $LnCoO_3$ systems. The apparent volcano shaped correlation between the enthalpy of formation of the rare earth oxide and activation energy on $LaCoO_3$ systems has been taken (25) to imply that the ease of oxygen removal seems to control the oxidation activity. Spin state populations probably affect the reactivities through

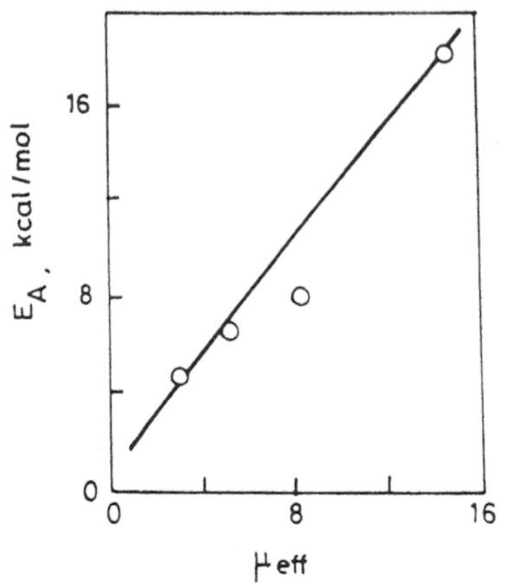

Fig. 5. The variation of activation energy for oxidation of CO against magnetic moment of $LnCoO_3$ compounds (Ln = La, Nd, Gd and Ho).

the alteration of the B-O bond strength as well as through the activation of CO.

The catalytic behavior of $La_{1-x}Sr_xCoO_3$ (x = 0.0 to 0.4) calcined at 1123 K showed high activity for the oxidation of CO (28). The catalytic activity increased with Sr^{2+} content up to x = 0.02 and then it decreased (29). The reducibility of $La_{1-x}Sr_xCoO_3$

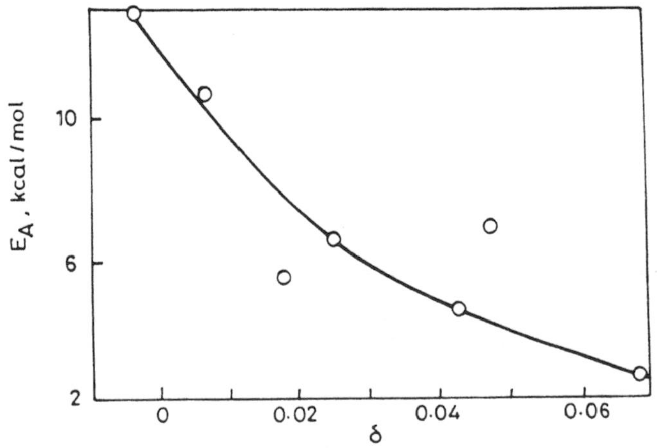

Fig. 6. The variation of activation energy for oxidation of CO against oxygen nonstoichiometry parameter δ in $LnCoO_{3-\delta}$ (Ln = La, Pr, Nd, Sm, Eu, Gd, and Dy).

by CO increased with the increase in Sr^{2+} content, whereas the rate of reoxidation of the reduced catalyst decreased with the increase in the value of x (30). It was therefore concluded that the catalytic oxidation of CO proceeds through a redox mechanism and the rate at the stationary state shows a maximum at a certain value of x. However, for smaller values of x, one has to invoke additional contribution from a mechanism involving the adsorbed oxygen besides the redox mechanism in order to account for the differrence between the rates of catalytic oxidation and reduction. Two types of oxygen desorption from $La_{1-x}CoO_3$ were identified by the presence of a broad desorption peak (α) below 1073 K and a sharp one (β) around 1093 K (31). From the XPS data it was concluded (32) that α state is ascribable to the desorption of the adsorbed oxygen, while β peak may arise due to desorption of the lattice oxygen. Nakamura et al. (33) concluded that the reactivity of the lattice oxygen, desorption of oxygen and reduction tend to increase with the increase in the value of x probably due to the formation of unstable Co^{4+} and the high chemical potential of lattice oxygen. Co^{4+} is reduced to Co^{3+} with the formation of an oxygen vacancy. This vacancy creation causes easy diffusion of lattice oxygen from the bulk to the surface.

It is also assumed that lattice and adsorbed oxygen attain equilibrium ($O^{2-}_{(ads)} \rightleftharpoons O^{2-}_{(l)}$) at reaction conditions. It is therefore proposed (34) that oxygen species adsorbed in the

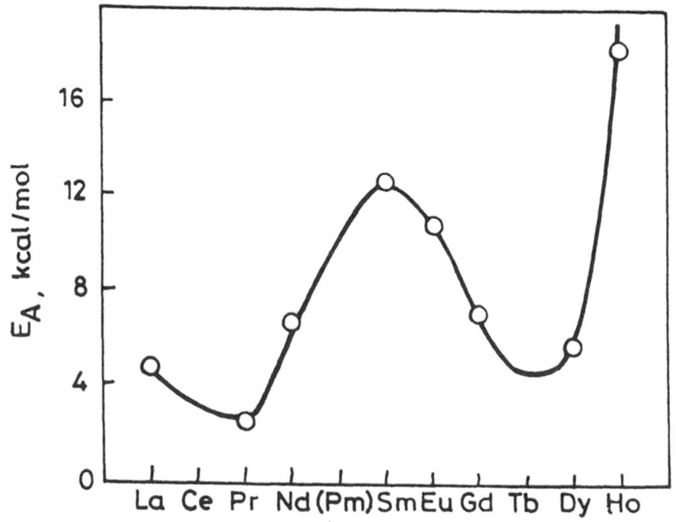

Fig. 7. Plot of activation energy for oxidation of CO against rare earth ion in $LnCoO_3$ compounds.

vacancies being labile, participates in the oxidation reaction and CO adsorbed on cations adjacent to these vacancies alone undergo oxidation. This postulate that CO is adsorbed on cations adjacent to oxygen vacancy (and hence of variable oxidation state) is substantiated by the multiple IR bands on adsorption of CO on these $LaBO_3$ oxides (35). Even though the lattice oxygen is participating in this oxidation reaction the adsorbed oxygen plays an important role and is substantiated by the same trend observed for oxygen adsorption, reducibility and total oxidation of CO (18,36). It is interesting to note that the catalytic activity for the oxidation of CO on high T_c superconducting oxide $YBa_2Cu_3O_{7-\delta}$ also decreased with increasing δ with a turning point at $\delta = 0.9$ showing that the oxygen vacancy of a particular type alone is responsible for this reaction (37).

Though one can obtain a number of correlations of catalytic activity of oxidation of CO on perovskites with a variety of physico-chemical properties of these systems every one of them is directly or indirectly related to the electronic configuration of the B site ion and the B-O bond strength. If one were to optimize the activity for CO oxidation on perovskites then the parameter that has to be modulated is the net charge density around the B-site ion. Since this can be achieved in a number of ways in perovskite systems, they appear to be promising catalysts for the oxidation of CO.

II. Reduction of NO

Shelef and Gandhi (38) have reported that ruthenium containing catalysts promoted the reduction of NO to N_2. In order to improve the stability of these catalysts they (39) have suggested that they could be used in combination with barium or lanthanum as barium ruthenate or lanthanum ruthenate. Voorhoeve et al. (7) identifying NO reduction reaction as a "reagent catalysis" (intrafacial) proposed a sequential mechanism involving lattice oxygen, with NO adsorption as the rate determining step (11). The NO adsorption can take place as nitrosyl or chelate complex type (40) or by dissociation with concurrent oxidation of the catalyst (41). Peña et al. (42) studied the interaction of NO with $LaFeO_3$ by IR spectroscopy and showed that IR bands assignable to interactions of NO with surface cations (dinitrosyl 1910, 1810 cm^{-1}), mono-nitrosyl (1775, 1725 cm^{-1}) and surface anions (bridge bidentate nitrates and nitrites < 1600 cm^{-1}) are observed. Because of the intrafacial nature of this interaction, it is natural to expect an effect of oxygen binding energy on the observed catalytic activity. $LaMnO_{3.5}$ was found to be more active in NO reduction than a sample with lesser oxygen content namely $LaMnO_{3.1}$ (7). This difference in reactivity could arise due to: (i) the differences in the Mn^{4+} content, (ii) the nature of the A ion, or (iii) the extent of A ion vacancies. The results on a series of systems with constant Mn^{3+}/Mn^{4+} ratio but varying A site ions showed that the molecular adsorption of NO which occurs on the B site ion is the same for all the systems studied, while the dissociative adsorption takes place on an oxygen vacancy. Since the formation of a number of oxygen vacancies is related to the binding energy of surface oxygen to the different type of A site ions, the catalytic activity in NO reduction is inversely related to the binding energy of oxygen (1).

The adsorption profile of NO with a series of $LaBO_3$ oxides (B = Cr, Mn, Fe, Co and Ni) showed two maxima at $LaMnO_3$ and $LaCoO_3$ (43). This type of twin peaked structure could result from the crystal field effects. The data are insufficient at present, to project that one would get the same type of twin peaked structure in the reactivity for the reduction of NO.

Though oxygen deficient perovskites could be used as catalysts for the reduction of NO, their activity is not comparable to that of Pt supported catalysts. This could have been due to either poisoning of the catalyst by the oxygen generated in NO reduction or to restricted mobility of lattice oxygen. $YBa_2Cu_3O_{7-\delta}$ systems are known to sustain considerable oxygen deficiency. The exploitation of this system for NO reduction has already been probed, since these systems are known to take up a considerable amount of NO in its lattice (44,45). As expected a MgO supported $YBa_2Cu_3O_{7-\delta}$ system was found to have higher activity for NO reduction than the Pt supported catalyst (45). Comparison of the catalytic activity of these two systems with that of a highly active perovskite system is shown in Fig. 8. These studies support the postulate that this reaction is intrafacial in nature and those cations with the possibility of variable valence (as in $YBa_2Cu_3O_{7-\delta}$ system; $Cu^+ \rightleftharpoons Cu^{2+}$) adjacent to the oxygen vacancy function as the active center for the activation of NO.

It is clear that the formulations for highly active catalysts for NO reduction should be based upon systems which can sustain the basic crystal structure but at the same time can tolerate a

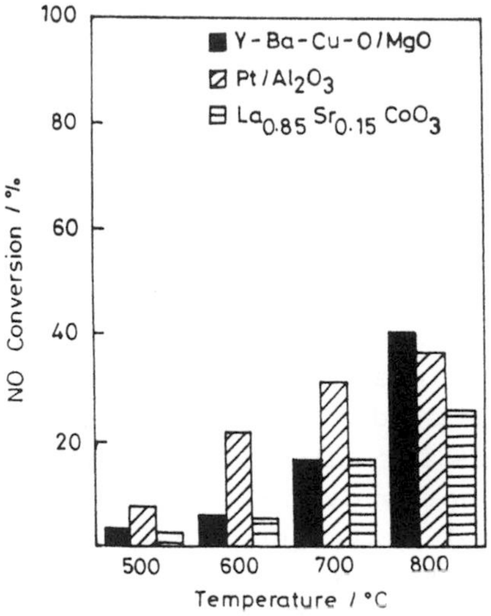

Fig. 8. Comparison of catalytic activity for NO reduction by superconducting oxide $YBa_2Cu_3O_{7-\delta}$ on MgO, Pt/Al_2O_3 and $La_{0.85}Sr_{0.15}CoO_3$ (catalyst weight used is 0.5 g).

considerable amount of oxygen vacancy. Among the known systems, only perovskite oxides can effectively satisfy this postulate and thus account for the emphasis on this series.

III. Simultaneous NO-CO Reaction

Libby (47) has suggested as early as 1971, that $LaCoO_3$ might be a good catalyst for treatment of auto-exhaust gases and would be cheaper than the noble metal based systems. Sorenson et al. (48) first examined $LaCoO_3$ as a catalyst for exhaust emission control and showed that high NO conversion could be achieved at high CO concentrations while high conversions in CO and hydrocarbons are achieved at low levels of CO. They have shown that NO reduction takes place with concurrent reduction of $LaCoO_3$ to an oxygen deficient phase, thus corroborating the postulate that NO reduction is intrafacial in nature. Johnson et al. (49) have evaluated $La_{0.7}Pb_{0.3}MnO_3$ as catalysts for auto-exhaust control with a view to develop promising systems which can resist poisoning by leaded gasoline exhaust. However, these systems were found to decompose with the precipitation of metallic lead, which complicated the interpretation of the observed behavior in terms of catalytic activity of the parent oxide. It is therefore necessary that in addition to choosing appropriate high active B ions, in ABO_3 compounds, the nature of the A ion should also be looked into so that the system is stable under the ambient auto-exhaust emission conditions as well as resistant to poisoning by lead sulphur or to other ambient physical conditions like temperature (sintering, stability) presence of acid (etching) and water vapor. The studies reported in literature seem to favor based on these considerations, Mn or Co as the B site ions and La and Sr as the A site ions. It is also possible some noble metals can be substituted at the B site so as to generate " three way" catalyst systems for auto-exhaust emission control. However, these systems have not provided either the required selectivity (N_2 from NO) or could resist degradation.

Mechanistically, however, the CO-NO reaction on perovskites seems to proceed through the formation of N_2O and surface isocyante species (43,50). If this possibility were to exist, then it is necessary that the systems have to be tailored in such a way that direct oxidation of CO becomes a facile process over the isocyante species formation. This may be possible with proper choice of the cations in the perovskite oxides with orbitals of suitable symmetry and energy which can activate the CO molecule directly so as to promote the oxidation reaction over the isocyanate species formation route.

Since the high T_c superconducting oxide phase couples the desirable features for NO reduction and CO oxidation, it is natural to expect that these systems will be examined in considerable depth for the catalytic reduction of NO by CO. The preliminary reports that are already available (51) show that this expectation is true and that the reaction: $NO + CO \rightarrow 1/2\ N_2 + CO_2$ could be carried out stoichiometrically by proper choice of temperature of the reaction and pretreatment of the initial catalytic phase chosen. The advent of a new series of perovskite oxides capable of sustaining the structure with cations of variable valency and considerable oxygen vacancy may lead to a remarkable progress in the search for alternative auto-exhaust emission control catalysts.

References

1. Voorhoeve, R.J.H., Perovskite-related oxides as oxidation-reduction catalysts, *Advanced Materials in Catalysis*, *J.J. Burton*

and R.C. Carter (Eds.), Academic Press Inc., New York, 1977, p.129.
2. Edwards, P.P., M.R. Harrison and R. Jones, Superconductivity returns to chemistry, *Chem. Brit.*, **23**: 962 (1987).
3. Reller, A. and T. Williams, Perovskites-chemical chameleons, *Chem. Brit.*, **25**: 1227 (1989).
4. Balasubramanian, M.R., R. Natesan and P. Rajendran, Correlation between catalytic activities and physico-chemical properties of perovskite oxides, *J. Sci. Ind. Res.*, **43**: 500 (1984).
5. Viswanathan, B., Solid state and catalytic properties of rare earth orthocobaltites-new generation catalysts, *J. Sci. Ind. Res.*, **146**: 151 (1984).
6. Tejuca, L.G., Properties of perovskite type oxides.II. Studies in catalysis, *J. Less Common Met.*, **146**: 261 (1988).
7. Voorhoeve, R.J.H., J.P. Remeika and L.E. Trimble, Defect chemistry and catalysis in oxidation and reduction over perovskite-type oxides, *Ann. New York Acad. Sci.*, **272**: 3 (1976).
8. Vrieland, E., The activity and selectivity of Mn^{3+} and Mn4+ in lanthanum calcuium manganites for the oxidation of ammonia, *J. Catal.*, **32**: 415 (1974).
9. Shimizu, T., Effect of electronic structure and tolerance factor and CO oxidation activity of perovskite oxides, *Chem. Lett.*, 1 (1980).
10. Goldschmidt, V.M., Geochemische verteilungsgesetze der Elemente VII, *Skrif. Nor. Vidensk Akad. Mat. Natur. Oslo*, **I(8)**: 7 (1926).
11. Voorhoeve, R.J.H., J.P. Remeika and L.E. Trimble , Nitric oxide and perovskite catalysts. Solid state and catalytic chemistry, *The Catalytic Chemistry of Nitrogen Oxides, R.L. Klimisch and J.G. Larson (Eds.)*, Plenum Press, New York, 1976, p. 215.
12. Tascon, J.M.D. and L.G. Tejuca, Adsorption of oxygen on the perovskite type oxide, $LaCoO_3$, *Z. Phys. Chem. N.F.*, **121**: 79 (1980).
13. Tascon, J.M.D. and L.G. Tejuca, Adsorption of CO on the perovskite type oxide, $LaCoO_3$, *Z. Phys. Chem. N.F.*, **121**: 63 (1980).
14. Tascon, J.M.D., J.L.G. Fierro and L.G. Tejuca., Kinetics and mechanism of CO oxidation on $LaCoO_3$, *Z. Phys. Chem. N.F.*, **124**: 249 (1981).
15. Goodenough, J.B., Metallic oxides, *Progress in Solid State*

Chemistry, H. Reiss (Ed.), Pergamon Press, Oxford, 1971, Vol. 5, p. 145.
16. Arakawa, T., A. Yoshida and J. Shiokawa, The catalytic properties of rare earth cobaltites and related compounds, *Mat. Res. Bull.*, **15**: 347 (1980).
17. Arakawa, T., A. Yoshida and J. Shiokawa, The catalytic activity of rare earth manganites, *Mat. Res. Bull.*, **15**: 268 (1980).
18. Fierro J.L.G., J.M.D. Tascon and L.G. Tejuca, Physico-chemical properties of $LaMnO_3$: readucibility and kinetics of oxygen adsorption, *J. Catal.*, **85**: 205 (1984).
19. Om Prakash, P. Ganguly, G. RamaRao, C.N.R. Rao, V.G. Bhide and D.S. Rajoria, Rare earth cobaltite catalyst-relation of activity to spin and valence state of cobalt, *Mat. Res. Bull.*, **9**: 1173 (1974).
20. Nagasubramanian, G., B. Viswanathan and M.V.C. Sastri, Orbital symmetry and its role in catalysis - Decomposition of nitrous oxide and oxidation of carbon monoxide, *Ind. J. Chem.*, **16A**: 642 (1978)
21. Wolfram, T., E.A. Kraut and F.J. Morin, d band surface states in transition metal perovskite crystals 1. Qualitative features and application to $SrTiO_3$, *Phys. Rev. B*, **7**: 1677 (1973).
22. Tascon, J.M.D., and L.G. Tejuca, Catalytic activity of perovskite type lanthanum metal oxides $LaMeO_3$, *React. Kinet. Catal. Lett.*, **15**: 185 (1980).
23. Viswanathan, B. and S. George, Oxidation of carbon monoxide on rare earth cobaltites - Role of spin state equilibrium, *React. Kinet. Catal. Lett.*, **27**: 321 (1985).
24. Yamazoe, N., S. Furukawa, Y. Teraoka and T. Seiyama, The effect of oxygen sorption on the crystal structure of $La_{1-x}Sr_xCo_{3-\delta}$, *Chem. Lett.*, 2019 (1982).
25. Viswanathan, B., and S. George, Kinetics and mechanism of carbon monoxide oxidation on rare earth orthocobaltites, *Ind. J. Tech.*, **22**: 348 (1984).
26. Richter, L., S.D. Badar and M.B. Brodsky, Ultraviolet, X-ray photoelectron and electron energy loss spectroscopy studies on $LaCoO_3$ and oxygen chemisorbed on $LaCoO_3$, *Phys. Rev. B*, **22**: 3059 (1980).
27. Kojima, I., H. Adachi and I. Yasumori, Electronic structure of the $LaBO_3$ (B = Co,Fe,Al) perovskite oxides related to catalysis, *Surf. Sci.*, **130**: 50 (1983).

28. George, S. and B. Viswanathan, Catalytic oxidation of CO on $La_{1-x}Sr_xCoO_3$ perovskite oxides, *React. Kinet. Catal. Lett.*, **22**: 411 (1983).
29. Nakamura, T., M. Misono and Y. Yoneda, Catalytic properties of perovskite type mixed oxides $La_{1-x}Sr_xCoO_3$, *Bull. Chem. Soc. Jpn.*, **55**: 394 (1982).
30. Nakamura, T., M. Misono and Y. Yoneda, Reduction-oxidation and catalytic properties of $La_{1-x}Sr_xCoO_3$, *J. Catal.*, **83**: 151 (1983).
31. Teraoka, Y., M. Yoshimatsu, N. Yamazoe and T. Seiyama, Oxygen sorptive properties and defect structure of perovskite type oxides, *Chem. Lett.*, 893 (1984).
32. Yamazoe, N., Y. Teraoka and T. Seiyama, TPD and XPS study of thermal behavior of adsorbed oxygen on $La_{1-x}Sr_xCoO_3$, *Chem. Lett.*, 1767 (1981).
33. Nakamura, T., M. Misono and Y. Yoneda, Reduction and oxidation and catalytic properties of perovskite type mixed oxide catalysts, *Chem. Lett.*, 1589 (1981).
34. Viswanathan, B. and S. George, On the nature of active species in the oxidation of CO on $LnCoO_3$ type perovskites, *Ind. J. Tech.*, **23**: 470 (1985).
35. Tejuca, L.G., C.H. Rochester and J.L.G. Fierro., Infrared spectroscopic study of the adsorption of pyridine, CO and CO_2 on the perovskite type oxides $LaMO_3$, *J. Chem. Soc. Faraday Trans.I*, **80**: 1089 (1984).
36. Fierro, J.L.G., J.M.D. Tascon and L.G. Tejuca, Surface properties of $LaNiO_3$: Kinetic studies of reduction and of oxygen adsorption, *J. Catal.*, **93**: 83 (1985).
37. Jiang, A., Y. Peng, Q.W. Zhou, P.Y. Geo, H.Q. Yuan and J.F. Deng, The catalytic oxidation of CO on superconductor $YBa_2Cu_3O_{7-\delta}$, *Catal. Lett.*, **3**: 235 (1989).
38. Shelef, M. and H.S. Gandhi, Ammonia formation in catalytic reduction of nitric oxide by molecular hydrogen, *Ing. Eng. Chem., Prod. Res. Dev.*, **11**: 2 (1972).
39. Shelef, M. and H.S. Gandhi, The reduction of nitric oxide in automobile emissions - stabilisation of catalysts containing ruthenium, *Plat. Met. Rev.*, **18(1)**: 2 (1974).
40. Voorhoeve, R.J.H., J.P. Remeika and L.E. Trimble, Perovskites containing ruthenium as catalysts for nitric oxide reduction, *Mat. Res. Bull.*, **9**: 1393 (1974).
41. Voorhoeve, R.J.H., J.P. Remeika, L.E. Trimble, A.S. Cooper, F.J. Disalvo and P.K. Gallagher, Perovskite like $La_{1-x}K_xMnO_3$

and related compounds: solid state chemistry and catalysis of the reduction of NO by CO and H_2, *J. Solid State Chem.*, **14**: 395 (1975).
42. Peña, M.P., J.M.D. Tascon and L.G. Tejuca, Surface interactions of NO with $LaFeO_3$, *Nouv. J. Chim.*, **9**: 591 (1985).
43. Tascon, J.M.D., L.G. Tejuca and C.H. Rochester, Surface interactions of NO and CO with $LaMO_3$, *J. Catal.*, **95**: 558 (1985).
44. Tabaka, K., H. Fukuda, S. Kohiki, N. Mizuno and M. Misono, Uptake of NO by $YBa_2Cu_3O_7$, *Chem. Lett.*, 799 (1988).
45. Arakawa, T. and G. Adachi, The direct reaction between nitric oxide and the superconductor, $YBa_2Cu_3O_{7-\delta}$, *Mat. Res. Bull.*, **24**: 529 (1989).
46. Shimada, H., S. Miyama and H. Kuroda, Decomposition of nitric oxide over Y-Ba-Cu-O mixed oxide catalysts, *Chem. Lett.*, 1797 (1988).
47. Libby, W.F., Promising catalyst for autoexhaust, *Science*, **171**: 499 (1971).
48. Sorenson, S.C., J.A. Wronkiewicz, L.B. Sis and G.P. Wirtz, Properties of $LaCoO_3$ as a catalyst in engine exhaust gases, *Bull. Am. Ceram. Soc.*, **53**: 446 (1974).
49. Gallagher, P.K., D.W. Johnson, J.P. Remeika, F. Schrey, L.E. Trimble, E.M. Vogel and R.J.H. Voorhoeve, The activity of $La_{0.7}Sr_{0.3}MnO_3$ without Pt and $La_{0.7}Pb_{0.3}MnO_3$ with varying Pt catalysts for the catalytic oxidation of CO, *Mat. Res. Bull.*, **10**: 529 (1975).
50. Olivan, A.M.O., M.A. Peña, L.G. Tejuca and J.M.D. Tascon, A comparative study of the interactions of NO and CO with $LaCrO_3$, *J. Mol. Catal.*, **45**: 355 (1988).
51. Mizuno, N., M. Yamato and M. Misono, Reactions of monoxides of carbon and nitrogen over the superconducting lanthanoid mixed oxide $YBa_2Cu_3O_7$, *J. Chem. Soc., Chem. Comm.*, 887 (1988).

Chapter 14

Partial Oxidation of Hydrocarbons and Oxygenated Compounds on Perovskite Oxides

T. Shimizu

Hakodate National College of Technology, Hakodate,
Tokura-cho 14-1, 042 Japan

I. Introduction

Research of perovskite oxide catalysts (ABO_3) with a rare-earth ion as an A-site and a transition metal ion as a B-site has been concentrated on the complete oxidation of hydrocarbons, particularly related to exhaust control, and revealed that they are potential catalysts for deep combustion of hydrocarbons (1). The complete oxidation activities have been reported to be mainly controlled by the physicochemical property of the B-site metal cations such as the electronic configuration of d-electron (2), the binding energy of B-O bond (3) and the stabilization energy of the crystal field (4), rather than the relatively small and less important effect of the rare-earth ion of the A-site (5) and also improved by the substitution of other metal cations for A- or B-site (6,7). Although it is difficult to find many investigations on the application of perovskite-type oxides to partial oxidation, in this chapter, the movement to the partial oxidation of hydrocarbons and oxygenated compounds using various perovskite oxides as catalysts is summarized.

The physical properties of perovskite oxides that are important in catalytic partial oxidation of hydrocarbons and oxygenated compounds are primarily the activation of oxygen in the bulk and the mobility of oxygen on the surface which are strongly affected by the valence states of A- or B-site metal cation in the perovskite structure.

Voorhoeve (8) proposed two types of catalytic processes for perovskite-type catalysts: one is the intrafacial and the other the suprafacial process. The intrafacial process, including the release of oxygen from the lattice or the reverse process, shows the redox behavior. The suprafacial process is the reaction on the surface where the adsorbed species strongly participates. The difference of lattice and adsorbed oxygen on the surface, is not yet sufficiently clear and should be distinguished for the discussion of the role of oxygen in the partial oxidation process.

Lattice oxygen has been usually shown to be reasonable for many useful partial oxidation reactions and its catalytic activites are often determined by thermodynamic properties of oxides (9). Particularly, the binding energy of oxygen is considered to be important because of the high reaction temperature. Consequently, the energy levels at the surface or bulk of oxygen atoms, which coordinate with the metal cations in a lattice, should be known. However, not even the relation between the physical properties of oxygen atoms and the rate of partial oxidation processes is well known.

On the other hand, the adsorbed oxygen is generally regarded as a nonreacting species in partial oxidation and weakly bonded to the B-site metal cation at the surface of perovskite oxides. The reactivity of adsorbed oxygens differs from that of the lattice oxygen because of their bonding strength and is correlated with high complete oxidation activity for hydrocarbons. However, this reactivity seems to be often characteristic of partial oxidation reactions for which the adsorbed oxygen is responsible. In such a case, the reaction proceeds at relatively low temperatures, as in the oxidative dehydrogenation of alcohols.

Furthermore, the substitution by other metal cations of the A- and/or B-site has brought about and controlled the formation of lattice defects and oxygen vacancies. Such defects or oxygen vacancies are advantageous for the formation of adsorption or absorption sites of oxygen and the amount of adsorbed oxygen has been also considered to be appreciably affected by the kind of vacancies as well as its reactivity.

From the extensive literature on catalyst applications of perovskite oxides only a few papers relevant for our purpose were reviewed here.

II. Partial Oxidation of Hydrocarbons

In partial oxidation of hydrocarbons for applications of perovskite oxides where the transfer of oxygen to the surface from bulk or gas phase was applied, one of the important problems, as mentioned above, is the source of an active state of oxygen species which contributes to partial oxidation under the given conditions. Whether this oxygen species is caused through the lattice or adsorbed state, as shown in Table 1, is considered to be strongly dependent on the reaction temperature, reactant composition and kind of catalysts. In this table, all of the reactions, except complete oxidation of methane and propane, was done in very diluted oxygen and reactant gas; however, the complete oxidation predominantly occurred with partial oxidation. At a higher temperature lattice oxygen mainly participates as observed in a methane coupling reaction, while at lower temperatures adsorbed oxygen exhibits larger activity for oxidative dehydrogenation of 1-butene or complete oxidation of methane or propane. At middle temperatures it is concluded

Table 1. Effect of Temperature and Surface Oxygen for Partial Oxidation of Hydrocarbon on Perovskite Catalysts

T (K)	Cat.	Oxygen[5]	Reactant	Product	Con.[6] (%)	Sel.[7] (%)	Re
1023	$BaTiO_3$	L	Methane	C_2 compounds	23.0	31.7	18
770	$LaCoO_3$	L	Toluene	Benzaldehyde	27.5	42.6	29
773	Ba-Bi-Te[1]	A<L	Propane	PAD[8] Acrolein	45.0	98.0	26
680	Ba-Bi-Te	A<L	Propane	Methanol Acetaldehyde	25.0	70.9	26
593	La-Sr-Fe[2]	A=L	1-Butene	Butadiene	88.0	59.0	28
873	La-Sr-Mn[3]	A	Methane	CO_2 Water	80.0	100	1
500	La-Sr-Mn[4]	A	Propane	CO_2 Water	5.5	100	16

1, $Ba_{1.85}Bi_{0.1}\square_{0.05}(Bi_{2/3}\square_{1/3}Te)O_6$; 2, $La_{0.75}Sr_{0.25}FeO_3$; 3, $La_{0.6}Sr_{0.4}MnO_3$; 4, $La_{0.4}Sr_{0.6}MnO_3$; 5, Contribution of surface oxygen: L and A refer to lattice and adsorbed oxygen, respectively; 6, Conversion; 7, Selectivity; 8, PAD: Propionaldehyde.

that the perovskite oxide containing very extensive vacancies and defects is active.

A. Oxidative Coupling of Methane

Oxidative coupling reaction of methane is the most important reaction to convert the natural gas to liquid hydrocarbon fuels or as one of the economically valuable reaction processes. Many alkali-promoted metal oxides that enhance this reaction have been examined under a variety of reaction conditions. However, it has been shown that the activity and selectivity of catalysts for this reaction are very dependent on the experimental conditions such as temperature, space velocity, CH_4/O_2 ratio, etc. (10-12).

Some perovskite-type oxides, which were shown in Table 2 with other alkali-doped metal oxides, have been also studied as catalysts for a methane coupling reaction. Imai et al. (13) have reported that C_2 selectivity of 47% and total methane conversion of 25% are obtained at 983 K on $LaAlO_3$ catalyst prepared by the mist decomposition method. They suggested that an amorphous phase of $LaAlO_3$, which can adsorb oxygen reversibly and in which aluminum ion dispersed highly into the amorphous lanthanum oxide, is responsible as the active site.

Furthermore, the study of the nature of oxygen species adsorbed on $LaAlO_3$ was carried out by these authors (14) using the delayed pulse technique and TPD method. They revealed that an adsorbed oxygen species, which is strongly bonded to the surface and stable even after outgassing at 823 K, is effective for the formation of C_2 compounds, while a gaseous or weakly adsorbed oxygen species is involved in the combustion reaction.

A similar assignment was suggested also by France et al. (15) using $LaMnO_3$ and alkali doped $LaMnO_3$. The selectivity to C_2 hydrocarbons was correlated with higher binding energies for oxygen at the surface. It was shown that the existence of strongly bonded oxygen on the surface is effective to obtain more selectively hydrocarbons from methane, in contrast to the weakly bonded oxygen necessary for deep oxidation of hydrocarbons.

In general, the behavior of oxygen species on the perovskite surface has been known to be reflected in the Temperature Programmed Desorption (TPD) spectrum. TPD spectra of $LaMO_3$ (M = Co, Ni and Mn) perovskite type oxide shows a large desorption peak in the temperatures higher than 773 K, but those with Cr or Fe do not show any oxygen desorption peak at this temperature range. In Sr-substituted perovskite oxides, for example, Sr-substituted $LaCoO_3$ and $LaFeO_3$ show two desorption peaks which increase with the increase of Sr-substitution at two temperature ranges: the peak of low temperature (below 700 K) is usually produced by desorption of weakly adsorbed oxygen and has been suggested (9) to be also associated with oxygen vacancy, while, that of high temperature (upper 773 K) is considered to be lattice oxygen.

Table 2. Comparison of Activity and Selectivity for the Oxidative Coupling of Methane on Perovskite Catalysts and Other Metal Oxides

Catalyst	S_{BET} (m_2/g)	Conv. (mol% CH_4)	C_2 Sel. (mol% C)	C_2 yield (mol% C)	$C_2^=/C_2$	Ref.
[Perovskites]						
$LaMnO_{3.0}$	3.4	10.1	11.0	1.1	0.6	15
$LaMnO_{3.1}$	6.8	13.6	37.2	5.1	0.8	15
$La_{0.9}K_{0.1}MnO_3$	1.9	19.2	41.8	8.0	1.2	15
$La_{0.9}Na_{0.1}MnO_3$	1.5	21.0	63.4	13.3	2.6	15
$La_{0.8}Na_{0.2}MnO_3$	0.9	14.1	71.6	10.1	2.1	15
$LaAlO_3$	2.0	25.1	46.8	11.8	0.6	13
$BaTiO_3$	0.3	23.0	31.7	7.3		18
$SrZrO_3$	0.8	19.5	60.0	11.7		18
$SrCeO_3$	0.3	18.7	47.3	8.9		18
$BaCeO_3$	0.2	26.2	30.8	8.1		18
[Metal Oxides]						
3%Li/MgO	1.0	37.0	49.8	18.4	1.6	17
7%Li/MgO	2.0	37.8	50.3	19.0	1.6	17
10%Li/Sm_2O_3		36.8	57.0	21.0		19
12.5%Li/ZnO		35.8	66.8	23.9	4.9	20
12.5%Na/ZnO		35.5	41.5	14.1	2.3	20
12.5%K/ZnO		33.5	29.6	9.9	1.1	20
MgO	7.2	12.0	39.1	4.7	0.3	10
Li_2CO_3/MgO	0.1	11.5	76.6	8.8	0.4	10
LiCl/Sm_2O_3	1.5	9.0	64.0	5.8	0.7	10
NaCl/MnO_2	0.6	14.0	78.2	11.0	0.9	10
$Na_{0.1}/MnO_2$	0.4	3.0	84.4	2.5	0.5	15
MnO_2	123.0	19.4	48.4	9.4	1.3	15
La_2O_3	6.5	19.1	44.6	8.5	0.7	15

On the other hand, in Sr-substituted $LaMnO_3$ (16), it was shown that the high-temperature peak shifts to a higher temperature and the peak height becomes lower with the increase of Sr-substitution. The same trend may be observed in alkali doped $LaMnO_3$ which is active for a methane coupling reaction. However, measurement of TPD on this catalyst is not conducted so that it is not possible to give an explanation for the TPD effects of alkali doping.

The relation between methane coupling activity and the TPD spectrum was studied by Tagawa et al. (14) for $LaAlO_3$ catalysts. They reported that the amount of oxygen molecules calculated from two desorption peaks of oxygen below 983 K is about 1.5 and 3.4 $\mu mol/m^2$ for lower temperature and higher temperature peaks, respectively, and total amounts (4.9 $\mu mol/m^2$) of oxygen molecules close to one half of monolayer converage (11 $\mu mol/m^2$) and correspond to active oxygen (5.7 $\mu mol/m^2$) required for the methane coupling and complete oxidation below 910 K. They stated that the oxygen species active for C_2 formation has to be reversibly adsorbed and to be stably bonded to the surface of the catalyst even after outgassing at 823 K. Although the quantitative difference between C_2 formation activity and the TPD spectrum or the amount of desorbed oxygen from the surface is not sufficiently clear in the paper by Tagawa et al. (14), it appears very sure that the adsorbed oxygen which shows TPD peak at higher temperature is much more active and highly selective for C_2 formation. Therefore, more active perovskite catalyst is necessary to have a large desorption peak at higher temperature range.

Alkali-metal-treated oxides (5-10 wt% for base metal oxides) have been reported (18-21) to be very active for methane coupling reaction at higher temperature than 1000 K. At these high temperatures, many metal oxides will probably be able to abstract a hydrogen atom from methane, producing a methyl radical which has been considered to be the crucial step in the coupling reaction. The main role of the alkali promoter in host oxides is concluded to increase the possibility of hydrogen abstraction by the more strongly charged lattice oxygen ions on the surface in addition to formation of active centers (17) and decrease the probability that methyl radicals will be further oxidized to CO and CO_2 on the surface (10).

The importance of the contribution of the physicochemical property of alkaline earth metal in the perovskite structure for methane coupling reaction was demonstrated by Nagamoto et al. (18) using perovskite-type oxides (ABO_3) containing Ba, Sr, Ca as a A-site and Ti, Zr, Ce as a B-site cation. The specific surface area of these catalysts, which were prepared by heating at 1423 K for 12 h, is very small. Perovskite oxides containing Ba or Sr in the A-site were stable against produced carbon dioxide at 1023 K and exhibited fairly good catalytic activity as shown in Table 2. According to them the catalytic activites for both C_2 and CO_2 production were closely related with the strength of the basicity of the A cation accelerated by the deviation from the equilibrium of the interatomic distance.

Production of the surface oxygen, with both sufficiently higher binding energy and a minimum number of weakly bonded oxygen sites in the perovskite structure by the doping of alkali as a guest under the formation of complete solid solution, is necessary to design more effective perovskite catalysts for a methane coupling reaction.

Although the decrease of surface area has been described to be necessary for increasing C_2 formation activity, however, coupling activity is concluded to be more greatly increased by the decrease of weakly adsorbed oxygen rather than the decrease of surface area by sintering at high temperature. Therefore, sintering beside a large surface area of perovskite oxides may be impossible, but will be more effective for increasing C_2 formation.

On the other hand, oxidative coupling of methane has also been achieved by the use of Ag/Bi_2O_3 as a catalyst with oxygen being electrochemically pumped by YSZ and the selectivity to C_2 hydrocarbons up to about 47% has been obtained (21). This reaction is also thought to proceed through the abstraction of hydrogen atoms from CH_4 by O^- ion of the surface (22) which is promoted by the oxygen electrochemically pumped on YSZ with perovskite oxide electrodes. The result of comparing perovskite oxides as oxygen catalytic electrodes has been reported by Isaacs and Olmer (23). $La_{0.5}Sr_{0.5}FeO_3$ was most active for oxygen reduction. The Sr-substituted $LaFeO_3$ as reported also by Yamamoto et al. (24), may be the most suitable electrode materials at the reaction temperature of methane coupling because of chemical stability with YSZ and high electronic and ionic conductivity. Furthermore, electrocatalytic effects of these perovskites are expected to advance fundamental studies of mechanisms because of the possibility of controlling the coupling reaction rate by applying different potentials.

B. Partial Oxidation of Other Hydrocarbons

The cation vacancy formed by the doping of Bi in the scheelite structure has been known to promote partial oxidation selectivity by the attraction of the proton from hydrocarbon (25). Conner and co-workers (26) tried to substitute Bi metal for the rare-earth-site of the perovskite catalysts with vacancies on the B-site such as $Ba_2(RE_{2/3}\square_{1/3}W)O_6$ discovered by Rauser and Kemmler-Sack (27) having perovskite structure, where RE is a trivalent rare-earth ion and □ represents a vacancy.

A series of defective perovskites with a type of $A_2(Bi_{2/3}\square_{1/3}B)O_6$, where A is Sr or Ba, and B is W, Mo, Te or U, were active for partial oxidation of propane. When $Ba_{1.85}Bi_{0.1}\square_{0.05}(Bi_{2/3}\square_{1/3}Te)O_6$ is used as a catalyst, the partial oxidation selectivity of about 30% for a total propane conversion of about 15% was obtained. The main partial oxidation products were methanol and acetaldehyde below 673 K. However, above 723 K propionaldehyde and acrolein were found. Other products were cracking components which increase with temperature and CO_x (CO and CO_2). Perovskite oxides with both Bi and vacancy in the A- and B-site were considered to be sensitive for partial oxidation of propane and similar to catalysts active for the partial oxidation of propylene.

In these catalysts, only $Ba_2(Bi_{2/3}\square_{1/3}Te)O_6$ was a single phase of perovskite, all others the mixed phase. The contribution of the second phase for partial oxidation may be important. Shimizu (28) studied the catalytic activities for oxidative dehydrogenation of 1-butene to butadiene and 2-butenes using $La_{1-x}Sr_xFeO_3$ ($0 \leq x \leq 1.0$) as catalysts and found that $SrFeO_{3-y}$ is present as a second phase above x = 0.136 and readily reduced by 1-butene, undergoing an incomplete structural change to the more oxygen deficient perovskite oxide by generating CO_2 gas. However, the $La_{1-x}Sr_xFeO_3$ phase was active for both oxidative dehydrogenetion and isomerization of 1-butene. After $La_{0.75}Sr_{0.25}FeO_3$ was reacted with 1-butene at 593 K for 3 h, subsequently reactant components (1-butene:oxygen:water:helium = 1:1:10:10) were introduced in a closed-circulation reaction system at the same temperature. The effect of reduction, as shown in Figure 1, was observed in the yield of butadiene which decreased to about one-half and the yield of 2-butenes which increased slightly. Therefore, it was concluded that the presence of $SrFeO_{3-y}$ as a second phase more strongly affects the isomerization activity than the $La_{0.75}Sr_{0.25}FeO_3$ which is the main phase.

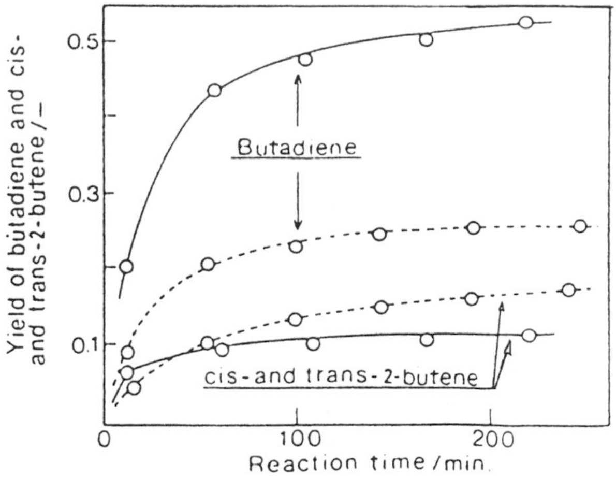

Fig. 1. Effect of reduction for the yield of butadiene and cis- and trans-2-butene. Catalyst: x = 0.25; Reaction temperature: 593 K.

On the other hand, aromatics were shown to be partly oxidized at a higher temperature as shown in Table 1. Madhok (29) reported the vapor phase partial oxidation of toluene to benzaldehyde, benzoic acid, maleic acid and carbon dioxide on $LaCoO_3$ at the temperature range of 623-873 K. Selectivity to benzaldehyde showed the maximum at 773 K. However, that to benzoic acid and maleic acid increased with increasing of temperature. For partial oxidation of benzene only CO_2 as final product was obtained by complete oxidation.

III. Partial Oxidation of Oxygenated Compounds

The oxidation of alcohol has generally been studied in order to develop processes which maximize the formation of partial oxidation while minimizing the formation of undesired by-products such as CO_2 and H_2O. Consequently, for example, the partial oxidation of ethanol to acetaldehyde has been investigated using mild oxidation catalysts such as Fe-molybdate (30), Mo-titanate (31), alumina supported Cu, Ag and Au (32).

Activity for the oxidative dehydrogenation of methanol, ethanol and 2-propanol to aldehydes has been examined at the relatively low temperature range of 473-673 K using various perovskite-type oxides. $SrVO_3$ in which 94% of the vanadium exists at the form of V4+ in cubic perovskite structure has been used by De and Balasubramanian (33) as a catalyst for the oxidative dehydrogenation of methanol and benzene to produce formaldehyde and maleic anhydride, respectively. Catalytic activity and selectivity were determined at the temperature of 553-673 K using air more than required for complete oxidation. In oxidative dehydrogenation of methanol to formaldehyde the selectivity of 90% and total methanol conversion of 44% were obtained at 593 K. On the other hand, for benzene oxidation the main product was CO_2 at same temperature.

Although $SrVO_3$ has been known to be easily reduced to $SrVO_{2.5}$ without a structural change in a mixture of an organic compound and air at reaction temperature (34), they concluded that $SrVO_3$ itself is not able to be reduced, in which the availability or mobility of the oxygen anion is not effective to the formation of formaldehyde.

On the other hand, Shimizu (35) studied the oxidative dehydrogenation of ethanol to acetaldehyde by use of a flow reactor in the temperature range of 473-723 K over $La_{1-x}Sr_xFeO_3$ ($0 \leq x \leq 1.0$) and $LaMeO_3$ (Me = Co, Mn, Ni and Fe). The ethanol oxidation to acetaldehyde occurred by introducing ethanol (15 vol%), oxygen (10-20 vol%) and helium as diluent gas. The formation activity of acetalaldehyde showed the maximum at x = 0.2 and then decreased with x values.

As shown in Figure 2, only water and acetaldehyde were the oxidation products, carbon dioxide and hydrogen were not detected

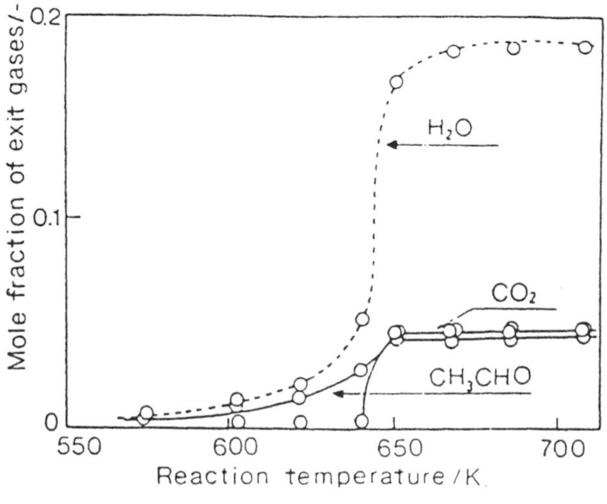

Fig. 2. Distribution of exit gases as a function of reaction temperature. Catalyst: x = 0.1

in the product in the range of 573-623 K. The formation of acetaldehyde proceeds in proportion to 0.5 order in the partial pressure of oxygen and ethanol, respectively; therefore, the overall rate is dependent on the partial perssure of both oxygen and ethanol below 623 K. It was concluded that ethanol dehydrogenation to acetaldehyde occurs between O_2 and ethanol molecules adsorbed on the catalyst. At temperatures higher than 623 K, the production of CO_2 occurred with a considerable amount of water in addition to acetaldehyde and showed no strong temperature dependence. The same trend was observed over the whole range of x in the $La_{1-x}Sr_xFeO_3$ ($0 \leq x \leq 1.0$) perovskite oxides. Acetone was obseved as a by-product in this temperature range but its amount was negligible compared to that of the other products.

Mole fractions of product in exit gases are plotted in Figure 3 against the inlet partial pressure of oxygen at 623 K. It was observed from this figure that the formation of acetaldehyde is appreciably dependent on the partial pressure of oxygen and proceeds through a maximum at about 0.1 atm whereas the formation of CO_2 and H_2O increases linearly with increasing oxygen partial pressure. The complete oxidation of ethanol and acetaldehyde

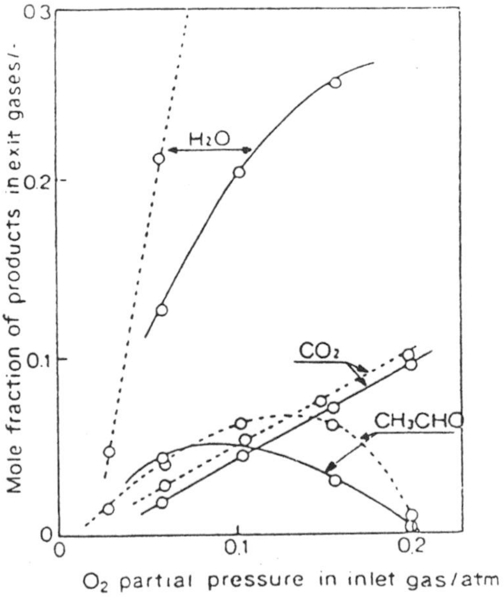

Fig. 3. Effect of inlet O_2 partial pressure for mole fraction of products. Catalyst: x = 0.2. Full and dotted line show before and after CO reduction (673 K), respectively.

occured at higher oxygen partial pressure. These results suggest that the acetaldehyde formation and the fast oxidation of ethanol to CO_2 and H_2O take place simultaneously. Similar results were reported for the catalytic oxidation of ethanol on Ru/SiO_2 (36). The activity of $LaMeO_3$ for the formation of acetaldehyde was in the following order: La-Co > La-Mn > La-Ni > La-Fe perovskite oxide. This order was similar to the one for the catalytic activity of CO oxidation, or the degree of surface reduction by CO, on these perovskite oxides.

The catalytic activity for the dehydrogenation of 2-propanol to acetone and hydrogen has been examined by Radha et al. (37) at the temperature range 473-573 K by the introduction of 2-propanol and nitrogen as a diluent gas in a flow reactor over the p-type La_2MnMO_6 (M = Co, Ni, Cu) perovskite oxides. They found that all the perovskites promote only the production of acetone and hydrogen in this reaction. The adsorption of acetone and hydrogen brought about the decrease of the conductivity of the catalyst by donation of an electron to the surface. They concluded from the changes in conductivity that the surface is predominantly covered with acetone under the reaction conditions and the desorption of acetone is the rate-controlling step of the reaction, involving electron transfer from the surface to adsorbed species.

In the above three reactions the complete oxidation did not occur. A very interesting aspect for the dehydrogenation of 2-propanol is that hydrogen is produced in the absence of oxygen even on the perovskite oxides usually used as complete oxidation catalysts at this temperature range. Further, the fact that the oxidation of alcohol to aldehyde proceeds at low temperature and oxygen pressure on these perovskite catalysts support that the oxygen weakly adsorbed on perovskite oxide contributes to the formation of aldehyde.

The degree of complete oxidation will be controlled by the amount of weakly adsorbed oxygen on the surface and the partial pressure of gas phase oxygen contained in a feed. If more or less oxygen with alcohol is fed in a reactor as observed in the case of the dehydrogenation of 2-propanol aldehyde alone may be produced.

In general, conductivity of perovskite oxides containing transition metal is strongly influenced by the oxygen pressure of atmosphere. This is explained by the electron donor and acceptor mechanisms from surrounding oxide ion. The changes in conductivity by the alcohol adsorption were studied to determine the complete oxidation mechanism and the behavior of sensors for alcohols. Arakawa et al. (38) studied the catalytic properties and activity of n-type $LnFeO_3$ (Ln = La-Gd) for the complete oxidation of methanol. Reaction was carried out by the flow method using a gaseous mixture of methanol (6 vol%) and oxygen (9 vol%) diluted with nitrogen and its activity was in the following order: Gd > Eu > Sm > Nd > Pr > La. They showed, from measuring binding energy of Fe $2p_{3/2}$ in X-ray photoelectron spectra at 653 K, that complete oxidation activity increases with the decrease of the radius of the rare-earth ion or relative magnitude of covalency for the Fe-O bond in $LnFeO_3$, that is, with increasing the ionic property of the Fe-O bond. As a possible mechanism, it was proposed that the reaction between the adsorbed methanol and oxygen is the rate-determining step in the overall reaction rate. The increase in conductivity by the adsorption of methanol was explained by a donor mechanism involving an electron transfer between the surface and methanol adsorbed at ionic state.

On the other hand, a series of the p-type perovskite oxides such as $LaCo_{1-x}Ni_xO_3$ and $LaCo_{1-x}Fe_xO_3$ were used by Ganguly (39) for the ethanol complete oxidation using oxygen as a carrier gas at the temperature range of 473-623 K. The measurement of electrical conductivity showed that the weakly adsorbed oxygen ions are involved in the oxygen transfer mechanism responsible for the oxidation.

Obayashi et al. (40,41) have shown that $LaNiO_3$ and similar perovskite oxides of the type of $(Ln,Sr)BO_3$, were Ln is the lanthanide metal and B is the transition metal, can be used as good ethanol sensors. In the presence of ethanol and air as a carrier gas at 423-673 K, the electrical resistivity of these oxides increases by the adsorption amount proportional to the concentration of ethanol with a comparatively short response time and it recovers to the initial value when the gas is removed. When $Ln_{0.5}Sr_{0.5}CoO_3$ is examined as a sensing element the response ratio increases and the response speed decreases with an increase in the atomic number of the Ln. Response speed is improved by the addition of In_2O_3 or PbO_2. It was shown by the above authors that the increase in resistivity is caused by an oxygen deficient composition at elevated temperatures. Furthermore, its degree and effect were related to the electronic configurations of the transition metal.

Rao et al. (42) showed that the increase in resistivity on the adsorption of ethanol is caused by the presence of high-spin ions in the $LaCo_{1-x}Ni_xO_3$ and $LaCo_{1-x}Fe_xO_3$ and therefore the changes in resistivity are connected with the spin state: particularly, the proportion of Co^{3+}.

IV. Conclusion

The perovskite oxides have successfully been studied as suitable catalysts to elucidate some relations between catalytic activity and solid state properties based on having a well-defined bulk structure and various compositions of cations at both A and B sites. However, the study of perovskite oxides as catalysts for partial oxidation has only just started. Although partial oxidation is based on the use of oxygen as co-reactant, the suppression of complete oxidation is necessary. As the systematic and extensive exploration of a variety of perovskite oxides for application as practical catalytic devices rapidly advances, improvements in relatively high temperatures of new catalytic devices such as methane coupling catalysts, sensors, electrode materials and combustion catalysts are anticipated to clarify catalytic reaction mechanisms in which oxygen transfer is almost always correlated with the binding energy of oxygen at the surface or in bulk. The discovery and designing of suitable selective and active perovskite oxides, if successful, would be of enormous industrial significance.

References

1. Arai, H., T. Yamada, K. Eguchi and T Seiyama, Catalytic combustion of methane over various perovskite-type oxides, *Appl. Catal.*, **26**: 265 (1986).
2. Voorhoeve, R.J.H., J.P. Remeika and L.E. Trimble, Defect chemistry and catalysis in oxidation and reduction over perovskite-type oxides, *Ann. N.Y. Acad. Sci.*, **272**: 3 (1976).
3. Shimizu, T., Oxygen adsorption and carbon monoxide oxidation over perovskite oxides, *Chem. Lett.*, 1 (1980).
4. Tascón, J.M.D. and L.G. Tejuca, Catalytic activity of perovskite-type oxides $LaMeO_3$, *React. Kinet. Catal. Lett.*, **15**: 185 (1980).
5. Nitadori, T., T. Ichiki and M. Misono, Catalytic properties of perovskite-type mixed oxides (ABO_3) consisting of rare

earth and 3d transition metals. The roles of the A- and B-site ions, *Bull. Chem. Soc. Jpn.*, **61:** 621 (1988).

6. Nitadori, T. and M. Misono, Catalytic properties of $La_{1-x}A'_x FeO_3$ (A'= Sr, Ce) and $La_{1-x} Ce_xCoO_3$, *J. Catal.*, **93:** 459 (1985).
7. Nakamura, T., M. Misono and Y. Yoneda., Reduction-oxidation and catalytic properties of $La_{1-x}Sr_xCoO_3$, *J. Catal.*, **83:** 151 (1983).
8. Voorhoeve, R.J.H., *Advanced Materials in Catalysis*, Academic Press, New York, 1977.
9. Seiyama, T., N. Yamazoe and K. Eguchi, Characterization and activity of some mixed metal oxide catalysts, *Ind. Eng. Chem. Prod. Res. Dev.*, **24:** 19 (1985).
10. Burch, R., G.D. Squire and S.C. Tsang, Comparative study of catalysts for the oxidative coupling of methane, *Appl. Catal.*, **43:** 105 (1988).
11. Keller, G.H. and M.M. Bhasin, Synthesis of ethylene via oxidative coupling of methane. I. Determination of active catalysts, *J. Catal.*, **73:** 9 (1982).
12. Otsuka, K., K. Jinno and A. Morikawa, Active and selective catalysts for the synthesis of C_2H_4 and C_2H_6 via oxidative coupling of methane, *J. Catal.*, **100:** 353 (1986).
13. Imai, H., T. Tagawa and N. Kamide, Oxidative coupling of methane over amorphous lanthanum aluminum oxides, *J. Catal.*, **106:** 394 (1987).
14. Tagawa, T. and H. Imai, Mechanistic aspects of oxidative coupling of methane over $LaAlO_3$, *J. Chem. Soc., Faraday Trans. 1*, **84:** 923 (1988).
15. France, J.E., A. Shamsi and M.Q. Ahsan, Oxidative coupling of methane over perovskite-type oxides, *Energy and Fuels*, **2:** 235 (1988).
16. Nitadori, T., S. Kurihara and M. Misono, Catalytic properties of $La_{1-x}A'_xMnO_3$ (A'= Sr, Ce, Hf), *J. Catal.*, **98:** 221 (1986).
17. Ito, T., J-X Wang, C.-H. Lin and J.H. Lunsford, Oxidative dimerization of methane over a lithium-promoted magnesium oxide catalyst, *J. Am. Chem. Soc.*, **107:** 5062 (1985).
18. Nagamoto, H., K. Amanuma, H. Nobutomo and H. Inoue, Methane oxidation over perovskite-type oxide containing alkaline-earth metal, *Chem. Lett.*, 237 (1988).
19. Otsuka, K., Q. Liu, M. Hatano and A. Morikawa, The catalysts active and selective in oxidation coupling of methane. Alkali-doped samarium oxides, *Chem. Lett.*, 467

(1986).

20. Matuura, I., Y. Utsumi, M. Nakai and T. Doi, Alkali-doped zinc oxide catalysts for the oxidative coupling of methane, *Chem. Lett.*, 1981 (1986).
21. Otsuka, K., S. Yokoyama and A. Morikawa, Catalytic activity and selectivity control for oxidative coupling of methane by oxygen pumping through yttria-stabilized zirconia, *Chem. Lett.*, 319 (1985).
22. Driscoll, D.J., W. Martir, J.-X. Wang and J.H. Lunsford, Formation of gas-phase methyl radicals over MgO, *J. Am. Chem. Soc.*, **107**: 58 (1985).
23. Isaacs H.S. and L.J. Olmer, Comparison of materials as oxygen catalytic electrodes on zirconia electrolyte, *J. Electrochem. Soc.*, **129**: 436 (1982).
24. Yamamoto O., Y. Takeda, R. Kanno and M.Noda, Perovskite-type oxides as oxygen electrodes for high temperature oxide fuel cells, *Solid State Ionics*, **22**: 241 (1987).
25. Aykan K., D. Halvorson, A.W. Sleight and D.B. Rogers, Olefin oxidation and ammoxidation studies over molybdate, tungstate, and vanadate catalysts having point defects, *J. Catal.*, **35**: 401 (1974).
26. Conner W.C., Jr., S. Soled and A. Signorelli, Propane oxidation over mixed metal oxides: perovskites, trirutiles and columbites, in *Proc. 7th Int. Congr. on Catalysis, K. Tanabe and T. Seiyama (Eds.)*, Kodansha-Elsevier, Tokyo, 1981, p. 1224.
27. Rauser G. and S. Kemmler-Sack, The variation of properties by incorporation of tantalum (V) in barium gadolinium uranium oxide ($Ba_2Gd_{0.67}UO_6$), *Z. Anorg. Allg. Chem.*, **429**: 181 (1977).
28. Shimizu T., Oxidative dehydrogenation of 1-butene over lanthanum strontium iron perovskite oxides, *Nippon Kagaku Kaishi*, 6 (1983).
29. Madhok K.L., Oxidation of toluene on lanthanum cobaltite perovskite ($LaCoO_3$) catalyst, *React. Kinet. Catal. Lett.*, **30**: 185 (1986).
30. Edwards, J., J. Nicolaids, M.B. Cutlip and C.O. Bennett, Methanol partial oxidation at low temperature, *J. Catal.*, **50** (1977).
31. Ono, T, Y. Nakagawa, H. Miyata and Y Kubkawa, Catalytic activity of MoO_3 and V_2O_5 highly dispersed on TiO_2 for oxidation reactions, *Bull. Chem. Soc. Jpn.*, **57**: 1205 (1984).
32. Bond, G.C., *Catalysis by Metals*, Academic Press, New York,

1962, p. 310.

33. De, K.S. and M.R. Balasubramanian, Cubic hypovanadate perovskite as an oxidation catalyst, *J. Catal.*, **81**: 482 (1983).
34. Kestigian, M., J.G. Dickinson and R. Ward, Ion-deficient phases in Ti and V compounds of the perovskite type, *J. Amer. Chem. Soc.*, **79**: 5598 (1957).
35. Shimizu, T., Activity of ethanol oxidation to acetaldehyde over $La_{1-x}Sr_xFeO_3$ and $LaMeO_3$ (Me = Co, Mn, Ni, Fe), *Appl. Catal.*, **28**: 81 (1986).
36. Gonzalez, R.D. and M. Nagai, Oxidation of ethanol on silica supported noble metal and bimetallic catalysts, *Appl. Catal.*, **28**: 57 (1985).
37. Radha, R. and C.S. Swamy., 2-propanol decomposition on La_2MnMO_6 (M = Co, Ni, Cu) perovskites, *Surface Technology*, **24**: 157 (1985).
38. Arakawa, T., S. Tsuchiya and J. Shiokawa, Catalytic properties and activity of rare-earth orthoferrites in oxidation of methanol, *J. Catal.*, **74**: 317 (1982).
39. Ganguly, P., Catalytic properties of transition metal oxide perovskites, *Ind. J. Chem.*, **15A**: 280 (1977).
40. Obayashi, H., Y. Sakurai and T. Gejo, Perovskite-type oxides as ethanol sensors, *J. Solid State Chem.*, **17**: 299 (1976).
41. Obayashi, H. and T. Kudo, Properties of oxygen deficient perovskite-type compounds and their use as alcohol sensors, *Nippon Kagaku Kaishi*, 1568 (1980).
42. Rao C.N.R., O. Prakash and P. Ganguly, Electronic and magnetic properties of $(LaNi_{1-x}Co_xO_3)$, $(LaCo_{1-x}Fe_xO_3)$, and $(LaNi_{1-x}Fe_xO_3)$, *J. Solid State Chem.*, **15**: 186 (1976).

Chapter 15

Photocatalysis on Fine Powders of Perovskite Oxides

T.R.N. Kutty* and M. Avudaithai

Department of Inorganic and Physical Chemistry
Indian Institute of Science, Bangalore 560 012, India

I. Introduction

Fine powders of semiconductor oxides have been widely used as photocatalysts for many reactions. Among the various photocatalytic reactions, water splitting has been given much importance, since it is a promising chemical route for solar energy conversion. Perovskite oxides, in particular $SrTiO_3$, have been commonly used as photocatalysts because some of them can decompose H_2O into H_2 and O_2 without an external bias potential (1). In turn, this is because the conduction band (CB) edges of some of the perovskite oxides are more negative than the H^+/H_2 energy level. Since the catalytic activity is related to the surface properties of the solids, fine powders rather than single crystals are used. Photocatalysis on fine powers can be conveniently discussed in three parts, viz. preparation, characterization and their catalytic activity. Presently, photo-decomposition of water using $SrTiO_3$ fine powders is discussed in greater detail, although other photocatalytic reactions on various perovskite oxides are also briefly dealt with.

---

* Also of Materials Research Centre, Indian Institute of Science, Bangalore 560 012.

II. Photodecomposition of Water

Wrighton et al. (1) observed the photoproduction of H_2 and O_2 in a photoelectrochemical cell having a $SrTiO_3$ single crystal photoanode at zero applied potential. Irradiation of the $SrTiO_3$ crystal, coated with alkali and in contact with water vapor, also resulted in H_2 formation (2). The rate of H_2 generation increased with the concentration of NaOH. A platinized $SrTiO_3$ crystal prereduced in H_2 atmosphere at high temperatures also produced H_2 and O_2 on illumination (3). There are also reports about the photocatalytic decomposition of water on the irradiation of $SrTiO_3$ powder suspensions (4-7). Lehn et al. (5) observed that $SrTiO_3$ particles alone could not catalyse water splitting and that a metal coating was essential. It was found that Rh/$SrTiO_3$ has higher activity than $SrTiO_3$ powders coated with other metals. In general, the intention of metal coating is to decrease the H_2 over-voltage of the semiconductor, which usually being very high, leads to low efficiency in the photoreduction of water. The over-voltage can be decreased by metal deposition. A metal-deposited semiconductor acts as a "short-circuited" photoelectrochemical cell, where oxidation and reduction reactions occur at two different sites, i.e., on the semiconductor and on the metal islets respectively of the same particle. Metal deposition also decreases the recombination of photoproduced electrons and holes by way of enhancing the rate of electron transfer to the metal. Surface modifications such as NiO mounting have also been attempted, where Ni metal is shown to exist at the NiO/$SrTiO_3$ interface (8). It is elucidated that H_2 evolution occurs on NiO/Ni islets whereas O_2 is produced on $SrTiO_3$ sites. The importance of the presence of Ni metal at the interface is suggestive of the transfer of electrons between the two oxides (9).

In all these investigations, commercially available $SrTiO_3$ powders have been used, which have particle size of ≥ 2 μm. It is of interest to prepare ultrafine particles of $SrTiO_3$, particulary in view of the recent interest on the size dependence of the band structure and consequently the electronic properties of the semiconductor crystallites (10). Ultrafine powders of binary oxides are not easy to realise, because of the high temperature required for the reaction between the refractory components. Although the low temperature preparation of $SrTiO_3$ through the metal alkoxide route has been known in literature (11), the preparation of the metal alkoxide is quite tedious and the resulting powders have strong tendency for agglomeration. Recently Thampi et al. (12) have studied the photocatalytic activity of $SrTiO_3$ powders prepared by the sol-gel technique. Kudo et al. (13) found that the photocatalytic activity was strongly dependent upon the calcination temperature of $SrTiO_3$ and the optimum calcination temperature is different for different reactions.

Investigations in our laboratory on the preparation of ABO_3 perovskites, where A = Pb, Ca, Sr, Ba; B = Ti, Zr, Sn have shown that fine powders of perovskites and their solid solutions can be obtained by the hydrothermal methods (14,15). We have also reported that $SrTiO_3$ prepared by this technique has higher photocatalytic activity in water splitting (16).

A. Experimental Procedures

Reactive gel of $TiO_2.xH_2O$ ($3 < x < 8$) and $Sr(OH)_2$ are the starting materials. $TiO_2.xH_2O$ gel is precipitated by mixing NH_4OH with aqueous $TiOCl_2$. A surface adsorbing polymer such as polyvinyl alcohol (PVA) can prevent agglomeration of the gel through steric effects of the adsorbed polymer molecule when present in low concentrations (0.01 %). The gel washed free of chloride ions and is suspended in $Sr(OH)_2$ solution so that the molar ratio of Sr/Ti is maintained at 0.99 to 1.02. The slurry is charged into a teflon-lined pressure vessel and heated at ~ 423 K. The product is separated by a centrifuge and purified by ultrafiltration. Particle size can be further decreased when 2-propanol-H_2O is used as the reaction medium. $SrTiO_3$, so obtained was characterized by wet chemical analysis and also by X-ray diffraction and electron diffraction. Coating of the $SrTiO_3$ fine powders has been carried out by the photodeposition method or the thermal deposition method. Details of the photolysis experiments are given in (16).

B. Characterization

X-ray diffractograms of $SrTiO_3$ powders show considerable line-broadening as compared to the sample prepared by the solid state reaction between $SrCO_3$ and TiO_2 (Figure 1). Half-widths of the reflections are used for calculating the average dimensions of the crystallites (17). The average crystallite diameter of particles prepared in H_2O medium is 40-80 nm and that from 2-propanol-H_2O medium is 8-20 nm.

Transmission electron microscopy indicated platelet morphology for the particles, prepared in water medium with 30-100 nm particle size, whereas the particles prepared in the presence of PVA are 20-60 nm and those from 2-propanol-H_2O medium are 5-20 nm. Electron diffraction revealed that the individual particle is single crystallite with cubic perovskite lattice structure. The lattice defect of these crystallites has been indicated by the elongation of the diffraction spots and associated streaks.

Thermogravimetric analysis of these powders have shown ≤ 0.45 % weight loss and the analysis of the evolved gases has indicated that only H_2O is lost during the heat treatment. Infrared spectra have shown low intensity absorption due to OH^- groups. Hydroxyl ion contents are low for the samples prepared in 2-propanol-H_2O medium. Nitrogen adsorption isotherms at 77 K of $SrTiO_3$ powder are indicative of the presence of transitional pores. The specific surface area obtained by the BET method varies from 46

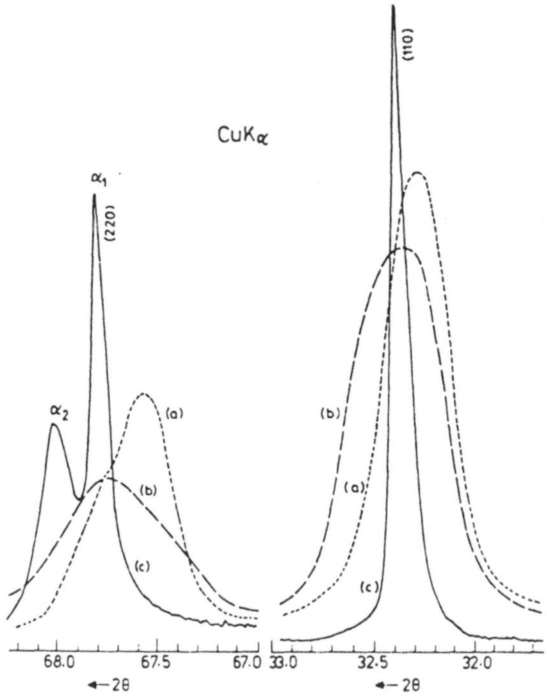

Fig. 1. X-ray diffraction traces of the (100) and (220) reflections of $SrTiO_3$ prepared hydrothermally (a) in aqueous medium, (b) in 2-propanol-H_2O medium, and (c) prepared by the ceramic method.

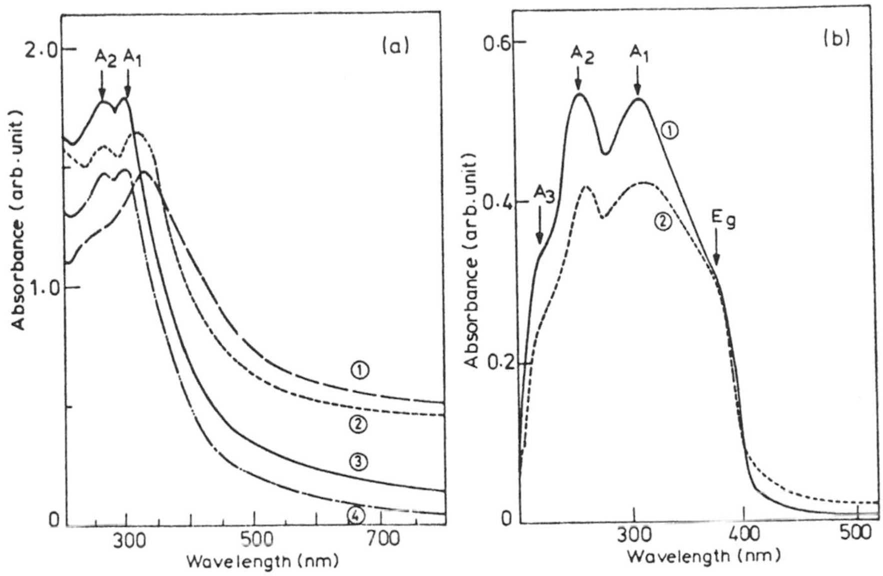

Fig. 2. Optical absorption spectra of $SrTiO_3$ at 298 K. (a) particle suspension in H_2O medium prepared: (1) in aqueous medium, (2) in aqueous medium containing PVA, (3) in 2-propanol-H_2O medium, and (4) in 2-propanol-H_2O medium containing PVA; (b) (1) single crystal, (2) polycrystalline ceramic sample of 50 μm grain size.

to 65 m^2/g. The observed surface areas are somewhat larger than those calculated from the average diameters from TEM studies, further suggestive of the presence of pores on these particles.

The relationship between the crystallite size and optical absorption spectra has been reported for colloids of II-VI semiconductors with an average diameter ≤ 5 nm (10). The absorption spectra of $SrTiO_3$ particle suspensions do not show any abrupt variation around 387 nm, whereas bulk crystal of $SrTiO_3$ has a band gap near 3.2 eV (387 nm) (Fig. 2). The optical absorption of $SrTiO_3$ suspensions continues to rise at lower wavelengths with two distinct maxima (designated as A_1 and A_2) which are at 350 and 270 nm for the particles prepared in water medium, whereas the samples prepared in 2-propanol-H_2O medium have the maxima at 300 and 265 nm. The size-dependent variation of the position of A_1 peak is more prominent than that of A_2. Fig. 2 also shows the A_2-A_1 splitting decreases with decreasing particle size. In comparison with the reported absorption spectra of $SrTiO_3$ single crystal (19), A_1 peak is attributed to (Γ15-Γ12) and A_2 to (Γ25-Γ12) transition. In the case of ultrafine particles, the band edge transition is strongly affected even in the 5-20 nm size range. This can be explained on the basis of a quantum size effect due to the confinement of holes and electrons in a small volume and that the band-gap transition is sensitive to the long range order and lattice defects (10).

C. Electron Paramagnetic Spectra

Understanding the reaction paths of photogenerated holes and electrons in semiconductor particles is necessary for improving the efficiency of H_2 production through catalysed photolysis of water. In this respect, we have reported the results of an EPR investigation of the trap-centers produced in TiO_2 particles during water photolysis at room temperature, under band-gap irradiation (19). Recently the behavior of electron and hole centers on $SrTiO_3$ particles during water photolysis has been further investigated. It is observed that, in contrast to TiO_2 particles, electron-hole trap centers prevail in $SrTiO_3$ particles even before the photolysis experiments and that the nature of the trap-centers varies with the method of preparation. The hole center signals of the hydrothermally prepared $SrTiO_3$ powders are strong and broad (Fig. 3). Although they overlap considerably, $g_x = 2.035$, $g_y = 2.048$ and $g_z = 2.021$ can be deciphered from the multi-structured derivative curves. $g = 1.999$ signal corresponds to the electron-center and the intensity increases tremendously on short-time annealing in H_2 at 1073 K accompanied by the appearance of the broad and shallow signals of $g = 1.968$, corresponding to Ti^{3+} - V_0 centers. Hole-center signals of chemically reduced $SrTiO_3$ powder exhibit pseudo-axial symmetry which is illustrated in Fig. 3 by way of the theoretical line-shape when $g_x \approx g_z < g_y$. Chemical reduction decreases the line-widths. Hole-center signals are modified with stronger intensity, accompanied by the appearance of $g = 2.008$ and 2.055 signals when these powders are dispersed in dilute HNO_3, followed by the washing with distilled water. Sr^{2+} has been detected in the filtrate by the AAS method, demonstrating that the stability of hole-centers is related to Sr-vacancies. This experiment also indicates that the hole-centers are located in the surface to subsurface regions and not in the bulk of the particle. Similar treatment of the powder in dilute acetic acid leads to well-resolved signals with lower intensities and line-widths. CH_3COO^- ions are well known hole scavengers, though less effective and

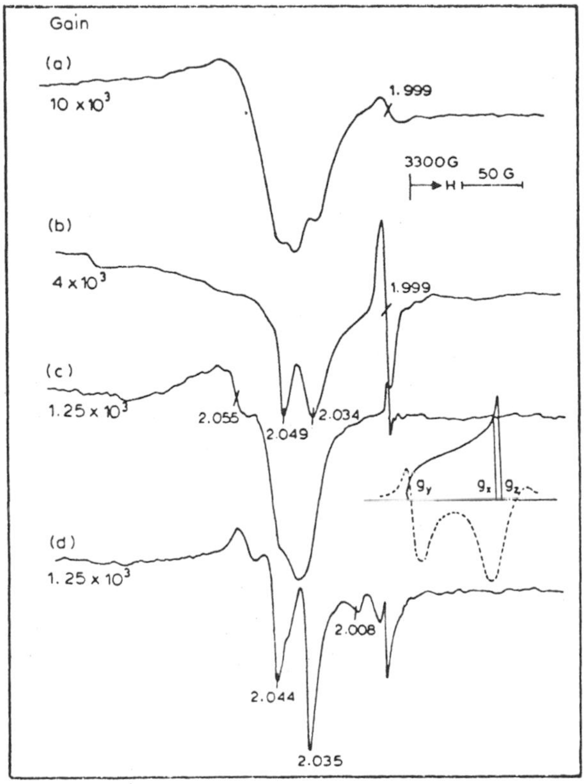

Fig. 3. EPR spectra of $SrTiO_3$ recorded at 77 K: (a) as-prepared sample, (b) annealed in H_2 at 1073 K, (c) dispersed in 0.5 M HNO_3 and washed with distilled water and recovered, (d) dispersed in diluted acetic acid. Inset shows the line shape expected for pseudo-axial centre with $g_x \approx g_z < g_y$.

their presence may decrease the concentration of the hole-centers.

On heating the acid-treated samples above 673 K, the signal intensities diminish, indicating that the center may be associated with the surface hydroxyls. Riederer et al. (20) have reported that OH_s^o radicals in semicrystalline matrix can have g_x, g_y = 2.008 and g_z ranges between 2.05 and 2.25, giving rise to the characteristic ascending absorption in the low-field region. The fact that the signals of $g_x = g_y = 2.008$ and $g_z = 2.055$ are more conspicous for samples with Ti/Sr > 1 indicates that OH_s^o may be associated with Ti^{4+} on the surface regions. This is also understandable from the chemistry of titanium where simple aquated Ti^{4+} ion does not exist because of high charge to ionic radius ratio. Consequently, $Ti(OH)_4$ is not stable whereas hydroxylated species such as $Ti(OH)^{3+}$, $Ti(OH)_2^{2+}$, $Ti(OH)_3^+$ are possible even in dilute aqueous acids. The latter can polymerize to various degrees giving rise to hydroxylated surface species. Therefore, the signals of g = 2.008 and 2.055 may originate from Ti^{4+}.. OH_s^o species where the radical formation may be assisted by the lighting conditions during handling the power in ambient atmosphere.

EPR signals corresponding to the electron-centers disappear on metal mounting except g = 1.999, whose intensity and line-width decrease perceptably (Fig. 4). After irradiating the aqueous suspensions and recovering the power, the intensities of the hole-center signals have diminished in comparison to those of the unirradiated powder. The signals of g = 2.088 and 2.055 have disappeared completely. This is indicative of the transference of holes to H_2O molecules as evidenced by the evolution of O_2. When $SrTiO_3$ particles suspended in aqueous solutions containing donor species such as disodium ethylene diamine tetraacetate (EDTA), NaH_2PO_2, Na_2SO_3 or triethanol amine (TEOA) are irradiated, the intensities of the hole-center signals decrease considerably.

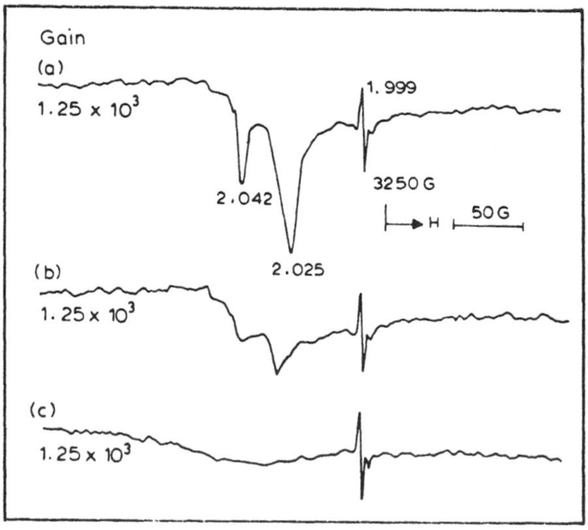

Fig. 4. EPR spectra of $SrTiO_3$ recovered after the band-gap irradiation for 4 h, dispersed in: (a) water, (b) 0.1 M EDTA, and (c) 0.5 M $H_2PO_2^-$.

D. Photocatalytic Activity

Suspensions of $SrTiO_3$ powders, without metal mounting, in water or in NaOH produced neither H_2 nor O_2 on illumination with uv light. However H_2 and O_2 are generated when these powders, after Pt- or Rh-mounting, are dispersed in NaOH on illumination but the yields are lower than those reported by Lehn et al. (5). The turn over numbers of H_2 and O_2 are expressed as the number of moles of the gas per mole of $SrTiO_3$. Fig. 5 shows the photoproduction of H_2 and O_2 with time. Oxygen turn over number is less than the expected $H_2:O_2$ = 2:1. The detection of O_2 suggests that H_2 arises from the photodecomposition of H_2O and not from the surface hydroxyl groups.

Sacrificial Systems

H_2 produced in NaOH solution gives a very low yield. Poor holes-scavenging ability of OH^- ions may be the reason for this. Therefore, hole-scavengers such as EDTA, TEOA and $H_2PO_2^-$ have been used so as to enhance the efficiency of H_2 production. Fig. 6 shows the amount of H_2 produced on band gap irradiation of $Pt/SrTiO_3$ suspended in EDTA, $H_2PO_2^-$ and TEOA solutions. $H_2PO_2^-$ is more effective than EDTA or TEOA. This is in contrast to that of

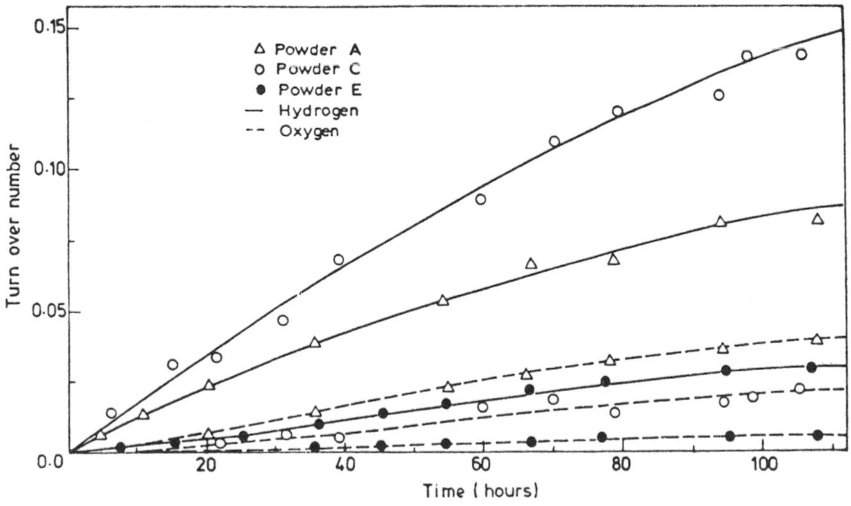

Fig. 5. Number of moles of gas per mole of $SrTiO_3$ [Turn over number] as a function of irradiation time for Pt-$SrTiO_3$ powder suspension in water.

TiO_2 where EDTA is better than other hole-scavengers (21). Fig. 7 gives the yield of H_2 with varying concentrations of $H_2PO_2^-$. The amount of H_2 produced increases up to 0.5 M and further increase of donor concentration has a deteriorating effect on catalytic activity. It is possible that enhanced adsorption of $H_2PO_2^-$ at the Pt/electrolyte interface at high $H_2PO_2^-$ concentration may block the electron transfer at the metal surface. Decrease in H_2 production can also be due to the strongly chemisorbed $H_2PO_2^-$ or its products of oxidation at the $SrTiO_3$ interface, which may weaken the binding of Pt particles on to $SrTiO_3$. Whereas, at lower concentrations of $H_2PO_2^-$, the adsorption density is low, which leads to a low scavenging rate for holes that arrive at $SrTiO_3$ interface during irradiation. Fig. 8(a) gives the relative rate of H_2 production with Pt-concentration, at constant $SrTiO_3$ and $H_2PO_2^-$ content. H_2 production rate increases with Pt, up to

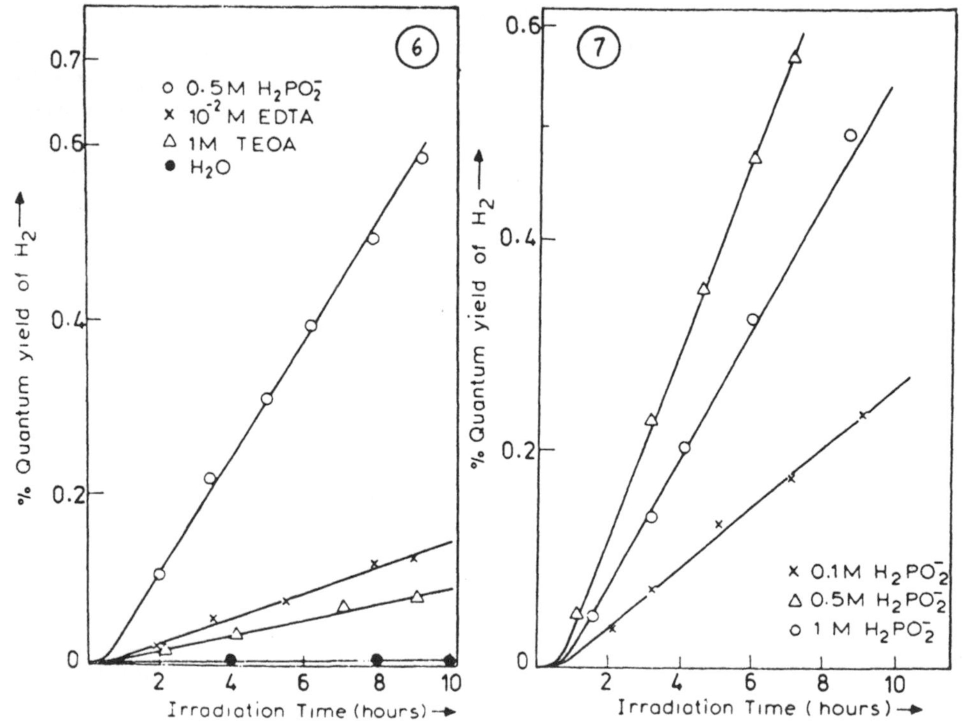

Fig. 6. Quantum yield of H_2 as a function of irradiation time with different hole-scavengers $[SrTiO_3] = 5 \times 10^{-3}$ M and $[Pt] = 1.5 \times 10^{-4}$ M.

Fig. 7. Quantum yield of H_2 with irradiation time for different concentration of $H_2PO_2^-$ $[Pt] = 1.5 \times 10^{-4}$ M, $[SrTiO_3] = 5 \times 10^{-3}$ M.

~4 wt%, and above this value, H_2 yield decreases. Although diminished over-voltage can be expected with Pt-deposition, steric hindrance by Pt-islets limits the adsorption density of donor molecules on $SrTiO_3$ at higher Pt-coverage. This leads to low hole scavenging rate, or in other words, high electron-hole recombination rate. Larger Pt coverage also reduces the effective intensity of light received by $SrTiO_3$ particles. Fig. 8(b) illustrates the effect of the varying dispersion density of $Pt/SrTiO_3$ particles at a given concentration of $H_2PO_2^-$ and $Pt:SrTiO_3$ ratio. H_2 production maximizes at an optimum dispersion density. At lower values, the light absorption increases with the particulate concentration, whereas above the optimun value, enhanced light scattering together with a reduction in adsorption density of donor molecules on $SrTiO_3$ diminishes the H_2 yield.

Formation of H_2 and O_2 observed on irradiating Pt or $Rh/SrTiO_3$ in water can be explained on the basis of the transfer of holes to the O_2/OH^- level because of the interband position of the energy levels arising from the hole centers (Fig. 9). Since the EPR signals of g = 2.088 and 2.055 assigned to Ti^{4+}.. OH_s^o disappear completely in a non-sacrificial system, in preference to other centers, the energy levels of this center have been shown closer to the O_2/OH^- level. Ti^{4+}.. OH_s^o centers may have shallower trap-depth, thereby increasing the hole transfer rate. Furthermore, these are located on the surface layers and thus possess more of surface state characteristics than of bulk trap

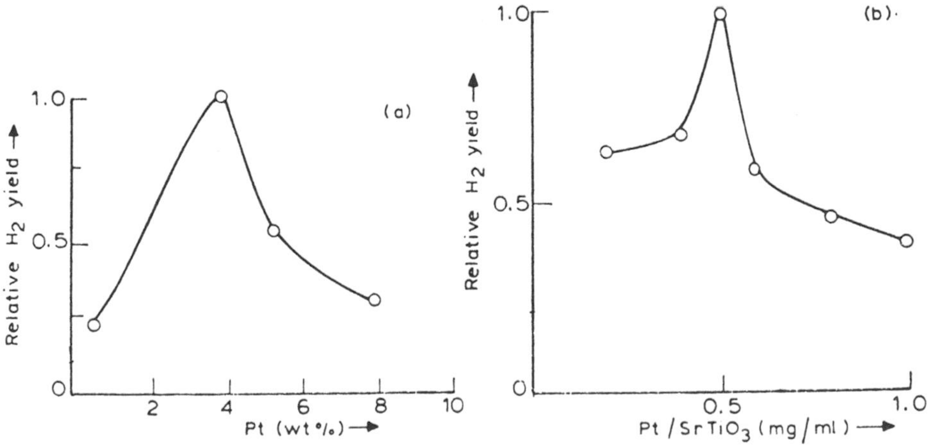

Fig. 8. Relative H_2 yield as a function of: (a) Pt concentration, (b) particle dispersion density, $[H_2PO_2^-]$ = 0.5 M, $[SrTiO_3]$ = 5 x 10^{-3} M, in the case of (b) [Pt] = 4 wt%.

states. $H_2PO_2^-$ is the efficient hole scavenger for $SrTiO_3$ because the energy states of the hole-centers in the band gap region are in proximity to the redox level of $H_2PO_2^-$ (Fig. 9). Transfer of photogenerated holes directly from valence band to O_2/OH^- level is scarcely probable since this involves dissipation of a large amount of energy. Holes from the valence band are at first transferred on to the hole-centers and then to the scavenger species. The latter step can be considered as a quasiisoenergetic process involving hole-centers in the subsurface regions and the surface states arising from chemisorbed scavenger molecules. The rate of hole-transfer depends on the trap depth in addition to the relative energy levels of hole-centers in the band gap region. Presence of more than one type of hole center ensures higher probability for isoenergetic charge transfer to favorably located surface states generated by chemisorbed donor species.

III. Other Photocatalytic Reactions

Perovskite oxide semiconductor powders exibit the photoactivity in the oxidation of alcohols, hydrocarbons, CO, etc. Also for the reduction of CO_2 and synthesis of NH_3, these powders are used.

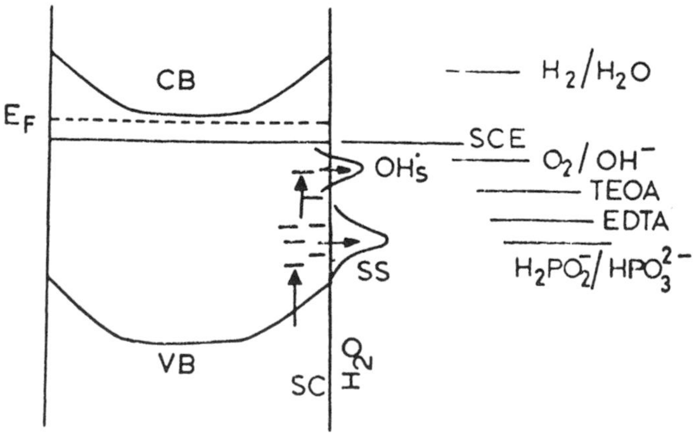

Fig. 9. Schematic diagram of the energy levels in $SrTiO_3$ particles with respect to the redox level of different hole-scavengers. SCE = Standard Calomel Electrode; SS = Surface State produced by chemisorbed donor species.

A. Photooxidation

The photocatalytic production of H_2 and acetone from mixtures of H_2O and 2-propanol in gas and liquid phases were studied on NiO supported $SrTiO_3$ (22). This reaction did not proceed without H_2O which indicated the coupling of the photodecomposition of H_2O and the photooxidation of 2-propanol. $SrHfO_3$ treated with sulfuration-oxidation cyclic process, i.e. heated in CS_2 atmosphere at 1173 K followed by oxidation in O_2 atmosphere at 873 K could absorb visible light. This powder photolysed 2-propanol and CH_3OH with visible light. The introduction of SO_2^- ions during this cyclic process is the reason for the visible light response (23).

The photooxidation of C_2H_6 with O_2 over $SrTiO_3$ and $BaTiO_3$ was reported (24). It was found that the photocatalytic activity depends on their source. The impurity in semiconductor photocatalyst exerts a great influence on activity since some impurity functions as a recombination center of photogenerated electron and holes. Alkane is photooxidized via alcohol and aldehyde, eventually to CO_2 and H_2O (25).

$LaCoO_3$, $BaTiO_3$ and $SrTiO_3$ powders were used for the photooxidation of CO (24,25). A strong photocatalytic effect was observed over $BaTiO_3$ and $SrTiO_3$, whereas the easily reducible oxide $LaCoO_3$ did not show any photo activity. The activity depended on their preparative source of the photocatalyst. Perovskite type compounds in the presence of phthalo cyanine accelarated the rate of photooxidation of HCOOH in the acid solutions with various degrees.

B. Photoreduction of CO_2

Photo assisted reaction of CO_2 with H_2 was investigated over platinized La-Ni-oxide and Pt/$SrTiO_3$ in the presence and absence of water vapor (26,27). CO was the product obtained in significant amount. CH_4 was the only side product in the absence of water vapor, whereas in the presence of water vapor, HCOOH and other minor products (CH_4, CH_2O and MeOH) were observed along with CO. A mechanism for the formation of different products was also discussed (26). Also the photoreduction of CO_2 by H_2O (28) over La-Ni-oxide was reported earlier.

C. Photosynthesis and Decomposition of NH_3

Photocatalytic synthesis of NH_3 from N_2 and H_2O was performed over $SrTiO_3$ and binary-wafered catalyst of $SrTiO_3$ and platinum black (29) where the copolymer of ethylene and vinyl alcohol was studied as a binder. The yield of NH_3 is larger in the case of binary-wafered catalyst than the semiconductor powder catalyst alone.

Photosynthesis of NH_3 and its photodecomposition were investigated over $BaTiO_3$ and $SrTiO_3$ based catalyst (30). Undoped $BaTiO_3$ and $SrTiO_3$ produced a small amount of NH_3 with no H_2 from H_2O and N_2. Effect of NiO and RuO_2 (which are mounted on these powderd catalyst) on this reaction was studied. NiO doping enhanced the decomposition of water on $SrTiO_3$ but the change in the yield of NH_3 is very small. On the other hand, the amount of NH_3 formation increased significantly for RuO_2-NiO-$BaTiO_3$ and RuO_2-NiO-$SrTiO_3$. These are attributable to the increased activites of NiO for H_2 evolution and RuO_2 for N_2 activation.

These findings also demonstrated that the thermodynamically unfavorable reaction of NH_3 formation from N_2 and H_2O can be produced under band-gap irradiation of these catalysts. As ammonia is accumulated, the photodecomposition of ammonia becomes important correspondingly.

D. Oxygen Isotope Exchange Reaction

Photocatalytic activity of $SrTiO_3$ for gas-solid oxygen isotope exchange (GSOIE) between $^{18}O_2$ and $SrTiO_3$ and dioxygen isotope equilibrium (GOIE) were studied by Sato and Kadowaki (24). The activity depended on the source of $SrTiO_3$.

IV. Conclusion

The catalytic conversion of light into chemical energy using colloidal particles is a fairly recent topic in solar energy conversion. The photocatalytic activity of semiconductors depends strongly on their preparation methods. Thus, increasing the efficiency of energy conversion also lies in the suitable method of preparation of the semiconductor fine particles. Optimum pretreatment of the photocatalyst and optimum reaction condition along with the selection of semiconductor material for the particular reaction are essential in increasing the efficiency.

Though the studies in this direction have not yet increased the efficiency of energy conversion much, a wealth of knowledge has been acquired in understanding the charge transfer processes in photocatalysis.

References

1. Wrighton, M.S., A.B. Ellis, P.T. Wolczanski, D.L. Morse, H.B. Abrahamson and D.S. Ginley, $SrTiO_3$ photoelectrodes. Efficient photo assisted electrolysis of water at zero applied potential, *J. Am. Chem. Soc.*, **98**: 277 (1976).
2. Wagner, F.T. and G.A. Somorjai, Photocatalytic hydrogen production from water on Pt-free $SrTiO_3$ in alkali hydroxide solutions, *Nature*, **285**: 599 (1980).
3. Wagner, F.T. and G.A. Somorjai, Photocatalytic and photoelectrochemical hydrogen production on $SrTiO_3$ single crystals, *J. Am. Chem. Soc.*, **102**: 5494 (1980).
4. Yoneyama, H., M. Koizumi and H. Tamura, Photolysis of water on illuminated strontium titanium trioxide, *Bull. Chem. Soc. Jpn.*, **52**: 3449 (1979).
5. Lehn, J.M., J.P. Sauvage and R. Ziessel; Photochemical water splitting. Continous generation of hydrogen and oxygen by irradiation of aqueous suspensions of metal loaded strontium titanate, *Nouv. J. Chem.*, **4**: 623 (1980).
6. Maglizzo, R.S. and A.I. Krasna, Hydrogen and oxygen photoproduced by titanate powders, *Photochem. Photobiol.* **38**: 15 (1983).
7. Lehn, J.M., J.P. Sauvage, R. Ziessel and L. Hilaire, Photochemical dissociation of water: Development and study of new heterogeneous catalysts. Part A water photolysis by UV irradiation of rhodium-loaded $SrTiO_3$ catalysts, *Israel J. Chem.*, **22**: 168 (1982).
8. Domen, K., S. Naito, T. Onishi, K. Tamura and M. Soma, Study of the photocatalytic decomposition of water vapor over a NiO-$SrTiO_3$ catalyst, *J. Phys. Chem.*, **86**: 3657 (1982).
9. Domen, K., A. Kudo, T. Onishi, N. Kosugi and K. Kuroda, Photocatalytic decomposition of water into hydrogen and oxygen over NiO-$SrTiO_3$ powder 1. Structure of the catalyst, *J. Phys. Chem.*, **90**: 292 (1986).
10. Brus, L.E., Electronic wave functions in semiconductor clusters: Experiment and theory, *J. Phys. Chem.*, **90**: 2555

(1986).

11. Turova, N.Y. and M.I. Yanovskaya, Oxides materials based metal alkoxides, *Inorg. Mater.*, **19**: 625 (1983).
12. Thampi, K.R., M. Subba Rao, W. Schwarz, M. Grätzel and J. Kiwi, Preparation of $SrTiO_3$ by sol-gel techniques for the photoinduced production of hydrogen and surface peroxides from water. *J. Chem. Soc., Faraday Trans. I*, **84**: 1703 (1988).
13. Kudo, A., A. Tanaka, K. Domen and T. Onishi, Mechanism of photocatalytic decomposition of water into H_2 and O_2 over NiO-$SrTiO_3$, *J. Catal.*, **111**: 296 (1988).
14. Kutty, T.R.N. and R. Balachandran, Direct precipitation of lead zirconate titanate by the hydrothermal method, *Mater. Res. Bull.*, **19**: 1479 (1984).
15. Vivekanandan, R., S. Philip and T.R.N. Kutty, Hydrothermal preparation of $Ba(Ti,Zr)O_3$ fine powders, *Mater. Res. Bull.*, **22**: 99 (1987).
16. Avudaithai, M. and T.R.N. Kutty; Ultrafine powders of $SrTiO_3$ prepared by hydrothermal method and their photocatalytic activity, *Mater. Res. Bull.*, **22**: 641 (1987).
17. Klug, H.P. and Alexander L.E., *X-ray Diffraction Procedures for Polycrystalline and Amorphous Materials*, John Wiley and Sons., Inc., New York, 1954, p. 530.
18. Cardona, M., Optical properties and band structure of $SrTiO_3$ and $BaTiO_3$, *Phys. Rev.*, **135A**: 132 (1965).
19. Avudaithai, M. and T.R.N. Kutty; EPR study of trapcenters produced on TiO_2 particles during water photolysis, *Mater. Res. Bull.*, **23**: 1675 (1988).
20. Riederer, H., J. Hüttermann, P. Boon and M.C.R. Symons, Hydroxyl radicals in aqueous glasses. Characterization and reactivity studied by EPR spectroscopy, *J. Mag. Res.*, **54**: 54 (1983).
21. Kutty T.R.N. and M. Avudaithai, Sacrificial photolysis of water on TiO_2 fine powders prepared by the hydrothermal method, *Mater. Res. Bull.*, **23**: 725 (1988).
22. Domen, K., S. Naito, T. Onishi and K. Tamaru, Photocatalytic hydrogen production from a mixture of water and 2-propanol on some semiconductors, *Chem. Lett.*, 555 (1982).
23. Yonemura, M., T. Sekine and H. Ueda, SO_2^- doping of $SrTiO_3$ and $SrZrO_3$ by cyclic (CS_2-O_2) processing, *J. Phys. Chem.*, **90**: 3003 (1986).
24. Sato, S. and T. Kadowaki, Photocatalytic activities of metal oxide semiconductors for oxygen isotope exchange and

oxidation reactions, *J. Catal.*, **106**: 295 (1987).

25. Van Damme, H. and W.K. Hall, Photocatalytic properties of perovskites for H_2 and CO oxidation-influence of ferroelectric properties, *J. Catal.*, **69**: 371 (1981).
26. Vijayakumar, K.M. and N.N. Lichtin, Reduction of CO_2 by H_2 and water vapor over metal oxides assisted by visible light, *J. Catal.*, **90**: 173 (1984).
27. Lichtin, N.N., K.M. Vijayakumar and B. I. Rubio, Photoassisted reduction of CO_2 by H_2 over metal oxides in the absence and presence of water vapor, *J. Catal.*, **104**: 246 (1987).
28. Lichtin, N.N. and K.M. Vijayakumar, Photocatalysed light driven reduction of CO_2 and N_2 on semiconductor surfaces, *J. Electrochem. Soc.*, **130**: 108C (1983).
29. Miyama, H., N. Fujii and Y Nagac, Heterogeneous photocatalytic synthesis of ammonia from water and nitrogen, *Chem. Phys. Lett.*, **74**: 523 (1980).
30. Li, Q-S., K. Domen and S. Naito, Photocatalytic synthesis and photodecomposition of ammonia over $SrTiO_3$ and $BaTiO_3$ based catalysts, *Chem. Lett.*, 321 (1983).

Chapter 16

Reduction of Sulfur Dioxide on Perovskite Oxides

D. Brynn Hibbert

Department of Analytical Chemistry, University of New South Wales, Kensington, New South Wales 2033, Australia

I. Introduction

Reactions of sulfur dioxide on perovskite oxides have been investigated from two distinct standpoints. First sulfur dioxide is a poison for reactions of carbon monoxide, nitric oxide and hydrocarbons (1-11). How to mitigate this poisoning and how to regenerate the catalyst is therefore of importance when designing pollution abatement catalysts for motor cars. However the reaction between sulfur dioxide and carbon monoxide, catalyzed by perovskites, itself may provide an important route for the removal of a major pollutant (12-16).

One difference between reactions of sulfur dioxide over perovskites and other gases is that sulfur dioxide is such an agressive agent. It is quite likely, for example, that no perovskite phase exists at all after a few minutes of reaction between sulfur dioxide and carbon monoxide (16,18). Even in small amounts when acting as a poison the activity of a catalyst can be permanently impaired.

II. Reactions at Perovskite Oxides Poisoned by Sulfur Dioxide

A. Experimental Studies

The effect of low levels of sulfur dioxide on reactions of perovskites has received attention with a view to characterize the extent of poisoning, to determine the possibility of reversing the effect and to establish the efficacy of adding platinum group metals to the catalyst (See Table 1). Because of the relevance of this work to the removal of pollutants from exhaust gases, typical reactions that have been studied are the oxidation of carbon monoxide and hydrocarbons, and the reduction of nitric oxide.

The oxidation of 1 - 2% carbon monoxide is a reaction that is poisoned by 4 to 50 ppm sulfur dioxide. The poisoning is typically characterized by a greater than 98% reduction in activity at the temperature of the reaction (473 - 573 K), or by a need to raise the temperature of the reaction by 50 K (6), 70 K (2) or even 100 K (5) to maintain the same activity. The extent of the interaction (or reaction) between sulfur dioxide and the catalyst is the subject of some debate. Studies of the coverage of sulfur dioxide show that increasing the temperature of the reaction leads to a greater adsorption of sulfur dioxide (1). For example on $La_{0.6}Sr_{0.4}CoO_3$ the adsorption rises from 0.14 of a monolayer at 473 K to 5.2 monolayers at 573 K. Other studies have shown that at least a monolayer is formed (3) and possibly more. In the light of adsorption studies of Hibbert and Campbell who observe XRD patterns of $SrSO_4$ after exposure of $La_{1-x}Sr_xCoO_3$ to 1% sulfur dioxide (15) (see also III. C) it is probable that sulfur dioxide reacts with the catalyst and so the poisoning effect, if it occurs, will be due to the inactivity of the reaction products rather than an adsorption phenomenon.

Sulphate formation has in fact been proposed by Voorhoeve (11) as the mechanism of poisoning. Katz et al. (6) see no evidence for a change in phase arising from exposure of $La_{0.7}Pb_{0.3}MnO_3$ to 0.09 % sulfur dioxide and 2 % oxygen at either 823 or 873 K for 60 hours.

In terms of the nature of the interaction most authors favor two surface structures in which sulfur is bonded to the transition element. For eample Huang et al. (8) propose $-SO_2$ and $-OSO_2$ on {100} $La_{0.5}Sr_{0.5}MnO_3$ in which there are two manganese sites Mn and MnO. From infrared studies on $La_{0.6}Sr_{0.4}CoO_3$ Li et al. (1) claim to show the presence of M-(O)S(O)-M and $M-SO_2$, the former being formed at high temperatures (673 K) and the latter at lower temperatures (473 K).

The oxidation of hydrocarbons appears to follow similar patterns. For example the oxidation of propene (6) was deactivated by sulfur dioxide. The reaction required an increase in the temperature of 35 K to reinstate the activity. Ethene

Table 1. Poisoning Effects of Sulfur Dioxide

Catalyst	Reaction[a]	$[SO_2]$ (ppm)	(% red. in activity) Remarks	Ref.
$La_{0.8}K_{0.2}Mn_{0.94}Ru_{0.06}O_3$	NO	50	(98) NH_3 formation reduced	(2)
$SrRuO_3$	NO	50	(98)	(2)
$La_{0.6}Sr_{0.4}Co_{1-x}M_xO_3$	CO	20-400	M not specified 0.14 monolayer SO_2 at 473 K	(1)
$La_{0.6}Sr_{0.4}CoO_3$	CH_4	"	5.2 monolayer SO_2 at 673 K. Activity restored if SO_2 removed and raise T	(3)
$LaCoO_3$	C_2H_4	4-20	(70 - 90)	(3)
"	CO	10	(97)	(3)
$BaCoO_3$	C_2H_4	5	(60)	(3)
"	CO	4-15	(75-87) H_2O improves activity	
$LaMnO_3$	C_2H_4	10	>(99.5)	(3)
"	CO	10	>(99.5)	(3)
$La_{0.5}Sr_{0.5}MnO_3$	CO	5	>(99.5)	(3)
$La_{0.9}Sr_{0.1}MnO_3$	CO	4	(66)	(3)
"	CO	30	(98)	(3)
$La_{0.7}Pb_{0.3}MnO_3$	CO	40	(79-98.3)	(3)
$La_{0.7}Pb_{0.3}MnO_3$	CO	10	(77) +0.4% H_2O, 773 K	
(+ 100 ppm Pt)	CO	20	(87)	(3)
"	CO	25-40	Activity increased by SO_2 at 673 K	(3)
$La_{0.7}Pb_{0.3}MnO_3$+ Pt	CO	50	(100)	(4)
$La_{0.7}Pb_{0.3}MnO_3$	CO	0.01%	(100) No XRD evidence for reaction with SO_2	(6)
"	C_3H_6	0.01%	(100)	(6)
Ba_2CoWO_6	CO	50	(100) Activity depends on gas mixture oxidizing or reducing	(5)
Ba_2FeNbO_6	CO	50	(100) Activity partially restored on removal of SO_2	

[a] CO = CO + O_2; NO = NO + H_2; CH_4 = CH_4 + O_2; C_2H_4 = C_2H_4 + O_2

oxidation (3) is also poisoned, but on $LaCoO_3$ and $BaCoO_3$ the effect is not as pronounced as with carbon monoxide oxidation.

The reduction of nitric oxide is also poisoned by sulfur dioxide (2,11). At 523 K 98% of activity of $La_{0.8}K_{0.2}Mn_{0.94}Ru_{0.06}O_3$ and of $SrRuO_3$ was lost in 90 minutes after exposure to 50 ppm sulfur dioxide. One useful side effect was that with the composition of the gas (0.13 % NO, 0.4 % H_2, 1.3 % CO, 3 % H_2O, 3% CO_2, 50 ppm SO_2 plus He) the water gas shift was eliminated by the poisoning and thus the formation of ammonia was considerably reduced (2). It was suggested that sulfur dioxide prevents the formation of NCO- on the catalyst. Voorhoeve (5) comments on the production of ammonia from nitric oxide over Ba_2CoWO_6 and Ba_2FeNbO_6. However no study of the effect of sulfur dioxide was performed.

It is clear that the exact condition and history of the catalyst used is of utmost importance. This has been noted by Voorhoeve et al. (5) who observe different effects on the oxidation of carbon monoxide on Ba_2CoWO_6 and Ba_2FeNbO_6 depending on whether the catalyst was in a reduced or oxidised form. Yung-Fang (19) asserts that some binary oxides such as CuO and spinels such as $CuCr_2O_4$ have better resistance to sulfur dioxide than perovskites.

B. Perovskites Containing Platinum-Group Elements

Following the discovery that $La_{0.7}Pb_{0.3}MnO_3$ rivalled platinum as a catalyst for the oxidation of carbon monoxide and the reduction of nitric oxide, Katz and coworkers (6,10) showed that the high activity and resistance to sulfur dioxide occurred only for $La_{0.7}Pb_{0.3}MnO_3$ single crystals grown from a molten flux in platinum crucibles. Analysis showed these crystals contained 50 to 75 ppm platinum and had ten times the activity of crystals grown free of platinum. Yung-Fang (3) observed similar effects and noticed that for some concentrations of sulfur dioxide (25 to 40 ppm) the addition of platinum to $La_{0.7}Pb_{0.3}MnO_3$ actually increased the performance. Yung-Fang proposed that in the catalyst as prepared platinum is alloyed with, or covered by, lead. Reaction with sulfur dioxide gives $PbSO_4$ which frees platinum to catalyze the oxidation of carbon monoxide etc. As the mechanism of poisoning by sulfur dioxide is considered to be the formation of sulfates it is fortuitous that the most readily formed sulfate is of an element (lead) not crucial to the catalytic activity of the perovskite. Other platinum-group elements that have been studied are ruthenium as $La_{0.8}K_{0.2}Mn_{0.94}Ru_{0.06}O_3$ and $SrRuO_3$ (2) and palladium (7).

III. Reduction of Sulfur Dioxide by Carbon Monoxide

A. Reactions and Thermodynamics of the SO_2/CO System

Concern over the increase in the emission of sulfur dioxide into the atmosphere from industrial plants burning fossil fuels has led to a growth in interest in reactions that will remove this pollutant. One such reaction is the reduction of sulfur dioxide to elemental sulfur by carbon monoxide.

$$SO_2 + 2\ CO \longrightarrow 1/2\ S_2 + 2\ CO_2 \qquad (1)$$

Reaction (1) has potentially many advantages: elemental sulfur is the byproduct, the process is dry, sufficient carbon monoxide may be present in the gas stream (for example in coal-fired power stations) and the process is one step. Sulfur dioxide may also be reduced by methane over transition metal sulfides (e.g. FeS, MoS_2 and WS_2) (20) but the availability of carbon monoxide suggests reaction (1) is to be preferred.

An important and totally undesirable side reaction is

$$CO + 1/2\ S_2 \longrightarrow COS \qquad (2)$$

and in the presence of water

$$CO + H_2O \longrightarrow CO_2 + H_2 \qquad (3)$$

$$H_2 + 1/2\ S_2 \longrightarrow H_2S \qquad (4)$$

As an example of other catalysts alumina-supported copper (21,22) has received some attention but in a single bed reactor no more than 70% removal of sulfur was possible and the formation of carbonyl sulfide was catalyzed by the metal sulfides formed.

Before discussing the effectiveness of catalysts the thermodynamics of the reaction need to be considered. For a stoichiometric mixture of sulfur dioxide and carbon monoxide in nitrogen and carbon dioxide at levels found in stack gases, above 673 K nearly all the elemental sulfur is found as S_2. Hibbert and Tseung (23) performed free energy minimization calculations for the system containing CO, CO_2, SO_2, SO_3, S_2, S_8, H_2O, H_2, H_2S, COS, CS_2, O_2 plus the condensed phases Cu, CuO, CuS, and C. They showed that under typical reaction conditions the stable form of copper was CuS. The optimum removal of sulfur dioxide occurs at about 773 K and with stoichiometric mixtures of the gases.

The first report of the use of a perovskite for the catalysis of carbon monoxide and sulfur dioxide was by Happel (12) who showed that $LaTiO_3$ was an active catalyst for reaction (1) that also suppressed the formation of carbonyl sulfide. The approach taken was to choose a transition metal that had the greatest difference between the enthalpy of formation of an oxide and a sulfide. It was reasoned that as sulfides are catalysts for the production of carbonyl sulfide an oxide that did not easily form its sulfide would be a good choice. The criterion has been criticized by Baglio (18) who argues, correctly in our opinion, that following extensive reaction of the starting oxide the active catalyst is considerably changed and so is not addressed by these considerations. The use of a perovskite to catalyze this complex reaction mixture with many different products must take account of the thermodynamic realities of both gases and solids in the system.

The work of Happel (12,13) was closely followed by that of Bazes (17) on $LaCoO_3$ and later by Hibbert and Tseung (14) and Hibbert and Campbell (15,16) who studied $La_{1-x}Sr_xCoO_3$. The rationale of the use of $La_{0.5}Sr_{0.5}CoO_3$ systems is from its efficacy in promoting reactions of carbon monoxide such as its reaction with oxygen (24) and with nitric oxide (9). It was thought that the reaction between carbon monoxide and sulfur dioxide was of a similar genre.

B. Experimental Studies in Flow Reactors

The reaction has been studied in traditional flow systems with carbon monoxide and sulfur dioxide around 2% and 1% in a nitrogen stream. Mostly steady state conversions have been determined although some dynamic studies have been done (14). Typical results show reaction commencing at about 773 K and nearly all the sulfur is removed by 923 K. For example $La_{0.5}Sr_{0.5}CoO_3$ is a catalyst promoting more than 90% removal of sulfur species above 823 K (Fig. 1).

It is important to have a measure of the gas phase sulfur remaining rather than simply the removal of sulfur dioxide.

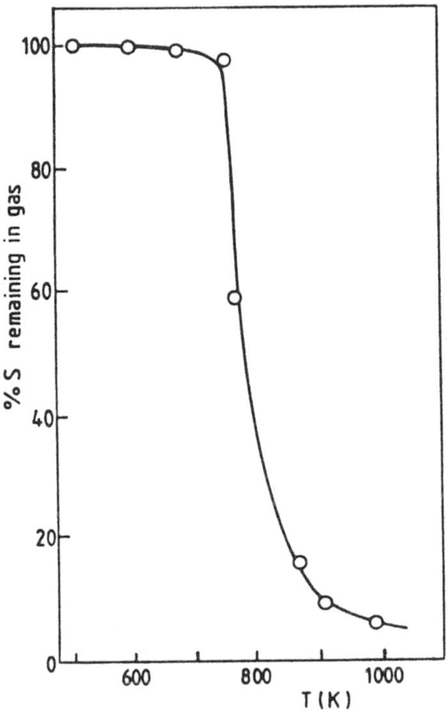

Fig. 1. The percentage of sulfur compounds remaining in the gas phase over $La_{0.5}Sr_{0.5}CoO_3$ as a function of temperature. Composition of reacting gases: CO = 2 %, SO_2 = 1 %, N_2 = 97 %. Contact time = 0.25 s. Reproduced with kind permission of Elsevier Science Publishers from J. Chem. Tech. Biotech. (Ref. 14).

Carbonyl sulfide (reaction 2) is a worse pollutant than sulfur dioxide and should not be formed in any amount. It is produced at higher concentrations of carbon monoxide following reaction between carbon monoxide and sulfur in the surface of the catalyst (as elemental sulfur or sulfide). The relative amounts of sulfur dioxide and carbon monoxide are therefore important. The parameter R_{CO} defines the composition of the gas as

$$R_{CO} = (P_{CO} - 2P_{O_2})/2P_{SO_2} \quad (5)$$

where P_{CO}, P_{O_2} and P_{SO_2} are the partial pressures of carbon monoxide, oxygen and sulfur dioxide respectively. A stoichiometric mixture of the gases thus has R_{CO} = 1. The efficiency of removal of sulfur from the gas phase over

$La_{0.5}Sr_{0.5}CoO_3$ is optimum at $R_{CO} = 1$ and becomes less good both as R_{CO} falls due to incomplete reaction of sulfur dioxide, and as R_{CO} increases due to the formation of carbonyl sulfide (Fig. 2). The region of efficient conversion is however better for $La_{0.5}Sr_{0.5}CoO_3$ than other catalysts.

Addition of up to four times as much oxygen as sulfur dioxide to the gas stream has a small effect of moving the minimum in the curve of remaining gas phase sulfur against R_{CO} to higher R_{CO}

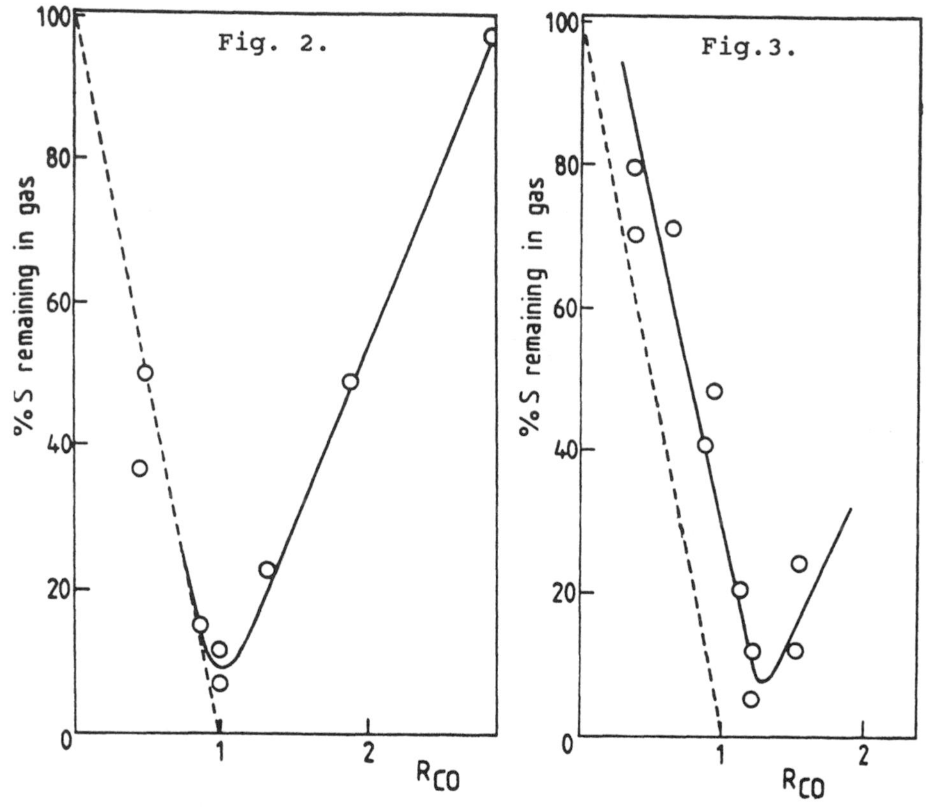

Fig. 2. The percentage of sulfur compounds remaining in the gas phase over $La_{0.5}Sr_{0.5}CoO_3$ as a function of inlet gas composition at 873 K. $R_{CO} = P_{CO}/2P_{SO2}$. Reproduced with kind permission of Elsevier Science Publishers from J. Chem. Tech. Biotech. (Ref. 14).

Fig.3. The percentage of sulfur compounds remaining in the gas phase over $La_{0.5}Sr_{0.5}CoO_3$ as a function of inlet gas composition at 873 K. $R_{CO} = (P_{CO}-2P_{O_2})/2P_{SO_2}$. P_{O_2} = 3 to 4 %. Reproduced with kind permission of Elsevier Science Publishers from J. Chem. Tech. Biotech. (Ref. 14).

(around 1.3) even after taking account of the oxygen which has a rapid reaction with carbon monoxide (Fig. 3).

Addition of both oxygen and water gives the optimum removal of sulfur, while water on its own leads to a further shifting of the minimum. As long as carbon monoxide was present to react with all added oxygen and have remaining carbon monoxide for reaction with sulfur dioxide up to 16% oxygen could be added without any deleterious effect on the reaction. However, oxygen plus sulfur dioxide poisoned the catalyst. At no time was sulfur trioxide observed.

In the presence of water, hydrogen sulfide, presumably from reaction (4) and reaction (3), was detected at levels an order of magnitude less than that of carbonyl sulfide. Sulfur species in the surface of the catalysts that could be removed by flowing carbon monoxide over a used catalyst was a maximum for cobalt oxide (4 mol S per mol oxide), was less for $La_{0.5}Sr_{0.5}CoO_3$ and decreased with increasing temperature of the reaction between sulfur dioxide and carbon monoxide (0.6:1 at 823 K to 0.16:1 at 973 K). This is consistent with the observation that at the start of an experiment with a fresh catalyst no sulfur at all was detected in the outlet stream of the reactor for several minutes.

C. Adsorption of SO_2, CO and Their Mixtures

As a precursor to a discussion of the mechanism of the reaction between sulfur dioxide and carbon monoxide it would be useful to understand the nature of the interaction between these gases and a perovskite surface. Some earlier studies on the adsorption of carbon monoxide and carbon dioxide have been reported by Fierro and Tejuca (25,26). In their study Hibbert and Campbell (15) used temperature programmed desorption mass spectrometry (TPDMS) and thermogravimetric analysis (TGA) to investigate the adsorption of carbon monoxide, sulfur dioxide and their mixtures on $La_{1-x}Sr_xCoO_3$ (x = 0.3, 0.5, 0.6, 0.7). The reaction at the surface was strongly correlated with the ease with which the perovskites lost oxygen during heating in vacuo, which increases with x. Adsorption of carbon monoxide alone at 923 K gave a decrease in mass of the catalyst for x = 0.3 but an increase for x = 0.7 (Fig. 4).

There is clearly a competition between the ability of carbon monoxide to scavenge surface oxygen from oxides having high oxygen concentration, and the tendency to adsorb on to oxygen vacancies on a surface. This observation is consistent with the finding of Voorhoeve (27) that prereduction of perovskite oxides leads to the better binding of carbon monoxide. Desorption of carbon monoxide could not be detected because of the high nitrogen background (also at m/e 28). However carbon dioxide, which showed a broad and complex desorption peak, was detected. Sulfur dioxide, on the other hand, caused a gain in weight of the catalyst for all x, and which increased with increasing temperature. At 773 K there is a linear relation between the increase in mass and decreasing x (Fig. 5).

One, two or three species were desorbed in the TPD-MS experiments, the number decreasing with temperature. For example for $La_{1-x}Sr_xCoO_3$ ($x = 0.7$) at 298 K three desorption peaks were seen with associated activation energies of 36, 54 and 55 kJ/mol. At 573 K a single peak was seen with Ea = 169 kJ mol^{-1}. The increase in weight at 773 K was not reversed by heating in vacuo at 923 K, nor could elemental sulfur or carbonyl sulfide be detected.

When the catalysts were exposed to both gases at 873 K there was a great increase in mass and yellow sulfur was seen to condense on the cooler parts of the apparatus. The trends in weight gain were similar to the sum of the effects when gases were separately adsorbed, but was of greater magnitude. Carbon dioxide and carbonyl sulfide were detected.

D. Structural Changes During Reaction

General Considerations

After Hibbert and Tseung (23) had shown that CuO was converted to CuS during the reaction of sulfur dioxide and carbon monoxide the possibility of the formation of a perovskite sulfide was investigated (14). For the perovskite ABX_3 with Goldschmidt radii of A, B and X respectively r_A, r_B and r_X, for stable six fold coordination of the B ion $r_B/r_X > 0.39$ and also the tolerance factor $t = (r_A + r_X)/\sqrt{2}(r_B + r_X)$ must lie between 0.75 and 1.00. For X = O (r_O = 132 pm) requires $r_A > 51$ pm and $r_B > 62$ pm. This is fulfilled by $La_{0.5}Sr_{0.5}CoO_3$ (r_{La} = 122 pm and r_{Co3+} = 63 pm giving t = 0.95). The larger sulfide ion (r_{S2-} = 174 pm) makes the requirements more stringent with $r_A > 83$ pm and $r_B > 68$ pm. Co3+ therefore just falls outside the conditions and Co4+ would have greater problems. $LaCoS_3$ has been synthesized by heating $LaCoO_3$ above 1423 K in CS_2, but has an hexagonal structure.

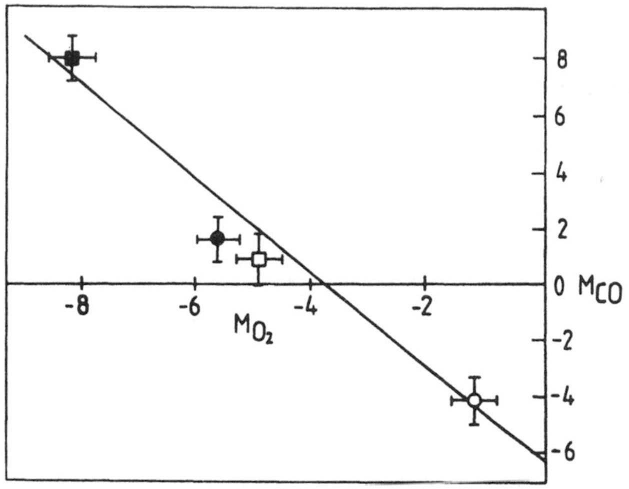

Fig. 4. Percentage mass change of $La_{1-x}Sr_xCoO_3$ on adsorption of carbon monoxide at 773 K (M_{CO}) plotted against the percentage mass change on heating to 923 K in vacuo. (○) x = 0.3; (●) x = 0.5; (□) x = 0.6; (■) x = 0.7. Reproduced with kind permission of Elsevier Science Publishers from Appl. Catal. (Ref.15).

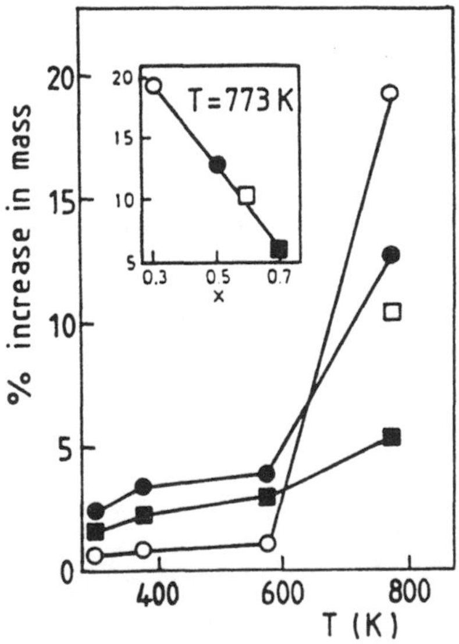

Fig. 5. Percentage mass change of $La_{1-x}Sr_xCoO_3$ on adsorption of sulfur dioxide as a function of temperature. (○) x = 0.3; (●) x = 0.5; (□) x = 0.6; (■) x = 0.7. Inset is the data for 773 K plotted against x. Reproduced with kind permission of Elsevier Science Publishers from Appl. Catal. (Ref.15).

X-Ray Diffraction (XRD)

After the flow reactor experiments of Hibbert and Tseung (14) XRD patterns of the used $La_{0.5}Sr_{0.5}CoO_3$ catalyst showed lines that were attributed to a mixture of perovskite, CoS_2, SrS and several unknown phases.

The structures of the perovskites were analyzed after the series of adsorption experiments (15). In the case of sulfur dioxide the perovskite phase could only be weakly detected and for x = 0.3 the structure seemed to be a mixture of La_2O_2S, Co_3S_4 and SrS. For x > 0.3, strontium sulfate was the predominant species. Little change in the XRD of any of the perovskites was observed after exposure to carbon monoxide. Following reaction (16) of both gases however no sample showed any evidence of a perovskite phase at all (Fig. 6). Instead the XRD was analyzed as a complex mixture containing compounds such as SrS, Co_3O_4, $LaS_{1.75\text{-}1.8}$, $La_2Sr_2O_5$, $La_2O_2SO_4$, La_2O_2S. This is a significant result as it indicates that although a perovskite may start off catalyzing the reaction, the substance that represents the steady state catalyst is certainly not a perovskite oxide. This conclusion was reached in an earlier paper of Baglio (18) who demonstrated that for higher concentrations of sulfur dioxide (10%) and carbon monoxide (20%) the perovskite $LaCoO_3$ decomposed and that the active catalyst was a mixture of La_2O_2S and CoS_2.

X-Ray Photoelectron Spectroscopy (XPS)

The chemical shift of the sulfur 2p peak may be used to distinguish between sulfur in different oxidation states. Of

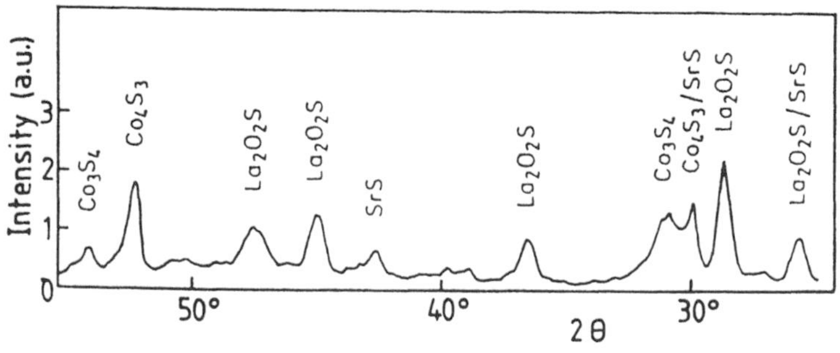

Fig. 6. Powder XRD of $La_{0.7}Sr_{0.3}CoO_3$ after exposure to sulfur dioxide and carbon monoxide at 923 K. Reproduced with kind permission of Elsevier Science Publishers from Appl. Catal. (Ref. 16).

interest in the work of Hibbert and Campbell (16) was the state of any sulfur taken up by a perovskite catalyst during the course of a reaction. Fig. 7 gives the XPS of a $La_{0.5}Sr_{0.5}CoO_3$ catalyst after exposure to carbon monoxide and sulfur dioxide at 923 K. The three possible species are SO_4^{2-}, SO_3^{2-} and S^{2-}.

It is quite clear that elemental sulfur is not present on the surface. In the reducing environment sulfide is the predominant species, but is of note that XPS picks up sulfate when this is not seen in the XRD. Another change in XPS between fresh and used $La_{1-x}Sr_xCoO_3$ was the disappearance of a shoulder on the low binding side of the oxygen 1s peak. The observation of this second oxygen peak which separates at low values of x has been reported by Yamazoe (28). This peak is attributed to adsorbed oxygen that is facilitated by increasing lanthanum content.

E. Mechanism of Reaction

Any mechanism must take account of the reactions that clearly take place with the perovskite oxide catalyst and show the separate and combined roles of carbon monoxide and sulfur

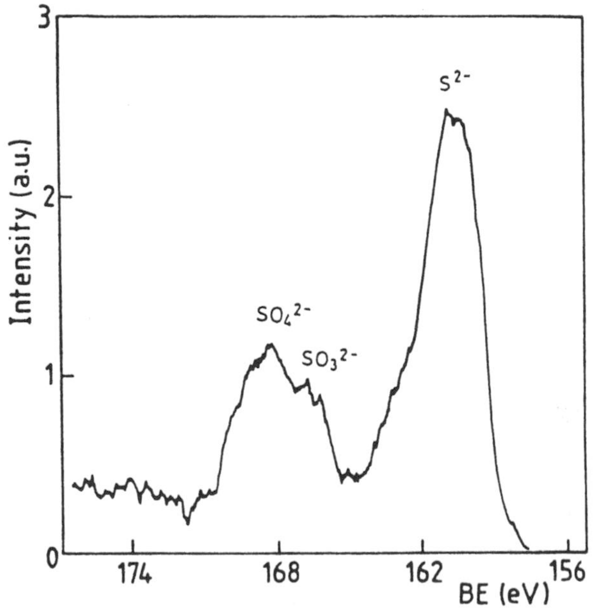

Fig. 7. XPS of $La_{0.5}Sr_{0.5}CoO_3$ after exposure to sulfur dioxide and carbon monoxide at 923 K. Reproduced with kind permission of Elsevier Science Publishers from Appl. Catal. (Ref. 16).

dioxide. With both gases being capable of reducing the oxide lattice parallel paths in which the gases react with the catalyst and in which they react with each other are required.

Although it is possible to write a step-wise reaction between adsorbed carbon monoxide and adsorbed sulfur dioxide that sequentially loses oxygen, it is not a probable reaction. Happel (13) in a kinetic isotope study of the reaction on lanthanum titanate proposes a mechanism in which surface sites and adsorbed oxygen act as intermediates. There is no direct reaction. The rate determining step is considered to be the adsorption of carbon monoxide or its reaction with adsorbed oxygen. No account is taken of any reaction leading to sulfides and the reaction is clearly more complex than a simple intrafacial one.

The reaction of each gas separately with the catalyst may be expressed as

$$CO\ (g) \longrightarrow CO\ (ads) \tag{6}$$

$$CO\ (ads) + \text{catalyst} \longrightarrow CO_2\ (ads) + \text{"CO reduced catalyst"} \tag{7}$$

$$CO_2\ (ads) \longrightarrow CO_2\ (g) \tag{8}$$

Reaction (7) would predominate on $La_{1-x}Sr_xCoO_3$ with $x = 0.3$ when there is a net gain in weight, and reaction 8 for $x = 0.7$ when carbon dioxide is evolved with loss of weight. If sulfur dioxide on its own is presented to the catalyst sulfate is produced.

$$SO_2\ (g) \longrightarrow SO_2\ (ads) \tag{9}$$

$$SO_2\ (ads) + \text{catalyst} \longrightarrow SO_4^{2-} + \text{"}SO_2\text{ reduced catalyst"} \tag{10}$$

This is the only reaction when $La_{1-x}Sr_xCoO_3$ is exposed to sulfur dioxide alone. It is not thought to play a major role when carbon monoxide is present. It is interesting to note that sulfate production is the preferred theory of the poisoning effect of sulfur dioxide (See II).

In the proposed scheme therefore adsorbed sulfur arises from the reaction between adsorbed sulfur dioxide and the "CO reduced catalyst" as suggested by Happel (12). However, also in the presence of adsorbed sulfur the "CO reduced catalyst" of reaction (7) must react to yield the final working catalyst which we designate "sulfided catalyst." When the catalyst is fully reacted adsorbed sulfur may then be desorbed as S_2.

Sulfides are catalysts for the preparation of carbonyl sulfide (29) and thus in the presence of excess carbon monoxide:

$$CO(ads) + \text{"sulfided catalyst"} \longrightarrow COS + \text{"CO reduced catalyst"} \quad (11)$$

The relationships between the species and reactions is summed up in Fig. 8 in which connections between species are indicated in a diagram showing increasing reduction of the surface.

Baglio (18) gives a more definite scheme in which carbonyl sulfide is an intermediate in a cycle involving CoS_2 and CoS as reactants and La_2O_2S as catalyst for reaction (13)

$$CoS_2 + CO \longrightarrow CoS + COS \quad (12)$$

$$COS + 1/2\ SO_2 \longrightarrow 3/4\ S_2 + CO_2 \quad (13)$$

$$CoS + 1/2\ S_2 \longrightarrow CoS_2 \quad (14)$$

Overall this scheme gives reaction (1).

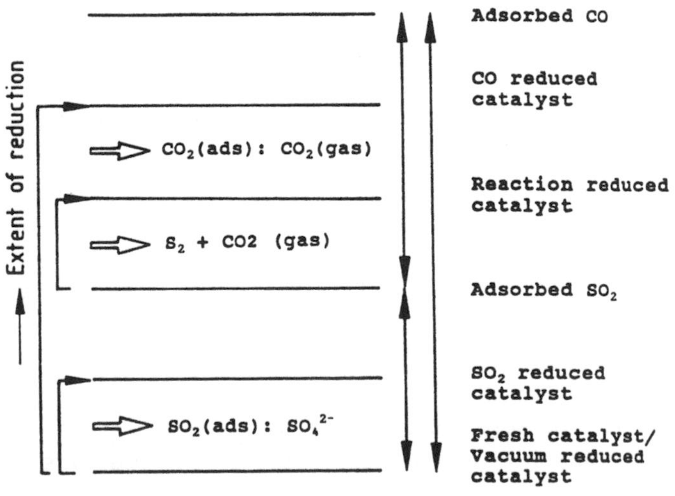

Fig. 8. Schematic of the relative levels of reduction found in the system $La_{1-x}Sr_xCoO_3/SO_2/CO$. Species at the ends of double arrows may react to give the products shown. Reproduced with kind permission of Elsevier Sci. Publ. from Appl. Catal. (Ref. 16).

IV. Conclusions

Sulfur dioxide is a poison for perovskite oxides used as catalysts for the oxidation of carbon monoxide. The perovskite is converted to mixtures of sulfates, sulfides and oxysulfides of the transition and lanthanide elements depending on whether the conditions are oxidizing or reducing. When low levels of sulfur dioxide are present its impact may be assuaged by the presence of platinum-group elements.

Perovskites are catalysts for the reaction between sulfur dioxide and carbon monoxide, a reaction of potential industrial importance that could serve to remove sulfur dioxide from stack gases and thus abate "acid rain." Structural and mechanistic studies have shown that after a brief induction perod no perovskite remains. It may therefore be argued that perovskites are in fact not catalysts for the reduction of sulfur dioxide, and technically this is probably the case. However it may transpire that the only route to the most efficient mixture of sulfur and oxygen species that is responsible for catalysis may be through a precursor perovskite.

References

1. Li, W., Q. Huang, W-J. Zhang, B-X. Lin, and G-L. Lu, Improving the resistance of perovskite type oxidation catalyst, *Catalysis and Automotive Pollution Control* (A. Crucq and A. Frennet, Eds.), Elsevier Science Publishers, Amsterdam, 1987, p. 405.
2. Trimble, L.E., Effect of SO_2 on nitric oxide reduction over Ru-containing perovskite catalysts, *Mat. Res. Bull.*, **9**: 1405 (1974).
3. Yung-Fang, Y.Y., The oxidation of hydrocarbons and CO over metal oxides IV Perovskite-type oxides, *J. Catal.*, **36**: 266 (1975).
4. Gallagher, P.K., D. W. Johnson, E. M. Vogel, and F. Schrey, Effects of the Pt content of $La_{0.7}Pb_{0.3}MnO_3$ on its catalytic activity for the oxidation of CO in the presence of SO_2, *Mat. Res. Bull.*, **10**: 623 (1975).
5. Voorhoeve, R.J.H., L. E. Trimble, and C. P. Khattak, Exploration of perovskite-like catalysts: Ba_2CoWO_6 and Ba_2FeNbO_6 in NO reduction and CO oxidation, *Mat. Res. Bull.*, **9**: 655 (1974).

6. Katz, S., J. J. Croat, and J. V. Laukonis, Lanthanum lead manganite catalyst for carbon monoxide and propylene oxidation, *Ind. Eng. Chem., Prod. Res. Dev.*, **14**: 274 (1975).
7. Kim, J. B., W. Y. Lee, H. K. Rhee, and H. I. Lee, A study on automobile exhaust gas control by perovskite-type oxide catalysts, *Hwahak Kongha*, **26**: 535 (1988) (*Chem. Abs.* **110**: 218205g).
8. Huang, M., H. Yang, Q. Wang, P. Lin, and J. Yong, The active centres of lanthanum strontium manganate ($La_{0.5}Sr_{0.5}MnO_3$), *Chehua Xuebao*, **4**: 311 (1983).
9. Libby, W.F., Promising catalyst for auto exhaust, *Science*, **44**: 499 (1971).
10. Croat, J. J., G. G. Tibbetts, and S. Katz, Rare earth manganates: surface segregated platinum increases catalytic activity, *Science*, **194**: 318 (1976).
11. Voorhoeve, R. J. H., D. W. Johnson, J. P. Remeika, and P. K. Gallagher, P.K., Perovskite oxides: materials science in catalysis, *Science*, **195**: 827 (1977).
12. Happel, J., M. A. Hnatow, L. Bajars, and M. Kundrath, Lanthanum titanate catalyst - sulfur dioxide reaction, *Ind. Eng. Chem., Prod. Res. Dev.*, **14**: 154 (1975).
13. Happel, J., A. L. Leon, M. A. Hnatow, and L. Bajars, Catalyst composition optimisation for the reduction of sulfur dioxide by carbon monoxide , *Ind. Eng. Chem., Prod. Res. Dev.*, **16**: 150 (1977).
14. Hibbert, D. B. and A. C. C. Tseung, The reduction of sulfur dioxide by carbon monoxide on a $La_{0.5}Sr_{0.5}CoO_3$ catalyst, *J. Chem. Tech. Biotech.*, **29**: 713 (1979).
15. Hibbert, D. B. and R. H. Campbell, Flue gas desulfuriation: catalytic removal of sulfur dioxide by carbon monoxide on sulfided $La_{0.5}Sr_{0.5}CoO_3$. I. Adsorption of sulfur dioxide, carbon dioxide and their mixtures, *Appl. Catal.*, **41**: 273 (1988).
16. Hibbert, D. B. and R. H. Campbell, Flue gas desulfurisation: catalytic removal of sulfur dioxide by carbon monoxide on sulfided $La_{0.5}Sr_{0.5}CoO_3$. II. Reaction of sulfur dioxide and carbon dioxide in a flow system, *Appl. Catal.*, **41**: 289 (1988).
17. Bazes, J. G., L. S. Caretto, and K. Nobe, Catalytic reduction of sulfur dioxide with carbon monoxide on cobalt oxides, *Ind. Eng. Chem., Prod. Res. Dev.*, **14**: 264 (1975).
18. Baglio, J. A., Lanthanum oxysulfide as a catalyst for the

oxidation of CO and COS by SO_2, *Ind. Eng. Chem., Prod. Res. Dev.*, **21**: 38 (1978).

19. Yung-Fang, Y. Y., The oxidation of CO and C_2H_4 over metal oxides. V. SO_2 effects, *J. Catal.*, **39**: 104 (1975).
20. Mulligan, D. J., and D. Berk, Reduction of sulfur dioxide with methane over selected transition metal sulfides, *Ind. Eng. Chem., Prod. Res. Dev.*, **28**: 926 (1989).
21. Querido, R., and W. L. Short, Removal of sulfur dioxide from stack gases by catalytic reduction to elemental sulfur with carbon monoxide , *Ind. Eng. Chem., Prod. Res. Dev.*, **12**: 10 (1973).
22. Quinlan, C. W., V. C. Okay, and S. R. Kittrell, S.R., Kinetics and yields for sulfur dioxide reduction by carbon monoxide, *Ind. Eng. Chem., Prod. Res. Dev.*, **12**: 107 (1973).
23. Hibbert, D. B., and A. C. C. Tseung, Catalyst participation in the reduction of sulfur dioxide by carbon monoxide in the presence of water and oxygen , *J. Chem. Soc., Faraday Trans. I*, **74**: 1981 (1978).
24. Nakamura, T., M. Misono, and Y. Yoneda, Catalytic properties of perovskite-type mixed oxides, $La_{1-x}Sr_xCoO_3$, *Bull. Chem. Soc. Jpn.*, **55**: 394 (1982).
25. Fierro, J. L. G. and L. G. Tejuca, Surface interactions of carbon dioxide and pyridine with $LaCrO_3$, *J. Chem. Tech. Biotech.*, **34A**: 29 (1984).
26. Fierro, J. L. G. and L. G. Tejuca, Kinetics of carbon monoxide adsorption on the perovskite-type $LaCrO_3$, *J. Colloid Interface Sci.*, **96**: 107 (1983).
27. Voorhoeve, R. J. H., L. E. Trimble, S. Nakajima, and E. Banks, Catalysis under transient conditions, (A.T. Bell and L.L. Hegedus, Eds.), *ACS Symposium Series 178*, Wiley, New York, p. 253.
28. Yamazoe, N., S. Furukawa, Y. Teraka, and T. Seiyama, The effect of oxygen sorption on the crystal structure of $La_{1-x}Sr_xCoO_3$, *Chem. Lett.*, 2019 (1982).
29. Ihara Chemical Industries, Carbonyl sulfide, *Fr. Patent* 2 146 664, 6th April 1973.

Chapter 17

Decomposition of N_2O on Perovskite-Related Oxides

C.S. Swamy[1] and J. Christopher[2]

[1]Department of Chemistry, Indian Institute of Technology, Madras 600 036, India

[2]R&D Centre, Indian Oil Corporation, Ltd., Faridabad 121 007, India

I. Introduction

Mixed metal oxides crystalizing in a perovskite-related structure have long been of interest to solid state chemists and physicists because of their technologically important physical properties. The ready availability of a family of isomorphic solids with controllable physical properties makes these oxides suitable for basic research in catalysis. These mixed metal oxides are more advantageous and are better catalytic materials than simple oxides because: (i) the crystal structure can accomodate various metal ions and can stabilize unusual and mixed valence states of active metal ion; (ii) appropriate formulation of these oxides leads to easy tailoring of many desirable properties such as valence state of transition metal ion, distance between active sites, binding energy, diffusion of oxygen in the lattice, magnetic and conducting properties of the solid; (iii) the catalytic activity can be correlated to solid state properties since many of their solid state properties are thoroughly understood; (iv) the surface of these oxides can be regenerated by suitable activation procedure.

In this section, the catalytic activity of mixed metal oxides having perovskite and related structure for the decomposition of N_2O has been reviewed.

II. Decomposition of N_2O Model Reaction

Decomposition of N_2O is one of the simple test reactions to evaluate the catalytic activity of oxides (1). The advantage of this reaction is that it follows an electronic mechanism (as shown below) which is very important to correlate the catalytic activity with electronic properties of the catalyst.

$$N_2O + e^- \Rightarrow N_2O^-_{ad} \quad (1)$$

$$N_2O^-_{ad} \Rightarrow N_2 + O^-_{ad} \quad (2)$$

$$O^-_{ad} + O^-_{ad} \Rightarrow O_2 + 2e^- \quad (3)$$

$$O^-_{ad} + N_2O \Rightarrow O_2 + N_2 + e^- \quad (4)$$

The kinetic results for the decomposition of N_2O can follow any of the following rate expressions reported by Cimino et al. (2) corresponding to no, strong, and weak, inhibition by oxygen respectively.

$$\frac{-dP_{N2O}}{dt} = k_1 P_{N2O}$$

$$\frac{-dP_{N2O}}{dt} = \frac{k_2 P_{N2O}}{P^{1/2}_{O2}}$$

$$\frac{-dP_{N2O}}{dt} = \frac{k_2 P_{N2O}}{1 + bP^{1/2}_{O2}}$$

If step 1, involving the adsorption of N_2O, is rate determining, then the kinetics of decomposition will obey eqn. (5). If step 3 and/or 4 involving desorption of oxygen are/is rate limiting, then the kinetics will follow eqn. (6). If adsorption of N_2O and desorption of oxygen are competing for the rate limiting step, then the kinetics will obey eqn. (7). Typical kinetic plots corresponding to no inhibition, strong, and weak inhibitions are shown in Fig. 1.

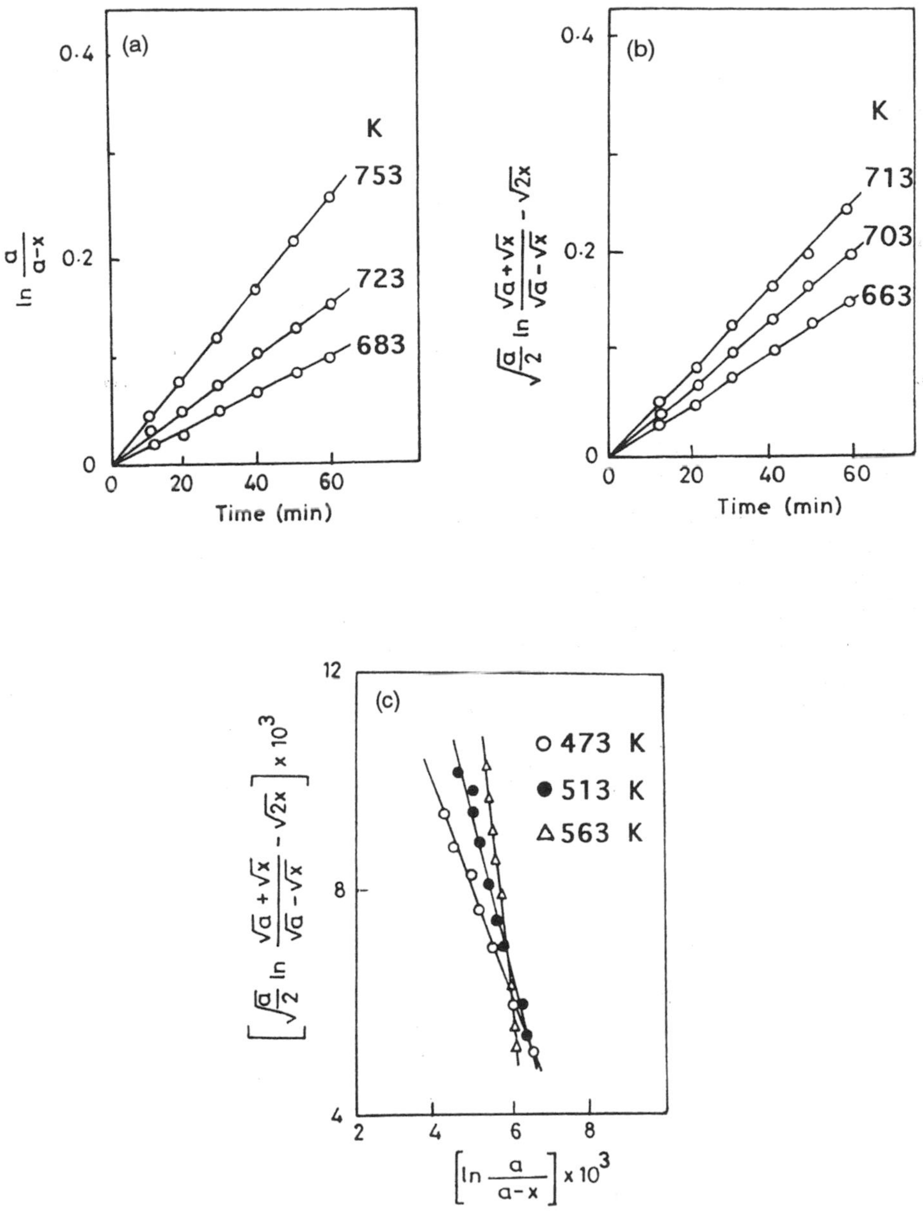

Fig. 1. Kinetic plots for the decomposition of N_2O (P = 50 torr) on: (a) $LaFeO_3$ (no inhibition); (b) $LaCoO_3$ (strong inhibition); (c) $LaNiO_3$ (weak inhibition).

III. Decomposition of N_2O on Perovskites

Nagasubramanian et al (3) studied the decomposition of N_2O on $MTiO_3$ (M = Ca, Sr or Ba) to evaluate the catalytic activity of titanates. The importance of localized d-orbitals occupancy of proper symmetry, depending upon whether desorption of oxygen (at 200 torr) or adsorption of N_2O (at 50 torr) is the rate determining step, has been discussed. Based on changes in electrical conductivity and Seebeck coefficients in various atmospheres, it has been proved that the decomposition of N_2O on titanates is suprafacial in nature, i.e. catalyst surface acts as a template providing orbitals of suitable or proper symmetry and energy for the adsorption of N_2O (4).

Since the desorption of oxygen is controlled by the availability of empty d_{z^2} orbitals on the active transition metal ions, Ti^{4+} with its empty d-orbitals effectively decreases the d-level occupancy of the transition metal ions at shorter distances and facilitates the desorption of oxygen. As M^{2+}-Ti^{4+} distance increases, cation-cation interaction decreases and the desorption oxygen is slowed down, which in turn, increases the activation energy. Such a correlation between E_a and cell parameter is shown in Fig. 2 for the series $MTiO_3$ (M = Mn, Co and Ni) (5). Also, a

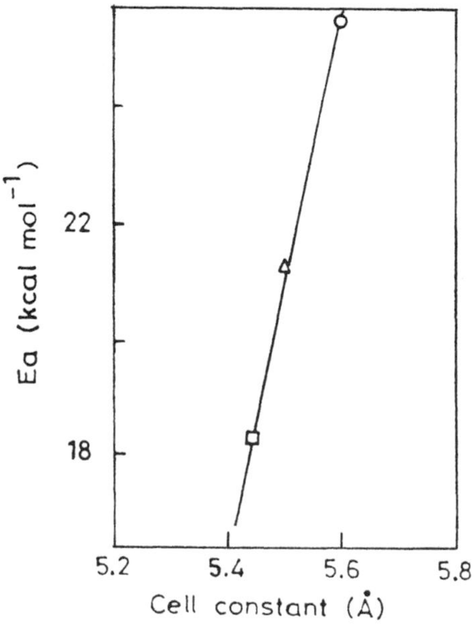

Fig. 2. Correlation between unit cell parameters and E_a for N_2O decomposition: (○), Mn^{2+}; (△), Co^{2+}; (□) Ni^{2+}.

linear correlation between E_a for decomposition and $E\sigma$ for conduction, observed in this series reveals that a system with $E\sigma$ 1.6 eV has maximum activity as this value corresponds to the value of one of the stimulation energy terms of oxygen molecules (6).

It has been reported by Nagasubramanian et al. (7) that a plot of ΔH and rate at 200 torr for the oxides $MTiO_3$ (M = Mn, Mg, Ca, Sr and Ba) showed a volcano shaped correlation (Fig. 3). However, at 50 torr such correlation was not observed. At high pressures of N_2O, the metal-oxygen bond strength may decide the desorption rate of adsorbed oxygen. A similar kind of correlation was reported for the decomposition of N_2O on simple oxides (8).

Louisraj et al (9) studied the role of oxygen species on the kinetics of N_2O decomposition over $LaMnO_3$. It has been found that the adsorption of oxygen species from N_2O decomposition is quite different from that obtained from the adsorption of molecular oxygen which normally inhibits the reaction. It has been shown to accelerate the decomposition rate depending upon the nature of the adsorbed oxygen species.

In perovskite oxides, it is well known that the "B" ion plays an important role in governing solid state properties and the A ion plays a modifying role (10, 11). Muralidhar et al. (12) studied the decomposition of N_2O on $LaMO_3$ (M = Cr, Ni) to study the effect of the B ion on catalytic activity. A plot of activity vs d-orbital occupancy (Fig. 4) showed a twin peak pattern with maxima for $LaCoO_3$ and $LaMnO_3$ similar to that reported for

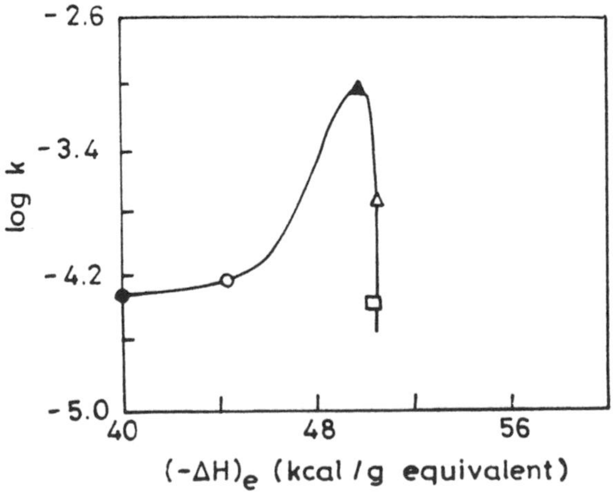

Fig. 3. Volcano-shaped correlation for $MTiO_3$ perovskites. M = Mn (●); Mg (○); Ca (▲); Sr (△); Ba (□).

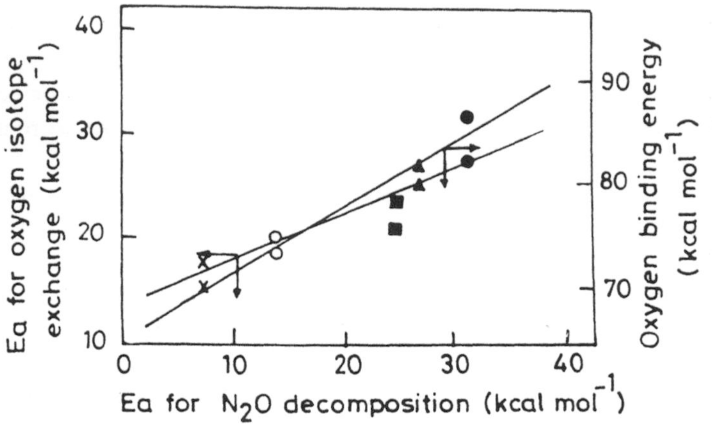

Fig. 4. Correlation of oxygen bong energy with catalytic activity for the oxides: (●), $LaCrO_3$; (▲), $LaFeO_3$; (■), $LaMnO_3$; (o), $LaCoO_3$; (×), $LaNiO_3$.

oxidation of propylene (13) and CO (14) indicating that perovskites containing low spin M^{3+} ions were highly active.

A linear correlation between E_a for O_2 isotope exchange and E_a for N_2O decomposition (Fig. 5) for the series $LaMO_3$ indicate that the desorption of oxygen is rate determinig (15). Also, a linear correlation between oxygen binding energy and catalytic activity (Fig. 5) emphasizes the importance of surface oxygen

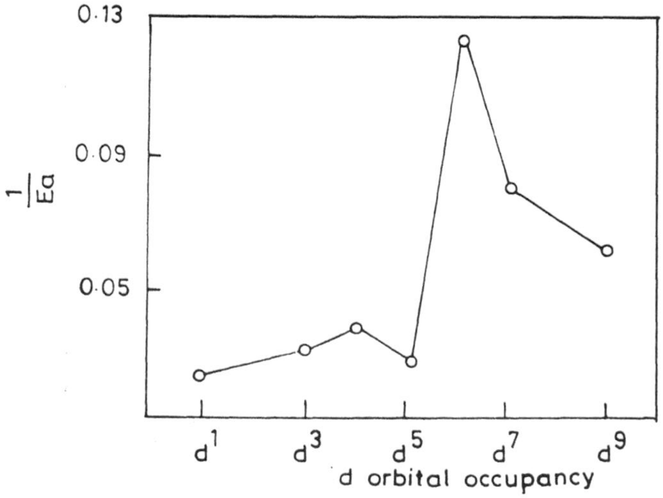

Fig. 5. Plot of d-orbital occupancy versus $1/E_a$ for the series $LaMnO_3$ (M = Ti ⟶ Cu).

Table 1. E_a Values for the Series $LnMnO_3$

Catalyst	$LaMnO_3$	$NdMnO_3$	$SmMnO_3$	$GdMnO_3$
E_a (kcal/mol)	25.42	12.42	9.90	6.46

bond energy in determining the catalytic activity of perovskites (15).

The role of rare earth at the A site was studied by carrying out decomposition of N_2O on $LnMO_3$ (Ln = La, Nd, Sm and Gd) (16). In that system, the Mn-O-Mn cluster has been considered to be the active site for the reaction. A linear correlation between E_a and lattice parameters "a" and "c" was observed. As the lattice parameter increases, Mn-O bond length increases thus increasing the d-electron density around Mn ion. As one proceeds from La → Ga, f-electrons are being filled successively and one can expect a relative increase in the net electron density around Mn. As a result the residual time of O^- on Mn sites can be expected to decrease and desorption of oxygen becomes facilitated. This is reflected in the decrease of E_a from La to Gd (Table 1).

Similarly, the effect of rare earth on catalytic activity was studied by Ramanujachary et al. (17) over $LnNiO_3$ (Ln =La, Sm, Nd). Over all these catalysts, a weak inhibition by product oxygen (eqn. 3) was observed. A multicenter adsorption of nitrous oxide has been envisaged to explain the kinetic data. Figure 6

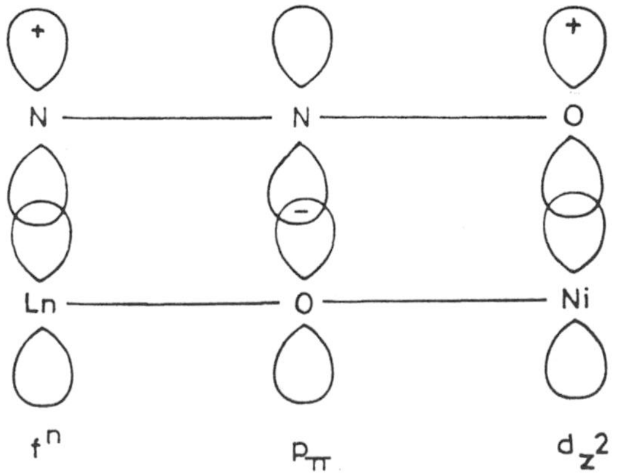

Fig. 6. Molecular orbital scheme for the multipoint adsorption of N_2O on $LnNiO_3$.

shows the molecular orbital scheme proposed for the adsorption of N_2O on $LnNiO_3$. In the case of $LaNiO_3$, La has an empty f orbital, hence $f-p_z$ interaction between La and O^{-2} ion will be weak, while $d_{z2}-p_z$ interaction between Ni^{3+}-O will be strong. Such a situation demands N_2O chemisorption to be rate determining. As the rare earth is changed to Sm, Ln-O interaction increases and consequently the Ni^{3+}-O interaction decreases resulting in the increase in the electron density around Ni^{3+}. In such a case, O^- is held strongly by Ni^{3+} on the surface and increased oxygen inhibition was found.

Louisray et al. (18) studied the effect of the valence state on the catalytic activity of $La_{1-x}Sr_xMnO_3$ (x = 0 to 1). The plot of E_a vs x or Mn^{4+} (Fig. 7) showed maximum activity at x = 0.41 corresponding to 50 % Mn^{4+}. Mn^{3+}/Mn^{4+} cluster is reported to be an active site for the decomposition of N_2O and this cluster acts as electron acceptor and donor site.

A linear plot generally observed between E_a and ln A in all the series indicates that the substitution of metal ions at A and B sites alter the energies of the active sites and hence composition effect was observed. Larsson (19) constructed a histogram of the partition of the data and observed that the value of E_a spaced regularly with an approximate interval of 3.5 kcal mol^{-1}. This empirical observation fits the suggestion that

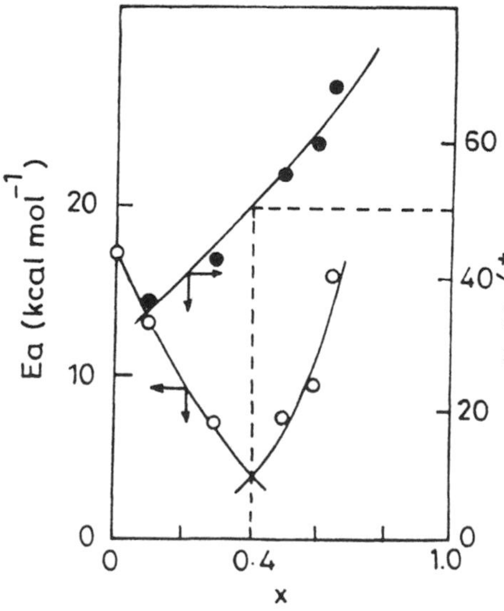

Fig. 7. Plot of E_a for N_2O decomposition or Mn^{4+} percentage versus x for the series $La_{1-x}Sr_xMnO_3$.

the change in E_a for a series of related catalysts for one and the same reaction occurs in a stepwise manner. Using this model, a quantitative description of the compensation effect has been made for the series $LnMnO_3$ (Fig. 8). According to the compensation law, i.e. the linear relation between E_a and ln A, the lines of Arrhenius plots for all the catalysts cut each other at the same temperature (iso catalytic temperature) which is found to be 900$^+$- 25 K for $LnMnO_3$ series.

IV. Decomposition of N_2O on Double Perovskites

The double perovskites ($A_2B_2O_6$) with their highly adaptable and stable structure (20) are suitable in the study of model catalytic reactions on solid state catalysts for a clear understanding of the role of solid state chemistry in catalysts. Sastri et al (21) studied the effect of pressure on kinetics of decomposition of N_2O on La_2TiMO_6 (M = Cu, Ni and Zn). Decomposition of N_2O on La_2TiCuO_6 obeyed no inhibition, weak inhibition and strong inhibition at 50, 100 and 200 torr

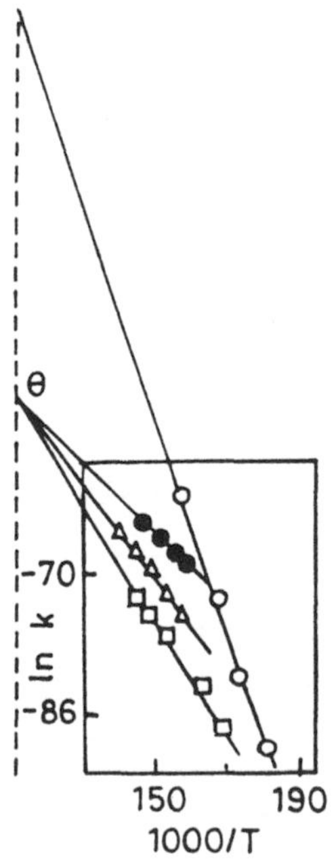

Fig. 8. Plot of ln k vs 10^3/T for $AMnO_3$ (A = La (O); Nd (Δ); Sm (□); Gd (●) showing isocatalytic temperature (Θ).

respectively. The activity of these catalysts follow the order $La_2TiNiO_6 > La_2TiCuO_6 > La_2TiZnO_6$ (22,23) which is not surprising because it is generlly known that catalysts containing Ni^{2+} are more active than Cu^{2+} which in turn will be more active than Zn^{2+}. The order of activity has been explained based on the conductivity and magnetic studies. At 200 torr, as mentioned earlier, the desorption of O^- as O_2 is rate determining which can be pictorially represented as

$$
\begin{array}{ccccccc}
O^- & O^- & & O & O & & O_2 \\
| & | & \xrightarrow{\;1\;} & \vdots & \vdots & \xrightarrow{\;2\;} & + \\
S\!\!-\!\!\!-\!\!\!-\!\!S & & & -S\!\!-\!\!\!-\!\!\!-\!\!S- & & & -S\!\!-\!\!\!-\!\!\!-\!\!S-
\end{array}
$$

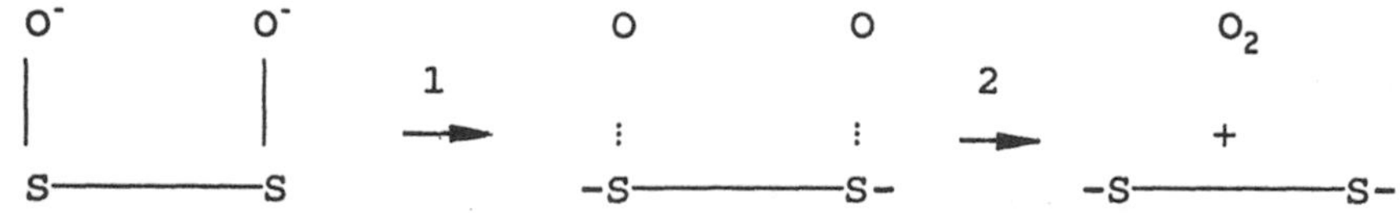

(adjacent sites)

Step 1 is determined by the electron acceptor property of the catalyst while step 2 involves decoupling of spins of $p\pi$ electron of oxygen atom to form O_2 molecule. This kind of decoupling will be facile on catalysts in which the spins of electrons are more ordered i.e. μ_{eff} of the catalyst is high. As we see from Table 2, La_2TiNiO_6 has the highest μ_{eff} and as a result the E_a is low for this catalyst.

Table 2 indicates that the E_a for the reaction is small when the catalyst has high E_a for conduction. When the compound has high E_a for conduction, the mobility of charge carriers must be low. Since charge carriers are positive holes (in the present series) and when the mobility becomes high, the net positive charge on the surface, at any moment, will be less. As a result the approaching N_2O molecules will feel less positive charge when

Table 2. Physicochemical and Catalytic Properties of La_2TiMO_6

Catal.	E_a for Reaction (kcal/mol)	E_a for Conduction (eV)	μ_{eff} (B.M.)	S_{BET} (m^2g^{-1})	Lat. par. (nm)
La_2TiNiO_6	3.5	0.34	3.3	4.1	0.7812
La_2TiCuO_6	21.5	0.28	1.84	1.6	0.7872
La_2TiZnO_6	40.7	0.22	0	3.5	0.7876

E_a for conduction is low. This will result in larger numbers of N_2O molecules getting adsorbed as N_2O^- and this would hinder the movement of O^- ions and hence desorption of O^- becomes difficult. The effect of such a hindrance is reflected in an increase in E_a for the reaction.

Gowri et al. (24) compared the catalytic activity of La_2TiMO_6 (M = Ni, Cu and Zn) for two different reactions viz. N_2O decomposition and isopropyl alcohol (IPA) decomposition. As discussed in the previous section, in the decomposition of N_2O, desorption of oxygen involving donation of electrons to the surface is rate limiting. Product inhibition studies and electrical conductivity measurements towards IPA decomposition showed that desorption of acetone involving donation of electron to the catalyst is rarely limiting. A positive correlation as shown in Fig. 9 between E_a for IPA reaction and E_a for N_2O reaction confirms that both reactions follow a similar kind of mechanism. In these series, the high activity of La_2TiNiO_6 has been attributed to its high tendency to accept electrons. Ramanujachary et al. (25) however attributed the activity of $La_{1.793}TiCoO_6$ to the presence of the mixed valence state of cobalt.

Based on % conversion, a reverse trend in activity has been reported for the system La_2MnMO_6 (M = Ni, Cu and Zn) (26). It has been reported that n-type oxides are less active than p-type oxides for nitrous oxide decomposition (27). But in the present series, n-type oxide La_2ZnMnO_6 showed the highest conversion which may be due to highly isolated Mn^{3+} ions present in the inert oxide matrix, which are present as a result of slight non-

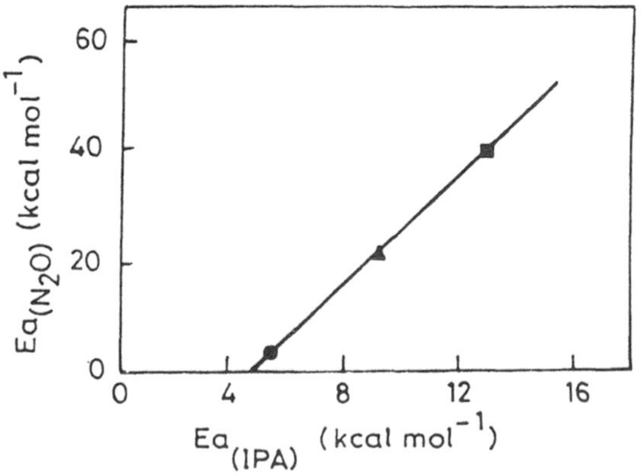

Fig. 9. Correlation between E_a for N_2O decomposition and IPA decomposition: (■), La_2TiZnO_6; (▲), La_2TiCuO_6; (●), La_2TiNiO_6.

stoichiometry of the compound. The observed activity has been explained based on clusters of the type Mn-O-M which are active sites in the series due to availability of the d-electron on Mn-ions. A similar type of cluster could not be attempted for the system La_2TiMO_6 (22, 23) because of non-availability of d-electrons on Ti^{4+}.

The low activity observed for $La_2Ni_{0.5}Mo_{0.5}O_6$ in the series $La_2Ni_{1.5}M_{0.5}O_6$ (M = Mo, W and Re) (28) has been attributed to the reduction of Mo^{6+} to Mo^{5+} with corresponding increase of Ni^{2+} to Ni^{3+} (29). Owing to the greater difficulty with which Ni^{3+} can donate electron to the N_2O molecules compared to Ni^{2+}, the activity is less.

V. Decomposition of N_2O on K_2NiF_4 Type Oxides

The catalytic activity of Ln_2CuO_4 (Ln = La-Gd) (30) and Ln_2NiO_4 (Ln = La, Pr, Nd) (31), oxides having perovskite related structure, has been studied to establish the role of A ion on catalytic activity. A molecular orbital scheme has been visualized to interpret the participation of rare earth ions on catalytic activity similar to the one proposed for $LnNiO_3$ (Fig. 8).

On Ln_2NiO_4, the oxide ion produced during decomposition is held by Ni^{2+}. When La(f°) is the rare earth ion, the electron density around Ni^{2+} is minimised by $f-p_z-d_{z2}$ interaction resulting in strong inhibition by oxygen. On proceeding to Pr^{3+} (f^2), the f-electrons contribute to redistribution of charge density resulting in relatively high electron density around Ni^{2+}. This facilitates the desorption of oxygen and hence obeys no inhibition kinetics. The weak inhibition kinetics observed by $Nd^{3+}(f^3)$ may be due to the competition between adsorption of N_2O and desorption of oxygen for the rate limiting step. Since the oxidation potential is very low for Nd^{3+}(2.43 eV) compared to La^{3+} (2.50) and Pr^{3+}(2.49), Nd^{3+} can independently offer a site for single point attachment of the N_2O molecule.

A linear correlation between E_a for reaction and E_a for conduction on Ln_2CuO_4 indicates suprafacial nature of this reaction on these systems. In the present case, 3d orbitals of cation and p-orbitals of oxygen interact to give rise to various bonding and antibonding levels. The antibonding levels appear to be of d-character if the cations involved are transition metal ions and they suffer usual crystal field splitting depending upon the nature and geometry of the coordination. The position of the Fermi level appears between these levels and this governs the activity of the system in electron transfer reactions.

Comparing the activity of La_2CuO_4 and La_2NiO_4, d_{z2}-p_z interaction is strong in La_2CuO_4 due to the filled d_{z2} orbital than Ni^{3+}(half filled). Hence f_o-p_z-d_{z2} interaction leads to increase in electron density around La^{3+} than in La_2NiO_4. This makes the N-O fission easy. Thus, the orthorhombic distortion along with high charge carriers make La_2CuO_4 more active than La_2NiO_4. The isocatalytic temperature reported for the series Ln_2CuO_4 is 785 K (19).

Oxide solid solutions can be utilised as the best means for studing these factors in a controlled manner. The solid solution technique enables a systematic tailoring of structural and

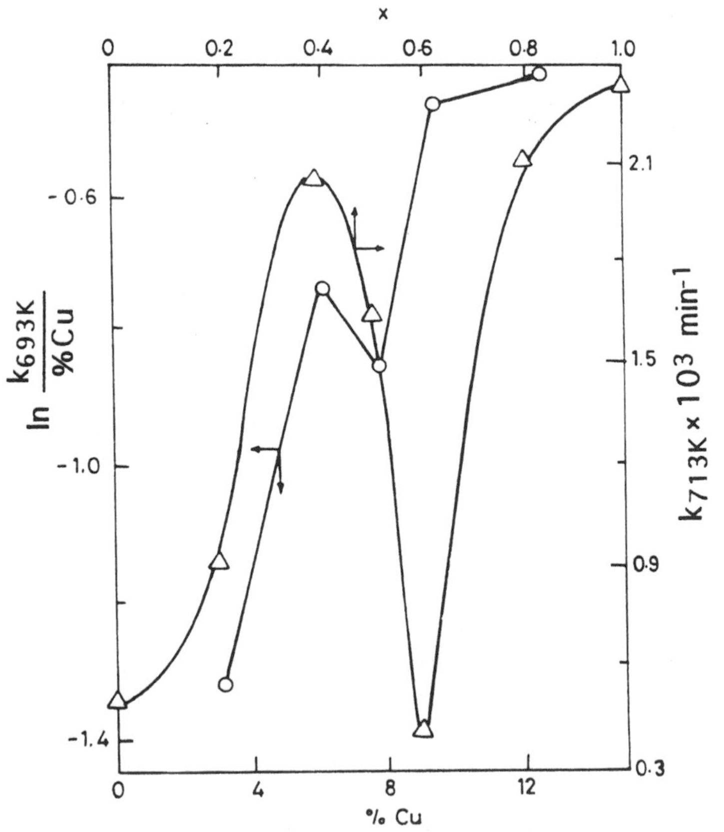

Fig. 10. Variation of the rate constant with composition: (O), $La_2Zn_{1-x}Cu_xO_4$; (△), $La_2Cu_{1-x}Ni_xO_4$.

influence on surface reactivity. A systematic study has been made of their electronic properties, thus allowing a study to be carried out to establish the role of B ions using N_2O decomposition on $La_2Cu_{1-x}Ni_xO_4$ ($0 \le x \le 1.0$) (32) and $La_2Cu_xZn_{1-x}O_4$ ($0 < x < 1.0$ (33). Initial substitution of Ni in the former case and Cu in the latter case increase the activity as shown in Fig. 10. Further substitution decreases the activity and increases again. Similar kinds of observation have been reported for sample oxides like $Ni_xMg_{1-x}O$, $Cr_xAl_{2-x}O_3$ and $Co_xMg_{1-x}O$ (34). The activity trend observed in the present series has been explained based on solid state polaron hopping mechanism proposed by Austin and Mott (35). Both series showed a linear correlation between E_a and lnA indicating that the substitution at the B site results in heterogenity of the surface, thereby confirming that the energetics of the active component are different in these series.

A parallel variation between E_a and magnetic susceptibility with composition in the series $La_2Cu_xNi_{1-x}O_4$ indicated a requirement of proper electronic occupancy at the emerging orbitals for the decomposition of N_2O. Similarly, a plot of the rate constant and magnetic susceptibility showed a linear relationship indicating the importance of the spin coupling process as the rate determining step for the decomposition. This is also in accordance with the observation that on this system the kinetic data obeyed first order kinetics.

Kameswari et al. (33) carried out decomposition of N_2O on $La_2Cu_{0.5}M_{0.5}O_4$ (M = Co, Ni and Zn) to understand the mutual interaction of two different active metal ions of the same concentration and valency in determining the physico-chemical and catalytic properties. As the desorption of oxygen requires empty d-orbitals at the surface, the non-transition metal ion, Zn^{2+} cannot function as the center for oxygen desorption. Hence, it followed strong inhibition kinetics. The rest of the ions (Co^{2+}, Ni^{2+} and Cu^{2+}) have the capacity to do so and hence obeyed first order kinetics. It has been observed that at 50 torr, the rate is governed by the electronic factor while at 200 torr, the geometric factor controls the overall rate. Christopher and Swamy (36) carried out the decomposition of N_2O on $La_{1.85}Sr_{1.05}CuO_{4-y}$, La_2CuO_4 and $CuAlO_2$ to study the effect of the valence state on catalytic activity. The high activity of $La_{1.85}Sr_{1.05}CuO_{4-y}$ has been attributed to the presence of mixed valent copper, namely Cu^{2+}/Cu^{3+}, higher surface concentration of copper, and anion vacancies in the system.

References

1. Viswanathan, B., Decomposition of nitrous oxide on mixed oxides, *Indian Chemical Manufac.* **9:** 1 (1980).
2. Cimino, A., V. Indovina, F. Pepe and F.S. Stone, Kinetics of the decomposition of nitrous oxide catalized by nickel ions in a magnesium oxide matrix, *Gazz. Chim. Ital.*, **103:** 935 (1973).
3. Nagasubramanian, G., B. Viswanathan and M.V.C. Sastri, Orbital symmetry and its role in catalysis: decomposition of nitrous oxide and oxidation of carbon monoxide, *Indian J. Chem.* **16A:** 642 (1978).
4. Nagasubramanian, G., M.V.C. Sastri and B. Viswanathan, Suprafacial catalysis: is nitrous oxide a model reaction?, *J. Indian Chem. Soc.* **56:** 158 (1979).
5. Nagasubramanian, G, B. Viswanathan and M.V.C. Sastri, Studies on catalytic decomposition of nitrous oxide on titanates, *Indian J. Chem.* **16A:** 645 (1978).
6. Nagasubramanian, G., M.V.C. Sastri and B. Viswanathan, Energy resonance hypothesis in heterogeneous catalysis, *Ind. J. Chem.* **16A:** 243 (1978).
7. Nagasubramanian, G., M.V.C. Sastri and B. Viswanathan, Catalytic activity of ternary oxides of titanates and volcano relationships, *Ind. J. Chem.* **16A:** 242 (1978).
8. Vijh, A.K., Sabatier-Balandin interpretation of the catalytic decomposition of nitrous oxide on metal oxide semiconductors, *J. Catal.* **31:** 51 (1973).
9. Louisraj, S., B. Viswanathan and V. Srinivasan, Role of adsorbed oxygen species in kinetics of catalytic decomposition of nitrous oxide, *Indian J. Chem.* **21A:** 689 (1982).
10. Ramadass, N., Preparation and characterization of La_2TiMO_6 (M = Co, Ni, Cu, Zn) perovskites, *Mater. Sci. Eng.* **36:** 231 (1978).
11. Rao, C.N.R., P. Ganguly, K.K. Singh and R.A. Mohanram, A comparative study of the magnetic and electrical properties of perovskite oxides and the corresponding two-dimensional oxides of K_2NiF_4 structure, *J. Solid State Chem.* **72:** 14 (1988).
12. Muralidhar, G., K.M. Vijayakumar, C.S. Swamy and V. Srinivasan, Catalytic properties of rare-earth transition metal oxide perovskites, in *Preprints of the 7th Canadian Symposium on Catalysis*, Edmonton, Alberta, 1980, p.289.

13. Kremenic, G., J.M.L. Nieto, J.M.D. Tascon and L.G. Tejuca, Chemisorption and catalysis on $LaMO_3$ oxides, *J. Chem. Soc., Faraday Trans. I* **81**: 939 (1985).
14. Nitadori, T., T. Ichiki and M. Misono, Catalytic properties of perovskite-type mixed oxides (ABO_3) consisting of rare earth and 3d transition metals. The roles of the A-and B-site ions, *Bull. Chem. Soc. Jpn.* **61**: 621 (1988).
15. Muralidhar G. and V. Srinivasan, Surface oxygen bond energy and kinetics and catalysis on perovskite oxide catalysts, *Int. J. Chem. Kinet.* **14**: 435 (1982).
16. Louisraj, S., B. Viswanathan and V.Srinivasan, Decomposition of nitrous oxide on rare earth manganites, *J. Catal.* **65**: 121 (1980).
17. Ramanujachary, K.V., K.M. Vijayakumar and C.S. Swamy , *J. Bangladesh Acad. Sci.* **5**: 49 (1981).
18. Louisray, S., B. Viswanathan and V. Srinivasan, The activity of Mn^{3+} and Mn^{4+} in lanthanum strontium manganite for the decomposition of nitrous oxide, *J. Catal.* **75**: 185 (1982).
19. Larsson, R., On the catalytic decomposition of nitrous oxide over metal oxides, *Catalysis Today* **4**: 235 (1989).
20. Le Flem G., G. Demazeau and P. Hagenmuller, Relations between structure and physical properties in K_2NiF_4-type oxides, *J. Solid State Chem.* **44**: 82 (1982).
21. Sastri, V.R., R. Pitchai and C.S. Swamy, Kinetics of the decomposition of N_2O on lanthanum titanates, *Indian J. Chem.* **18A**: 213 (1972).
22. Sastri, V.R., R. Pitchai and C.S. Swamy, Catalytic activity of La_2TiMO_6 (M = nickel, copper and zinc), *Indian J. Chem.* **19A**: 738 (1980).
23. Sastri, V.R., R. Pitchai and C.S. Swamy, Nitrous oxide decomposition on La_2TiMO_6 (M = nickel, copper and zinc), *Curr. Sci.* **48**: 441 (1979).
24. Gowri, D.K., K.V. Ramanujachary and C.S. Swamy, Comparison of the catalytic activities between isopropyl alcohol decomposition and nitrous oxide decomposition over La_2TiMO_6 (where M = Ni, Cu and Zn), *Curr. Sci.* **50**: 226 (1981).
25. Ramanujachary, K.V., N. Kameswari and C.S. Swamy, *J. Bangladesh Acad. Sci.* **8**: 55 (1984).
26. Kameswari, N., B. Rajasekhar, R. Radha and C.S. Swamy, Catalytic decomposition of nitrous oxide on $La_2Ni_{1.5}M_{0.5}O_6$ (M = Ni, Cu and Zn), *Curr. Sci.* **54**: 229 (1985).

27. Dell, R., F.S. Stone and P.F. Tilley, The decomposition of nitrous oxide on cuprous oxide and other catalysts, *J. Chem. Soc.* **49:** 201 (1952).
28. Kameswari, N. and C.S Swamy, Decomposition of nitrous oxide on $La_2Ni_{1.5}M_{0.5}O_6$ (M = molybdenum, tungsten and rhenium), *Curr. Sci.* **54:** 383 (1985).
29. Subramanian M.S. and G.V. Subba Rao, Synthesis and electrical properties of perovskite oxides $LnB_{0.75}B'_{0.25}O_3$, *J. Solid State Chem.* **31:** 329 (1980).
30. Ramanujachary, K.V., N. Kameswari and C.S. Swamy, Studies on the catalytic decomposition of N_2O on rare earth cuprates of the type Ln_2CuO_4, *J. Catal.* **86:** 121 (1984).
31. Ramanujachary, K.V., K.M. Vijayakumaar and C.S. Swamy, Studies on the catalytic decomposition of N_2O on rare earth nickelates, *React. Kinet. Catal. Lett.* **20:** 233 (1982).
32. Ramanujachary K.V. and C.S. Swamy, Studies on the decomposition of N_2O on the solid solution $La_2Cu_{1-x}Ni_xO_4$ ($0 \leq x \leq 1.0$), *J. Catal.* **93:** 279 (1985).
33. Kameswari, U., J. Christopher and C.S. Swamy, Studies on the catalytic decomposition of N_2O over the solid solution $La_2Cu_xZn_{1-x}O_4$ ($0 \leq x \leq 1.0$), *React. Kinet. Catal. Lett.* **41:** 381 (1990).
34. Pomonis, P. and J.C. Vickerman, An interpretation of the activity behavior of model oxide solid solution catalysts on the basis of a solid state polaron hopping mechanism, *J. Catal.* **55:** 88 (1978).
35. Austin, I.G. and N.F. Mott, Polaron in crystalline and non-crystalline materials, *Adv. Phys.* **18:** 41 (1968).
36. Christopher J. and C.S. Swamy, Catalytic decomposition of N_2O on ternary oxides containing copper in different valence states, *J. Mol. Catal.* **62:** 69 (1990).

Chapter 18

Perovskite Oxides as Solid State Chemical Sensors

T. Arakawa

Department of Industrial Chemistry, Faculty of Engineering, Kinki University in Kyushu, Iizuka, Fukuoka 820, Japan

I. Introduction

The principle that a significant change in electrical resistance may be caused by adsorption of a gas on the surface of semiconductor oxides has long been implicated by Brattain and Bardeen (1) or Heiland (2). Subsequently, the original measurements were made on zinc oxide and tin oxide (3,4). Since then, many workers have made efforts to obtain good devices of gas detection, which are so-called "sensors." At present, it is well known that a SnO_2 sensing element is one of most commercially available gas detectors (5) because it shows a high sensitivity at lower operating temperatures. The search for new materials for use in gas sensors has attracted continued interest. The application of these gas sensors is limited by stability in ambient atmosphere. Recently, a number of perovskite type oxides (ABO_3) were used as gas sensor materials because of the stability in thermal and chemical atmospheres. Some of these perovskite oxides are semiconductors (6) and in some cases oxygen anion conductors can be obtained by the appropriate combination of A and B site metals and by doping a third metal element (7,8). Thus, the electrical and magnetic properties (9), as well as the catalytic features (10,11), and the crystal structures (12) have been discussed by many workers. In this chapter, a correlation between physicochemical and gas sensing properties of perovskite oxides has been described.

II. Structures of Gas Sensor and Characterization of Gas Sensor Materials

The simple structure of gas sensors is shown in Fig. 1. The layer of semiconductor oxides used as a gas sensor is covered with ceramic insulating tube (1.2 mm x 4.0 mm) assembled with a heater, lead wires, lead pins, insulating substrates and stainless steel net. The gas sensors were placed in the center of the test box (a flow-type or a closed-type) containing controlled fresh air. A pure test gas (i.e. alcohol, CO, H_2, hydrocarbon) was injected directly into the box by a syringe and stirred by a fan. In order to obtain electrical characteristics of the sensors, the fixed D.C. current was applied to the sensor and the change of sensor resistance was monitored by an electronic recorder. In general the sensitivity of the sensor was influenced by the operating temperature, the concentration of gases and materials used. The sensitivity is defined as follows: (i) the resistance ratio R_{air}/R_{gas} (R_{air} and R_{gas} are the resistances in air and in an atmosphere containing a constant gas concentration, respectively); (ii) the response ratio $(R_{air}- R_{gas})/R_{air}$ x 100 %, where R_{air} and R_{gas} are the resistances in air and after exposure to a flammable gas, respectively.

When a flammable gas was introduced into the system, a conductivity change in the thin oxide film was observed. An

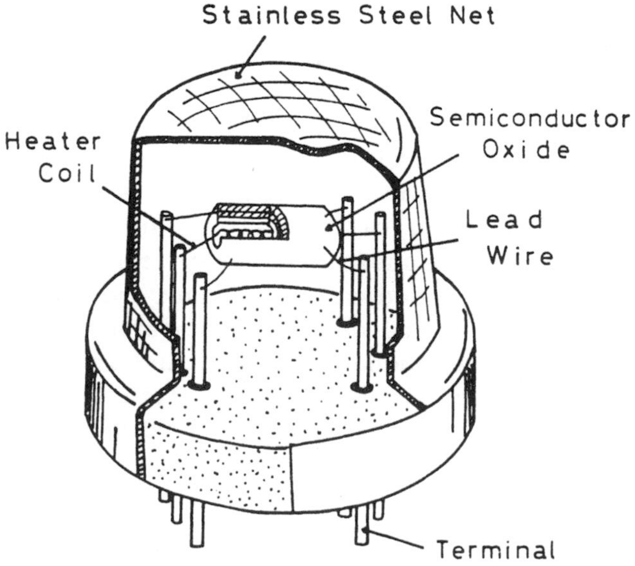

Fig. 1. Sketch diagram of a typical sensor structure.

isolated example of conductivity change in flow system is shown in Fig. 2. For the $SmCoO_3$ thin film, the conductivity decreases immediately after the injection of methanol and then is restored to its initial value. Since the conductivity changes and the catalytic reactions under adsorption of reducing gas occur simultaneously on the surface of gas sensors, it is necessary to consider the catalytic reaction on the surface of Perovskite oxides used as gas sensor materials.

III. Perovskite Oxides as Gas Sensor Materials

Since the gas sensing function of semiconducting oxides utilizes mainly the change of the resistivity of semiconducting oxides under adsorption of flammable gases, it may be essential to understand various physical parameters including the electrical conductivity for developing the semiconductor oxide gas sensors using perovskite oxides.

Most rare earth and transition metal mixed oxides of the type $LnMO_3$ (Ln = rare earth, M = transition metal) have perovskite structure (13,14) and are semiconductors (15), although $LaNiO_3$ and $LaTiO_3$ display metallic conductivity (16). For rare earth orthochromites, manganites and ferrites, the conductivity in each series of perovskites decreases with increasing atomic

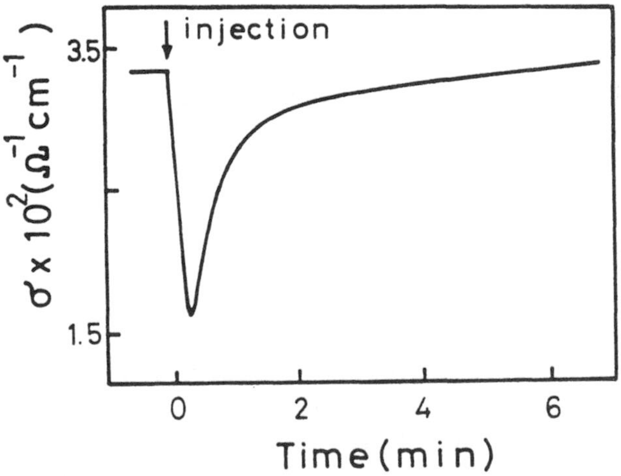

Fig. 2. The change of electric conductivity of $SmCoO_3$ with time after injection of CH_3OH at 513 K. The carrier gas was N_2 containing ca. 50 ppm O_2 at a flow rate of 40 cm^3/min.

number of rare earth (17). The electrical transport properties of these oxides can be explained in terms of the localized behavior of the d-electrons. The $LaCoO_3$ behaves as a semiconductor below 398 K because $LaCoO_3$ contains predominantly Co^{III} with a random distribution of Co^{3+} (18). In the temperature range 398 < T < 1210 K, the conductivity increases much more rapidly with an increase in the temperature and was metallic above 1210 K, since the transition from localized to itinerant e_g electrons occurs in the form of a first-order phase transition (15).

The practical uses of gas sensors are limited by the stability of the perovskite phase in reducing atmosphere at high temperatures. The stability of a series of perovskite-type oxides $LaBO_3$ (B = V, Cr, Mn, Fe, Co, Ni) at 1273 K in gas mixtures of CO_2/H_2 was investigated by Nakamura et al. (19). The sequence of stability of these oxides was found to be $LaCrO_3$ > $LaVO_3$ > $LaFeO_3$ > $LaMnO_3$ > $LaCoO_3$ > $LaNiO_3$.

Obayashi et al. (20,21) employed $(Ln,M)BO_3$ (Ln = rare earth, M = alkaline-earth metal, and B = transition metal) for detecting ethanol in the exhaled air. In the presence of ethanol vapor, the electrical conductivity of these oxides drastically changed. This was the first example that perovskite oxides applied to the gas sensor. However, no attempt has been made to establish a correlation between physicochemical and gas sensing phenomena of perovskite oxides. This relationship has been described in this article.

A. Rare Earth Cobaltates and Manganates

The activity for methanol sensing is shown in Fig. 3 (22). A response ratio of 20% is attained for $LnCoO_3$ and $LnMnO_3$ and the relative sensitivity attains 1 for $LnCrO_3$ and $LnFeO_3$. It is clear that the activities among these compounds are influenced by the transition metal ions and the rare earth ions. The activity of the rare earth cobaltates is the highest. Based on the classification of perovskites by Googenough (23), the cobaltates are located on the boundary between itinerant and localized e_g electrons. Rao and Bhide (24) have suggested that the catalytic properties of these cobaltates are connected with the spin states, and particularly with the proportion of high-spin state Co^{3+}. It is noteworthy that the activity of rare earth cobaltates is the greatest for $SmCoO_3$, which has the highest room-temperature ratio of high-spin to low-spin cobalt ions.

There are many investigations of the correlation between electrical conductivity and magnetic properties. Both the ferromagnetic Cúrie temperature and the electrical conductivity reach a maximum for $0.3 < x < 0.4$ in $La_{1-x}Sr_xMnO_3$ (25). There is a linear correlation between resistivity and Cúrie point in $La_{1-x}Pb_xMnO_3$ ($x = 0.31$) (26). The $LnMnO_3$ (Ln = La - Eu series) has a paramagnetic Cúrie temperature which is dependent on the kind of rare earth incorporated. The change of activity with Θ_p is shown in Fig. 4 (27). The activity decreases linearly with Θ_p, except for $LaMnO_{3+x}$. In addition, a linear relation between Θ_p and the c axis Mn^{3+}-Mn^{3+} separation exists in $LnMnO_{3+x}$ (27). The results in Fig. 4 are consistent with the activity that increases with the ease of exchanging oxygen atoms with the solid, since the activity of methanol sensing increases with Θ_p and Θ_p increases with the concentration of mobile holes in the Mn^{3+}:d^4 band. This, in turn, appears to correlate with the length of the Mn^{3+}-O-Mn^{3+}. The catalytic activity of $LnMnO_3$ may also be correlated with the cell size. The values of E_a for the decomposition reaction of N_2O on $LnMnO_3$ increase with an increase in the value of cell constant (28). The (100) face containing Mn-O-Mn group is active for the decomposition of N_2O. As the value of the cell constant increases, the Mn-O bond length increases, thereby weakening the Mn-O bond and increasing the d electron density of the Mn ion. The order of activity of $LnMnO_3$ for CO oxidation reaction was La > Pr < Nd > Sm < Eu > Gd (29). This order appears to correlate well with the size of the Ln^{3+} ions.

In order to increase the catalytic activity, partial substitutions of the Ln ion by a second ion having a different valency have been tried in simple perovskite oxides (30-32). The substituted manganates and cobaltates are the most widely studied compounds because these perovskite oxides have higher catalytic activities. Parkash et al. (33) found that $LaCoO_3$ was less active for CO oxidation than its substituted $La_{1-x}Sr_xCoO_3$ counterparts, which contain a mixture of Co^{3+} and Co^{4+} ions. The reducibility of catalysts $La_{1-x}Sr_xCoO_3$ in the oxidation of CO, CH_4 and C_3H_8, the ease of oxygen desorption and the diffusivity of oxygen in the bulk of these oxides were found to increase markedly with an increase in the value of x (34). Raj et al. (35) also found that the catalytic activity reached a maximum at the composition $x = 0.41$ (Mn^{4+} = 50 %). This suggests that Mn^{3+} and Mn^{4+} become indistinguishable with regard to their catalytic activity. Thus the replacement of the Ln ion by a second ion changes the M^{4+}/Mn^{3+}

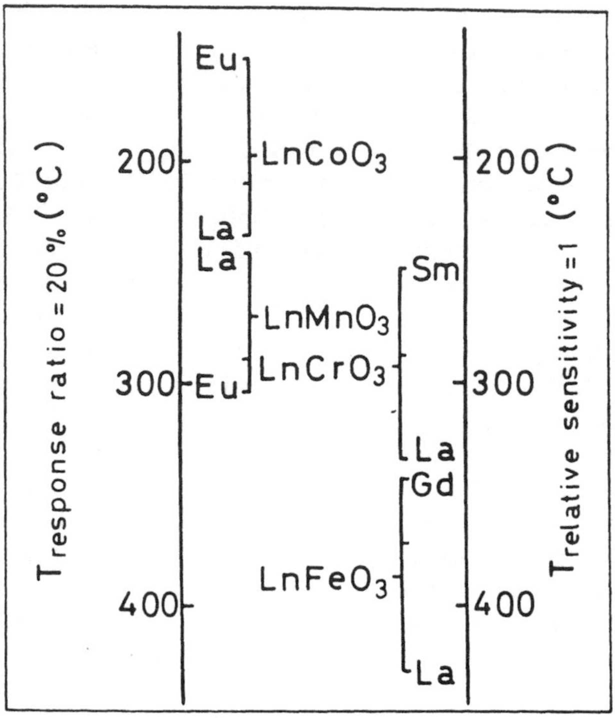

Fig. 3. Activity sequence for methanol sensing of various perovskites.

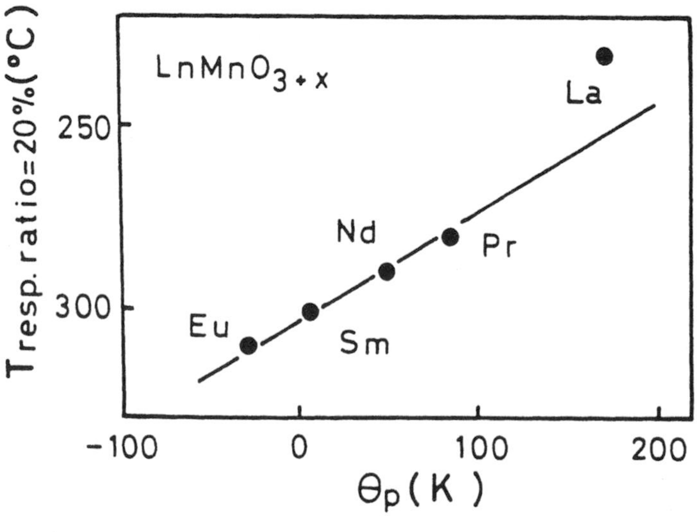

Fig. 4. Relationship between Θ_p and activity.

ratio. In the $Ln_{1-x}A_xMO_3$ systems, the introduction of alkaline earth ions results to the creation of a mobile hole. However, the gas sensing activity may not be appreciably influenced by the number of holes, but the depth of the trap centers of holes in these perovskite oxides (36). If Ln and A ions are fixed, M ions have a role in determining the gas sensing activity. In $Sm_{0.5}Sr_{0.5}MO_3$ system (20), the activity of $Sm_{0.5}Sr_{0.5}FeO_3$ compound is higher than that of $Sm_{0.5}Sr_{0.5}CoO_3$ homologue if the activity is compared by the response ratio alone. However, the $Sm_{0.5}Sr_{0.5}CoO_3$ compound shows the best trade-off characteristics concerning both the response ratio and response time.

B. Orthoferrites and Orthochromites

Orthoferrites can be either a p- or a n-type semiconductors as indicated by Seebeck coefficient measurements (37,38). The n-type $LnFeO_3$ compounds exhibited a remarkable activity for methanol sensing above the Néel temperature (T_N) of these oxides (39). The sensing activity appeared at the temperature ranges in which these oxides shows paramagnetic behaviors. The variation of the activity of n-type $LnFeO_3$ oxides with the atomic number of the rare earth elements is shown in Fig. 5, where the activity is compared also with that of the T_N. The activity increases as the radius of rare earth ion decreases, while T_N decreases as the radius of the rare earth ion decreases. Orthochromites exhibit p-type conduction and they are antiferromagnetic with Néel temperature (40). T_N of orthochromites is lower than T_N of orthoferrites. It is expected that the activity of orthochromites increases with the decrease of the radius of the rare earth ion as the case of orthoferrites since T_N decreases ongoing from La to Sm, i.e. 282, 214, and 190 K for $LaCrO_3$, $NdCrO_3$ and $SmCrO_3$, respectively. But there is no correlation between activity and Néel temperature.

C. Perovskite Related Oxides

Perovskite sheets separated by AO sheets with rock salt structure are found in the $(SrO)_n(SrTiO_3)$ compounds (41). For n = 1 these oxides have the K_2NiF_4 structure, and many of these are known (42). Similar perovskite sheets are found in the isostructural La_2NiO_4, La_2CuO_4 and La_2MnO_4 compounds (43).

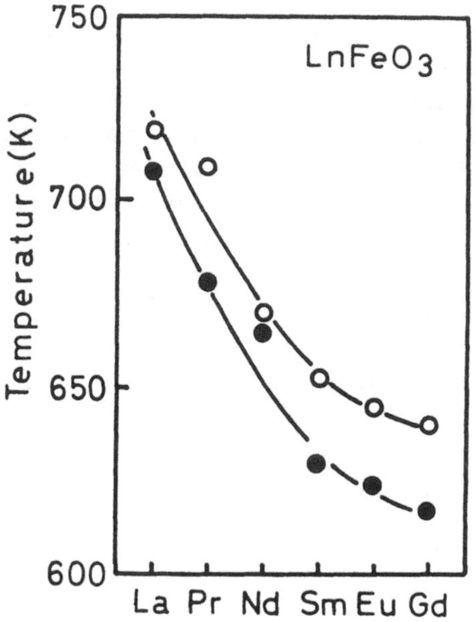

Fig. 5. Comparison of the catalytic activity (○) with the Néel temperature (●).

Recently, Bednorz and Muller discovered the facts that the Ba-La-Cu-O system of which a part of the La^{3+} ions are replaced by Ba^{2+} in La_2CuO_4, exhibits high-temperature superconductivity (44). The electrical properties of these oxides reveal drastic changes ongoing from Pr to Nd; chemisorption of methanol on La_2CuO_4 and Pr_2CuO_4 led to a decrease in the conductivity of the oxide, whereas on Sm_2CuO_4, Eu_2CuO_4 and Gd_2CuO_4 the reverse was observed.

D. Activities for Gas Sensing and Surface Reaction

We estimated $\Delta E = E_c - E_t$ from the conductivity changes of $LnMnO_3$ after the injection of methanol, by using equation 1,

$$\sigma = A\ \sigma_0 \exp\ (-(E_c - E_t)/kT) \tag{1}$$

where σ is the minimum conductivity after the adsorption of a reducing gas, σ_0 is the conductivity in a steady gas flow, A is a constant, E_c is the energy of the conduction band, and E_t is the energy of the surface state (46). The results are summarized in Table I (22). The sequence of ΔE is $LnCoO_3 < LnMnO_3 \approx LnCrO_3$

Table 1. ΔE* Values for Various Perovskite-Type Oxides

Compound	E (eV)	Compound	ΔE (eV)
$LaCoO_3$	0.14	$LaCrO_3$	0.28
$SmCoO_3$	0.13	$SmCrO_3$	0.22
$EuCoO_3$	0.09	$EuCrO_3$	0.14
$LaMnO_3$	0.25	$LaFeO_3$	0.98
$SmMnO_3$	0.25	$NdFeO_3$	0.74
$EuMnO_3$	0.27	$SmFeO_3$	0.55

* Absolute values determined according to Eq. 1.

< $LnFeO_3$, which coincides with that of the activity for gas sensing. The $LnCoO_3$, which shows the highest activity, has the lowest ΔE. Moreover, ΔE values correlate well with the binding energy of oxygen which coordinates to metal ions. The pattern in Fig. 6 may be understood based on the strength of the metal-oxygen bond in various perovskites being an important factor in methanol sensing.

An electronic theory of catalysis on semiconductors was developed by Weisz (47) and others (48). Parravano et al. (49) showed that the activation energies for the rate of CO oxidation on NiO doped with Li, Ga or Cr were strongly dependent on the type of doping although the rate of reaction was little affected. For a number of catalysts, a relationship between the activation energy for the oxidation reaction of SO_2 and the activation energy for electrical conduction has been established (50). A good relationship between E_a and E_σ in these catalysts suggests that a charge-transfer process may be involved in the rate-determining step.

When the reducing gases were adsorbed on p-type $LnMO_3$ (M = Mn, Cr, Co), the conductivity decreased (22). These phenomena may be represented by the following equations,

$$CO\ (g) + O^{2-}\ (s) \longrightarrow CO_2\ (s) + 2\ e^- \quad (2)$$

$$H_2\ (g) + O^{2-}\ (s) \longrightarrow H_2\ (s) + 2\ e^- \quad (3)$$

$$CH_3OH\ (g) + 3\ O^{2-}\ (s) \longrightarrow CO_2 + 2\ H_2O\ (s) + 6\ e^- \quad (4)$$

where (s) denotes a surface species and (g) denotes a gaseous species. As an electron is liberated upon adsorption, the chemisorption of the reducing gases results in the loss of a conductivity hole by interaction with this free electron. These equations mean that the catalytic reaction occurs on the surface of perovskite oxides. Moreover, Fig. 6 demonstrates that O2-(s) and lattice oxygen in LnMO3 attains equilibrium fairly rapidly at the reaction temperature as follows,

$$O^{2-}(s) \Longleftrightarrow O_L^{2-} \quad (O_L^{2-} : \text{lattice oxygen in } LnMO_3) \qquad (5)$$

IV. Oxygen Sensors

Another application of perovskite oxides to gas sensors has been the measurement of oxygen pressure in combustion control

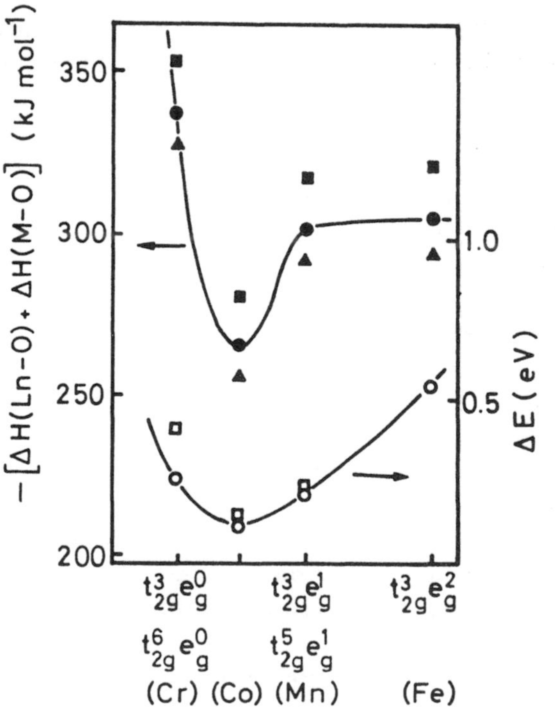

Fig. 6. $-[\Delta H(Ln\text{-}O) + \Delta H(M\text{-}O)]$ and ΔE as a function of the electron configuration of the M ion for $LnMO_3$. La (□, ■); Sm (○, ●); Eu (▲).

systems. Recently, the requirements for control of the fuel-air mixture to maintain a fixed concentration of excess oxygen, which is called "lean-burn control," have received increasing attention in order to increase the fuel efficiency as well as to reduce the air pollutant components in exhaust gases for boiler systems or vehicle engines.

There has been a continuing search for materials of lean-burn oxygen sensors. These oxygen sensors are divided into two groups, i.e. solid electrolyte type such as yttria-stabilized zirconia and semiconductor type such as titania (51) or $Co_{1-x}Mg_xO$ (52). The sensors using yttria-stabilized zirconia have been developed for the use in a lean-burn oxygen sensors. The semiconductor type sensors utilizing electrical conductivity changes due to oxygen adsorption or desorption has received considerable interest in recent years because of the simple structure and lower cost.

The change in the electrical conductivity of semiconducting oxides as a function of oxygen partial pressure at higher temperatures is well known. This change can be represented by the following equation,

$$\sigma = \sigma_0 \exp (E_A/kT)(P_{O2})^{1/n} \qquad (6)$$

where k is the Boltzmann's constant, T is the Kelvin temperature and E_A an activation energy. The sign and value of n depend on the semiconductive nature of metal oxides in the oxygen partial pressure, i.e. n = 4 or 6 for p-type semiconductors and n = -4 or -6 for n-type semiconductors. TiO_2, which has n = -4 and E_A of 1.6 eV in Eq. 6, was the first oxygen sensor of semiconductor type for control of the stoichiometric fuel-air mixture in car engine systems (53). However, TiO_2 cannot be used for sensing elements in the lean-burn region because the temperature coefficient of resistivity is too high, although titania has the advantage in chemical stability. On the other hand, a lean-burn oxygen sensors using p-type semiconducting oxides, i.e. CoO (54) or $Co_{1-x}Mg_xO$, has the advantage in the lean-burn regions since the resistance of these oxides in the lean-burn regions is low. In the patent literature (55,56), p-type perovskite oxides have been used. Fig. 7 shows the change of resistance of $La_{0.35}Sr_{0.65}Co_{0.7}Fe_{0.3}O_{3-y}$ in the gas atmosphere which is typical for exhaust gases (57). A steep increase of resistance occurred at the stoichometric gas composition. The magnitude of the steep resistance change increased with the rising sensor operation temperature. Arai et al. (58,59) observed that $SrMg_{0.4}Ti_{0.6}O_3$ among

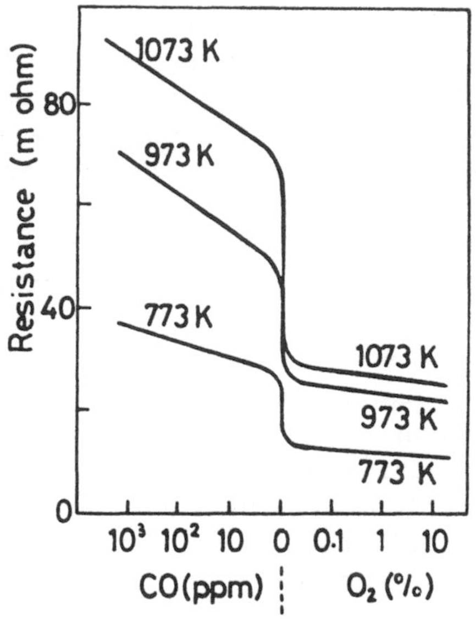

Fig. 7. Relationship between air-fuel ratio and sensor resistance.

$SrA_xTi_{1-x}O_3$ (A = Al, Mg) exhibits the largest change in the resistance at the stoichiometric point of the combustion.

In order to obtain an excellent lean-burn oxygen sensor it is necessary to continue a search for new materials with higher oxygen sensitivity and stability even at an extremely low oxygen partial pressure.

V. Conclusions

Since the use of perovskite oxides containg 3d transition metals in an alcohol sensor has attracted considerable interest, the study for the correlation of sensitivities with solid state properties has been advanced. Recently, the perovskite oxides were applied to the oxygen sensors in combustion control systems. The perovskite oxides which have unique optical properties can also be expected to produce the new chemical sensors in optoelectronics (60,61), because an optical method would have the advantage for practical applications in the future. Thus, the application of perovskite oxides to new sensing devices has just started. Further investigation of a wide variety of properties may stimulate to develop new practical chemical sensors, since various compositions of perovskite oxides are available.

References

1. Brattain, W. H., and J. Bardeen, Surface Properties of Germanium, *Bell Systems Tech. J.*, **32**: 1-41 (1953).
2. Heiland, G., Zum einfluss von adsorbiertem sauerstoff auf die electrische leitfahigkeit von zinkoxydkristallen, *Physik*, **138**: 459-464 (1954).
3. Seiyama, T., A. Kato, K. Fukkishi, and M. Nagatani, A new detector for gaseous components using semiconductive thin films, *Anal. Chem.*, **34**: 1502-1503 (1962).
4. Seiyama, T., and S. Kagawa, Study on a detector for gaseous components using semiconductive thin films, *Anal. Chem.* **38**: 1069-1073 (1966).
5. Taguchi, N., Gas-detecting device, *US Patent* 3,631,367 (1970)
6. Jonker, G. H., Semiconducting properties of mixed crystal with perovskite structure, *Physica*, **20**: 1118-1122 (1954).
7. Shimizu, Y., Y. Fukuyama, T. Narikiyo, H. Arai, and T. Seiyama, Perovskite-type oxides having semiconductivity as oxygen sensors, *Chem. Lett.*, 377-380 (1985).
8. Shimizu, Y., Y. Fukuyama, H. Arai, and T. Seiyama, Oxygen sensor using perovskite-type oxides, *ACS Symposium Series* **309**, (D.Schmetzle and R. Hammerle, Eds.), Am. Chem. Soc., Washington, DC, p.83.
9. Ramadass, N., ABO_3-type oxides. Their structure and properties. A bird's eye view, *Mat. Sci. Eng.*, **36**: 231-239 (1978).
10. Voorhoeve, R. J. H., D. W. Johnson, Jr., J. P. Remeika and P. K. Gallagher, Perovskite oxides: Materials science in catalysis, *Science*, **195**: 827-833 (1977).
11. Voorhoeve, R. J. H., Perovskite-related oxides as oxidation-reduction catalysts, *Advanced Materials in Catalysis* (J. J. Burton and R. L. Garten, Eds.), Academic Press, New York, 1977, p. 129.
12. Megaw, D. W., Crystal structures: a working approach, *Studies of Physics and Chemistry, No. 10* (W. B. Saunders Ed.), Philadelphia, 1973, p. 563.
13. Goodenough, J. B. and J. M. Longo, Crystallographic and magnetic properties of perovskite and perovskite-related compounds, *Landolt-Bornstein New Series*, **Vol. 4**, Springer-Verlag, Berlin and New York, 1970, p. 126.
14. Glazer, A. M., The classification of tilted octahedra in perovskite, *Acta Cryst.*, **B28**: 3384-3392 (1972).

15. Raccah, P. M., and J. B. Goodenough, First-order localized-electron collective = electron transition in $LaCoO_3$, *Phys. Rev.*, **155**: 932-948 (1967).
16. Goodenough, J. B. and P. M. Raccah, Complex versus band formation in perovskite oxides, *J. Appl. Phys.*, **36**: 1031-1032 (1965).
17. Rao, G. V. S., B. M. Wanklyn, and C. N. R. Rao, Electrical transport in rare earth ortho-chromites, -manganites and -ferrites, *J. Phys. Chem. Solids*, **32**: 345-358 (1971).
18. Bhide, V. G., D.S. Rajoria, Y. S. Reddy, G. R. Rao, G. V. S. Rao, and C. N. R. Rao, Localized-to-itinerant electron transitions in rare earth cobaltates, *Phys. Rev. Lett.*, **28**: 1133-1136 (1972).
19. Nakamura, T., G. Petzow, and L. J. Gauckler, Stability of the perovskite phase $LaBO_3$ (B = V, Cr, Mn, Fe, Co, Ni) in reducing atmosphere. I. Experimental results, *Mat. Res. Bull.*, **14**: 649-659 (1979).
20. Obayashi, H., Y. Sakurai and T. Gejo, Perovskite-type oxides as ethanol sensors, *J. Solid State Chem.*, **17**: 299-303 (1979).
21. Obayashi, H., and H. Okamoto, Properties of oxygen deficient perovskite-type compounds and their application as alcohol sensors, *Nippon Kagaku Kaishi*, 1568-1572 (1980).
22. Arakawa, T., H. Kurachi, and J. Shiokawa, Physicochemical properties of rare earth perovskite oxides as gas sensor material, *J. Mat. Sci.*, **20**: 1207-1210 (1985).
23. Goodenough, J. B., Comparisons of fluorides, oxides, and sulfides containing divalent transition elements, *Solid State Chemistry* (C. N. R. Rao, Ed.), Dekker, New York, 1974, p. 215.
24. Rao, C. N. R., and V. G. Bhide, Rare earth cobaltites, $La_{1-x}Sr_xCoO_3$ and related systems, *Am. Inst. Phys. Conf. Proc.*, **18**: 504-512 (1974).
25. Jonker, G. H., and J. H. van Santen, Ferromagnetic compounds of manganese with perovskite structure, *Physica*, **16**: 337-349 (1950).
26. Leung, L. K., A. H. Morrish, and C. W. Searle, Studies of the ionic ferromagnet $(LaPb)MnO_3$. II. Static magnetization properties from 0 to 800 K, *Can. J. Phys.*, **47**: 2697-2702 (1969).
27. Arakawa, T., A. Yoshida, and J. Shiokawa, Catalytic properties of rare earth cobaltites and related compounds,

Mat. Res. Bull., **15**: 269-273 (1980).
28. Ahamed, L. I., Lattice structure of the (100) free surface of barium titanate, *Surf. Sci.*, **12**: 437-453 (1968).
29. Yonghua, C., M. Futai, and L. Hui, Catalytic properties of rare earth manganites and related compounds, *React. Kinet. Catal. Lett.*, **37**: 37-42 (1988).
30. Voorhoeve, R. J. H., J. P. Remeika, and D. W. Johnson, Jr., Rare-earth manganites: Catalysts with low ammonia yield in the reduction of nitrogen oxides, *Science*, **180**: 62-64 (1963).
31. Webb, J. B., M. Sayer, and A. Manshingh, Polaronic conduction in lanthanum strontium chromite, *Can. J. Phys.*, **55**: 1725-1731 (1977).
32. Yo, C. H., E. S. Lee, and M. S Pyon, Study of the nonstoichiometry and physical properties of $Sr_xDy_{1-x}FeO_{3-y}$, *J. Solid State Chem.*, **73**: 411-417 (1988).
33. Parkash, O., P. Ganguly, G. R. Rao, C. N. R. Rao, D. S. Rajoria, and V. G. Bhide, Rare-earth cobaltite catalysts: Relation of activity to spin and valence states of cobalt, *Mat. Res. Bull.*, **9**: 1173-1176 (1974).
34. Nakamura, T., M. Misono, T. Uchijima, and Y. Yoneda, Catalytic activities of perovskite-type compounds for oxidation reactions, *Nippon Kagaku Kaishi*, 1679-1684 (1980).
35. Raj, S. L., B. Viswanathan, and V. Srinivasan, The activity of manganese (3+) and manganese (4+) in lanthanum strontium manganite for the decomposition of nitrous oxide, *J. Catal.*, **75**: 185-187 (1982).
36. Arakawa, T., T. Sudo, and J. Shiokawa, Surface state of the perovskite oxides $Sm_{0.5}A_{0.5}CoO_3$ (A = Ca, Sr, Ba) as gas sensor material, *Denki Kagaku*, **54**: 880-885 (1986).
37. Arakawa, T., S. Tsuchi-ya, and J. Shiokawa, Catalytic properties and activity of rare earth orthoferrites in oxidation of methanol, *J. Catal.*, **74**: 317-322 (1982).
38. Gaur, K., S.C. Verma, and H. B. Lal, Defects and electrical conduction in mixed lanthanum transition metal oxides, *J. Mat. Sci.*, **23**: 1725-1728 (1988).
39. Arakawa, T., S. Tsuchi-ya, and J. Shiokawa, Catalytic activity of rare earth orthoferrites and orthochromites, *Mat. Res. Bull.*, **16**: 97-103 (1981).
40. Tripathi, A. K., and H. B. Lal, Electrical transport in rare earth orthocromites, *Mat. Res. Bull.*, **15**: 233-242 (1980).
41. Ruddlesden, S. N., and P. Popper, New compounds of the K_2NiF_4 type, *Acta Cryst.*, **10**: 538-540 (1957).

42. Blasse, G., and A. Bril, Fluorescence of Eu^{3+}-activated sodium lanthanide titanates ($NaLn_{1-x}Eu_xTiO_4$), *J. Chem. Phys.*, **48**: 3652-3656 (1968).
43. Goodenough, J. B., Interpretation of the transport properties of Ln_2NiO_4 and Ln_2CuO_4 compounds, *Mat. Res. Bull.*, **8**: 423-431 (1973).
44. Bednorz, J. G., and K. A. Muller, Possible high T_c superconductivity in the Ba-La-Cu-O system, *Z. Phys. B*, **64**: 189-193 (1986).
45. Arakawa, T., S. Takeda, G. Adachi, and J. Shiokawa, Catalytic properties of compounds between the rare earths and copper oxide, *Mat. Res. Bull.*, **14**: 507-511 (1979).
46. Stetter, J. R., A surface chemical view of gas detection, *J. Colloid Interface Sci.*, **65**: 432-443 (1978).
47. Weisz, P. B., Electronic barrier layer phenomena in chemisorption and catalysis, *J. Chem. Phys.*, **20**: 1483-1484 (1952).
48. Wolkenstein, T., Théorie éléctronique de la catalyse sur les sémiconducteurs, *Adv. Catal.*, **12**: 189-264 (1960).
49. Parravano, G., Catalytic activity of lanthanum and strontium manganite, *J. Am. Chem. Soc.*, **75**: 1448-1451 (1953).
50. Rienacker, V. G., and J. Scheve, Untersuchungen an Cr_2O_3/ SnO_2-mischoxiden, *Z. anorg. allg. Chem.*, **328**: 201-218 (1964).
51. Ford Motor Co., Oxygen sensor and its operation, *European Patent*, G001-510 (1982).
52. Park, K., and E. M. Logothethis, Oxygen sensing with $Co_{1-x}Mg_xO$, *J. Electrochem. Soc.*, **124**: 1443-1446 (1977).
53. Tien, T. Y., N. L. Stadler, E. F. Gibbons, and P. J. Zacmanidis, TiO_2 as an air-to-fuel ratio sensor for automobile exhausts, *Am. Ceram. Soc. Bull.*, **54**: 280-282 (1975).
54. Logothethis, E. M., K. Park, A. H. Meitzler, and K. R. Land, Oxygen sensors using CoO ceramics, *Appl. Phys. Lett.*, **28**: 209-211 (1975).
55. Sekido, S., H. Tachibana, and Y. Ninomiya, Sensor element, *European Patent*, 55,104 (1984).
56. Williams D. E., B. C. Tofield, and P. McGeehin, Ceramics coatings, *UK Patent*, 2,138,321 (1984).
57. Ninomiya, Y., Y. Yamamura, K. Tachibana, and S. Sekido, Air-fuel ratio sensor (Oxygen deficient sensor), *National Technical Report*, **29**: 475-483 (1983).
58. Arai, H., C. Yu, Y. Fukuyama, Y. Shimizu, and T. Seiyama, Application of p-type semiconducting perovskite-type oxides

to a lean-burn oxygen sensor, *Proc. 2nd Int. Meeting on Chemical Sensors*, Bordeaux, 1986, p. 142.

59. Yu, C., Y. Shimizu and H. Arai, Investigation on a lean-burn oxygen sensor using perovskite-type oxides, *Chem. Lett.*, 563-566 (1986).
60. Van Damme, H., and W. K. Hall, Photocatalytic properties of perovskites for H_2 and CO oxidation. Influence of ferroelectric properties, *J. Catal.*, **69**: 371-383 (1981).
61. Thomas, H. M., Fuels from solar energy: How soon?, *Science* **222**: 151-153 (1983).

Index

Milton Keynes UK
Ingram Content Group UK Ltd.
UKHW051944071024
449327UK00026B/2165

9 780367 402525